Theory and Calculation
of Heat Transfer in Furnaces

Theory and Calculation of Heat Transfer in Furnaces

Yanguo Zhang
Department of Thermal Engineering
Tsinghua University, Beijing, P.R. China

Qinghai Li
Department of Thermal Engineering
Tsinghua University, Beijing, P.R. China

Hui Zhou
Department of Earth and Environmental Sciences
Columbia University, New York, USA

AMSTERDAM • BOSTON • HEIDELBERG • LONDON
NEW YORK • OXFORD • PARIS • SAN DIEGO
SAN FRANCISCO • SINGAPORE • SYDNEY • TOKYO
Academic Press is an imprint of Elsevier

Academic Press is an imprint of Elsevier
125 London Wall, London EC2Y 5AS, UK
525 B Street, Suite 1800, San Diego, CA 92101-4495, USA
50 Hampshire Street, 5th Floor, Cambridge, MA 02139, USA
The Boulevard, Langford Lane, Kidlington, Oxford OX5 1GB, UK

Notices
Knowledge and best practice in this field are constantly changing. As new research and experience broaden our understanding, changes in research methods, professional practices, or medical treatment may become necessary.

Practitioners and researchers must always rely on their own experience and knowledge in evaluating and using any information, methods, compounds, or experiments described herein. In using such information or methods they should be mindful of their own safety and the safety of others, including parties for whom they have a professional responsibility.

To the fullest extent of the law, neither the Publisher nor the authors, contributors, or editors, assume any liability for any injury and/or damage to persons or property as a matter of products liability, negligence or otherwise, or from any use or operation of any methods, products, instructions, or ideas contained in the material herein.

British Library Cataloguing-in-Publication Data
A catalogue record for this book is available from the British Library

Library of Congress Cataloging-in-Publication Data
A catalog record for this book is available from the Library of Congress

ISBN: 978-0-12-800966-6

For information on all Academic Press publications
visit our website at https://www.elsevier.com/

  **Working together
to grow libraries in
developing countries**

www.elsevier.com • www.bookaid.org

Publisher: Jonathan Simpson
Acquisitions Editor: Simon Tian
Editorial Project Manager: Naomi Robertson
Production Project Manager: Lisa Jones
Cover designer: Victoria Pearson

Typeset by Thomson Digital

Contents

5. Heat Transfer Calculation in Furnaces

6. Effects of Ash Deposition and Slagging on Heat Transfer

Foreword

Furnace technology is very commonly employed, industrially and in the home. The fuels available for furnaces include coal, oil, gas, and other combustible materials. More than one half of the energy consumed by humans comes from fossil fuel and biomass combustion; other energy sources include hydraulic, nuclear, wind, and solar power. All the thermal energy created by combustion is the result of a burning reaction between the furnace and combustible material, which takes place in its boiler or stove. The processes that occur in a furnace include not only combustion, but also flow, heat transfer, and mass transition. Heat transfer is the most important process in the furnace, and the focus of this book.

Heat transfer, one of the primary applications of thermal physics, covers heat conduction, convection, and radiation. Comprehensive understanding (and successful manipulation) of heat transfer is crucial for power engineering—including, of course, furnace engineering. The furnace is a reactor for combustion, in which the temperature is far higher than ambient, thus radiation is the dominant mode of heat transfer. The majority of this book concerns the behavior and principles of radiation heat transfer in furnaces.

There are normal modes of heat transfer in a furnace, such as the conduction in the water-cooled tube wall and convection between tubes and flue gas. These processes are not discussed in detail in this book because they are covered in basic heat transfer courses—instead, they are simplified to calculate convection heat transfer in the low-temperature area of the boiler flue gas passages after the furnace. The complete thermal calculation method is described in this book as a comprehensive boiler calculation process. Due to space limitations, there are a handful of other important topics related to heat transfer that are not introduced here (eg, numerical simulation of radiation heat transfer, heat transfer in industrial ovens, and waste-heat-recovery boilers).

Heat transfer in furnaces is a fundamental issue that extends the basic principles of heat transfer to specific thermal calculation for boilers. The information in this book is provided under two essential criteria: first, indispensable scientific rigor without theoretical exasperation, and second, delivering practical solutions to operational problems specific to boiler design. The information in this book also represents a typical, scientific progression from theory to practice, which helps the reader to understand how to simplify a real and complex problem to be solved according to theoretical physics; the basic approach taken by the authors involved both established engineering method and experimental application, with the ultimate goal of offering the reader a relatively simple and useful textbook as well as a valuable reference for future design of furnace technologies.

This book was written by Yanguo Zhang, Qinghai Li, and Junkai Feng first in Chinese. The current version, in English, was translated and edited by Yanguo Zhang, Qinghai Li, and Hui Zhou. Rongrong Cai also took part in the translation. The authors hope that this English edition will aid the successful and fruitful development of academic and engineering contacts and understanding between the Chinese community and non-Chinese communities which all have common interests in boiler design or application.

Yanguo Zhang acknowledges with gratitude the support received from Ganglian Ren (former Chief Engineer of Beijing Boiler Works, China), Dr Jin Sun, Professor Junfu Lv, and Ms Yi Xie. The authors would like to acknowledge Professor Junkai Feng for invaluable guidance, and Professor Buxuan Wang for his recommendation. The authors would also like to acknowledge the assistance of our editors, Ms Qian Yang, Ms Qiuwan Zhuang and Ms Xin Feng from Tsinghua University Press and the editors from Elsevier and typesetter Thomson Digital.

The authors' special thanks go to Tsinghua University Teaching Reformation Program for partially financial support.

The authors would also like to point out that this book cites formulae and other information from References [18] and [20] in chapter: Theoretical Foundation and Basic Properties of Thermal Radiation, Reference [16] in chapter: Heat Transfer in Fluidized Beds, and Reference [15] in chapter: Effects of Ash Deposition and Slagging on Heat Transfer. Although some other valuable literatures are referenced, it is difficult to cite them exactly in the text. We therefore list those literatures in the reference list. All of the references are greatly appreciated.

Yanguo Zhang

Department of Thermal Engineering, Tsinghua University

Preface

Energy, communication, and material are basic elements which push modern society forward in the processes of industrialization, electrification, and information development. Most energy and power for modern devices come from fossil fuels, which are combusted in furnaces to release heat by chemical reaction. In a boiler furnace, radiation is the dominant mechanism of transferring heat from flame and flue gas to the heating surface, combined with convection—the heat is delivered from the surface to the inner media by conduction of the tube wall. The physical and chemical processes in a furnace are a combination of combustion, heat transfer, and flows, all of which are limited by engineering factors. All devices related to combustion (including not only power plant boilers, turbines, and engines, but several other industry boilers and stoves) must satisfy environmental protection and economic demands.

This book was written based on a course on Heat Transfer in Furnaces taught by the authors at Tsinghua University, Beijing, for several years. The author would suggest that the reader first learn the basic scientific concepts of heat transfer. This book provides a connection between fundamental theories on the subject and real-world engineering applications, and the authors sincerely hope it will serve as a helpful reference for the reader during complex engineering design endeavors.

This book contains seven chapters in total. After a brief introduction to the essentials and basic principles of radiation in chapter: Theoretical Foundation and Basic Properties of Thermal Radiation, radiative characteristics of flame and flue gas (with walls) are examined in chapter: Emission and Absorption of Thermal Radiation and chapter: Radiation Heat Exchange Between Isothermal Surfaces. Chapter: Heat Transfer in Fluidized Beds describes the relatively novel concept of heat transfer in fluidized beds, which differs notably from heat transfer in stock boilers or pulverized coal boilers. Chapter: Heat Transfer Calculation in Furnaces provides thermal calculations for furnaces in three typical types of boilers. Chapter: Effects of Ash Deposition and Slagging on Heat Transfer illustrates the effects of ash deposition and slagging on the heat transfer of heating surfaces, and chapter: Measuring Heat Transfer in the Furnace discusses furnace heat transfer measurement, including flame emissivity and heat flux meters.

I strongly feel that this book contains unique and valuable characteristics, including clear and accurate depiction of relevant concepts, simple and fluent language, and a fascinating and practical extension of the authors' combined experience in engineering. I am happy to recommend it to the reader, and hope that students and practitioners of boiler technology will find this book inspiring and useful.

Academician of Chinese Academy of Sciences, Buxuan Wang
Department of Thermal Engineering
Tsinghua University

Symbols

A	area, m^2; ash content in fuel, wt.%
a	Constant; absorptivity; emissivity; fly ash fraction; width, m
b	depth, m
B	magnetic induction intensity; fuel supply rate, kg/s
c	constant; correction factor; light velocity, m/s; specific heat capacity, J/(kg·K), J/(Nm^3·K)
c_p	specific heat capacity at constant pressure, J/(kg·K)
C	carbon content in fuel, wt.%
d	diameter, m or mm
D	diameter, m; boiler capacity, kg/s or t/h
E	energy, J; emissive power (radiosity), W/m^2; electric field intensity, V/m
e	energy level; error, %
g	degeneracy of the energy level; mass fraction of flue gas flowing through superheated cold or hot sections
G	flue gas mass of unit fuel, kg/kg
h	Planck constant; heat transfer coefficient, W/(m^2·K); height, m
H	magnetic field intensity, A/m; heating area of furnace radiation, m^2; hydrogen content in fuel, %
I	radiation intensity, W/(m^2·sr); enthalpy of air or flue gas, kJ/kg
i	imaginary unit; enthalpy of working medium (water, steam), kJ/kg
K	extinction (attenuation) coefficient, m^{-1}; heat transfer coefficient, W/(m^2·K)
k	coefficient of radiant absorption, l/(m·Pa); Boltzmann constant
L	length, m
l	dimensionless length
m	mass, kg
M	moisture, %
n	refractive index
N	nitrogen content in fuel, wt.%
O	oxygen content in fuel, wt.%
P	pressure, Pa, MPa
Pr	Prandtl number
Q	heat flow, W; heating value, kJ/kg
q	heat flux, W/m^2; volumetric or sectional heat release rate (thermal load), W/m^3, W/m^2; heat percentage, %
R	radius, m; thermal resistance
r	radius vector, m; electrical resistivity, Ω·m; volume fraction
s	mean beam length, effective radiation layer thickness, m; Poynting vector, W/m^2; spacing, mm
S	sulfur content in fuel, wt.%

t	time, s; temperature, °C
T	absolute temperature, K
u	unknown variable in the integral equation; velocity, m/s
V	volume, m³; amount of air or flue gas, Nm³/kg
x,y,z	coordinates
x	effective configuration factor of the water wall
Z	number of tube rows
α	absorptivity; excess air coefficient for gas side; Lagrange factor
β	Lagrange factor; excess air coefficient for air side
δ	optical thickness
ε	emissivity; blackness; dielectric constant; ash deposition coefficient
φ	configuration factor; heat preservation coefficient
η	dimensionless coordinate
θ	polar angle, incident angle, rad; temperature, °C
Θ	dimensionless temperature
λ	wavelength, μm; thermal conductivity, W/(m·k)
μ	dimensionless concentration, kg/kg
ν	frequency, Hz; kinematic viscosity, m²/s
ξ	dimensionless coordinate; utilization coefficient
ρ	density, kg/m³; reflectivity; ash deposition coefficient
σ	Stefan-Boltzmann constant, W/(m²·K⁴)
τ	Transmissivity
χ	angle of refraction, rad
ω	circular frequency, s^{-1}; ratio; velocity, m/s
η	efficiency, %
θ	angle of inclination, degree; temperature, °C
δ	thickness, mm
ρw	steam mass flux, kg/(m² s)
τ	factor
ζ	fouling factor
ψ	thermal efficiency coefficient; effective coefficient
Ω	solid angle, sr
Δ	D-value
Π	coefficient

SUPERSCRIPTS

'	at inlet
"	at outlet
0	theoretical
-	averaged

SUBSCRIPTS

A	surface; section
a	absorption; air; adiabatic
ar	as received
ave	averaged

b	blackbody; bottom; boiler; bed
bd	blowdown
c	carbon black; corrected; convective; cold section
cal	calculation
co	coke
ca	cold air
d	diffuse; dispersed
daf	dry and ash free
ds	desuperheater spray
dw	down
e	emit; environment; outlet
f	front
fa	fly ash; furnace nose
fe	furnace exit
fl	flame
fw	boiler feed water
F	furnace; sectional
g	gas, flue gas
G	graybody
h	hot section
ha	hot air
H	section
H_2O	water
i	incident; inlet; inner diameter
i, j, k	surface numbers
I	projection
in	incoming
l	air leakage
lf	luminous flame
m	maximum; medium; average; modified
max	maximum
mh	manhole
min	minimum
ms	coal pulverizing system
n	normal direction; triatomic gas
N_2	nitrogen gas
o	outgoing; outer; outlet
p	projection; primary stage; platen
r, β, θ	spherical coordinates
r	radiation; rear
rb	refractory belt
rc	reversing chamber
roof	top surface of furnace
R	effective radiation
R_2O	CO_2 and SO_2
s	scatter; distance; secondary stage; side; surface; system; saturation
sys	system
ss	superheated steam

t	through; tube; water cooling wall
up	upward
V	volume
w	surface; wall
wch	water cooled combustion chamber hopper
x, y, z	components at *x, y, z* direction
zf	coal pulverizing system
λ	monochromatic wavelength
ν	monochromatic frequency
ϕ	solid angle
α	absorption
τ	transmission
ρ	reflectance
ut	utilized heat
ex	exhaust gas
ug	unburnt gas
uc	unburnt carbon
rad	radiation heat loss of boiler
ph	physical heat loss due to ash and slag
db	dense bed
fb	free board
sh	superheater
rh	reheater
eco	economizer
aph	air preheater
wm	working medium
os	outside, outer surface
is	inside, inner surface
DT	deformation temperature
ST	softening temperature
FT	fluid temperature
fh	fly ash
y	flue gas
k	air
ad	air dry basis
d	dry basis
daf	dry and ash free basis
gt	unburnt carbon
qt	unburnt gas
l	furnace
zs	corrected
r	burner
pj	averaged
hz	ash and slag

Chapter 1

Theoretical Foundation and Basic Properties of Thermal Radiation

Chapter Outline

All substances continuously emit and absorb electromagnetic energy when their molecules or atoms are excited by factors associated with internal energy (such as heating, illumination, chemical reaction, or particle collision). This process is called radiation. Radiation is considered a series of electromagnetic waves in classic physical theory, while modern physics considers it light quanta, that is, the transport of photons. Strictly speaking, radiation exhibits wave-particle duality, possessing properties of not only particles but also waves; this work

Theory and Calculation of Heat Transfer in Furnaces. http://dx.doi.org/10.1016/B978-0-12-800966-6.00002-8

1

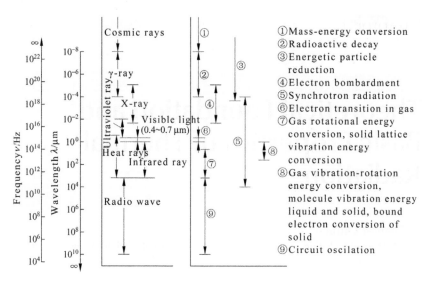

FIGURE 1.1   **Electromagnetic wave spectrum.**

considers these to be the same, that is, "radiation" refers simultaneously to both photons and electromagnetic waves.

At equilibrium, the internal energy of a substance is related to its temperature – the higher the temperature, the greater the internal energy. The emitted radiation covers the entire electromagnetic wave spectrum, as illustrated schematically in Fig. 1.1.

Thermal energy is the energy possessed by a substance due to the random and irregular motion of its atoms or molecules. Thermal radiation is the transformation of energy from thermal energy to radiant energy by emission of rays. The wavelength range encompassed by thermal radiation is approximately from 0.1 to 1000 μm, which can be divided into three subranges: the infrared from 0.7 to 1000 μm, the visible from 0.4 to 0.7 μm, and the near ultraviolet from 0.1 to 0.4 μm. Thermal radiation is a form of heat transfer between objects, characterized by the exchange of energy by emitting and absorbing thermal rays.

Consider, for example, two concentric spherical shells with different initial temperatures ($t_1 < t_2$) separated by a vacuum, as shown in Fig. 1.2. The temperature of sphere shell 2 increases as a result of heat exchange by thermal

FIGURE 1.2   **Radiation heat transfer between concentric sphere shells.**

radiation between the two shells, since there is no heat conduction or heat convection between them.

This chapter will briefly outline the essential characteristics of thermal radiation, and the fundamental parameters that describe thermal radiation properties. The description of the basic laws of thermal radiation and the general methods used in thermal radiation transfer calculation are emphasized, as these are the theoretical foundation for solving heat radiation transfer problems and conducting related engineering calculations.

1.1 THERMAL RADIATION THEORY—PLANCK'S LAW [1,2,23,24]

At the end of the 19th century, classical physics had encountered two major roadblocks: the problem of relative motion between ether and measurable objects, and the spectrum law of blackbody radiation, that is, the failure of the energy equipartition law. The solution to the first problem led to relativity theory, and the second problem was solved after the establishment of quantum theory. Quantum theory also solved the problems of blackbody radiation, photoelectric effect, and Compton scattering.

In quantum mechanics, a particle's state at a definite time can be described by wave function $\Psi(r)$, and the motion of the particle can be described by the change of the wave function with time $\Psi(r,t)$. The wave function $\Psi(r,t)$ satisfies the following Schrodinger equation:

$$i\hbar \frac{\partial}{\partial t}\Psi(r,t) = \hat{H}\Psi(r,t) \tag{1.1}$$

where $\hat{H}$ is the Hamilton operator, and $\hbar$ is a constant.

In classical mechanics, if the Hamilton of a system is known, its Hamilton equation can be obtained to determine the motion of the entire system. For a quantum system, as long as the Hamilton operator $\hat{H}$ is known, the motion of the whole system can be determined, including its energy level distribution and transition. Only quantum mechanics can strictly and accurately describe the generation, transmission, and absorption characteristics of radiation. Strict description of thermal radiation, quantum mechanics, statistical physics, and other basic theories are necessary, however, these theories are too complex for engineering calculation, particularly concerning solutions to complicated motion system equations. The basic theory of macroscopic thermal radiation is also rather difficult to describe. To this effect, it is necessary to reasonably simplify and approximate problems for the convenience of engineering application. This task is performed by professional engineering disciplines.

This section focuses on blackbody radiation theory (Planck's law), the theoretical basis of thermal radiation. During the heat transfer process, "blackbody" refers to an object which can absorb all radiant energy of various wavelengths projected onto its surface. Planck's law describes the behavior of the blackbody, the derivation of which requires some basic concepts and methods of quantum mechanics and statistical physics. The following section provides a simple introduction to Planck's law and blackbodies, to help readers to understand the basic theory and history of thermal radiation.

According to quantum mechanics, the energy of a photon with frequency v is:

$$e = hv \tag{1.2}$$

where $h = 6.6262 \times 10^{-34}$ Js is the Planck constant.

According to statistical physics, the distribution with the greatest chance of a system consisting of a large number of particles is called "the most probable distribution." The most probable distribution is typically used to describe the equilibrium distribution of an isolated system. The photon does not obey the Pauli Exclusion Principle, thus it is a Boson; neutrons, protons, and electrons are called Fermions, as they all do obey the Pauli Exclusion Principle.

The classical particles satisfy the classical Maxwell–Boltzmann distribution under the conditions of continuous energy and degeneracy. Bosons obey Bose–Einstein distribution (B–E distribution), and Fermions obey Fermi–Dirac distribution.

According to the basic principle of statistical physics, the statistical equation of B–E distribution is derived as follows:

$$N_i = \frac{g_i}{\exp(\alpha + \beta e_i) - 1} \tag{1.3}$$

where N_i is the number of particles with energy level of $e_i = hv_i$, g_i is the degeneracy of the energy level e_i, and α and β are Lagrange factors. For a photon, $\alpha = 0$, and the statistical equation the photon obeys is:

$$N_i = \frac{g_i}{\exp(\beta e_i) - 1} \tag{1.4}$$

Consider a cavity with volume V and surface temperature T. For the photon in V:

$$\beta = \frac{1}{kT} \tag{1.5}$$

where k is the Boltzmann constant, equal to 1.38×10^{-23}J/K. The degeneracy of the photon in V at energy level e_i is:

$$g_i = \frac{8\pi V v_i^2}{c^3} dv_i \tag{1.6}$$

Substituting Eq. (1.5) and Eq. (1.6) into Eq. (1.4) provides the number of photons in the frequency range from v_i to $v_i + dv_i$:

$$dN_i = \frac{8\pi V v_i^2}{c^3} \frac{dv_i}{e^{\frac{e_i}{kT}} - 1} \tag{1.7}$$

Substituting the photon energy $e_i = h v_i$ into Eq. (1.7) results in the following:

$$dN_i = \frac{8\pi V v_i^2}{c^3} \frac{dv_i}{e^{\frac{h v_i}{kT}} - 1} \tag{1.8}$$

Then, the energy of dN_i photons is:

$$de_i = h v_i dN_i = \frac{8\pi h V v_i^3}{c^3} \frac{dv_i}{e^{\frac{h v_i}{kT}} - 1} \tag{1.9}$$

The unit volume radiant energy density ranging from frequency v_i to $v_i + dv_i$ is:

$$u_i dv_i = \frac{de_i}{V} = \frac{8\pi h v_i^3}{c^3} \frac{dv_i}{e^{\frac{h v_i}{kT}} - 1} \tag{1.10}$$

where u_i is monochrome radiation intensity, this is Planck's law in the form of radiant energy density. For a specified frequency v_i, the subscript i can be removed. Eq. (1.10) can also be expressed by wavelength λ, because $v = c/\lambda$; thus, $dv = -\frac{c}{\lambda^2} d\lambda$, so radiant energy density ranging from wavelength λ to $\lambda + d\lambda$ is:

$$u_\lambda d\lambda = \frac{8\pi h c}{\lambda^5} \frac{d\lambda}{e^{\frac{hc}{\lambda kT}} - 1} \tag{1.11}$$

That is to say,

$$u_\lambda = \frac{8\pi h c}{\lambda^5} \frac{1}{e^{\frac{hc}{\lambda kT}} - 1} \tag{1.12}$$

Obviously, Planck's law is an inevitable result derived from the basic theory of quantum mechanics and statistical physics which indicates that the radiant energy density ranging from $\lambda \sim \lambda + d\lambda$ only relates to wavelength λ and cavity temperature T, denoted as $u_\lambda = f(\lambda, t)$. From this law, the radiant energy density of different wavelengths at the same temperature T can be obtained, then the spectral distribution of radiant energy density can likewise be obtained. Planck's law is the basis of the entirety of thermal radiation theory. Thus, a good understanding of the law is necessary for mastering the properties and laws of thermal radiation.

For simplicity, "radiation" in the following chapters actually refers to "thermal radiation" unless otherwise specified.

1.2   EMISSIVE POWER AND RADIATION CHARACTERISTICS

A blackbody is an ideal radiation absorber and emitter. Similar to an "ideal gas" in thermodynamics, the blackbody is an ideal concept which forms the criteria used to compare actual radiators. Similarly, Planck's law as derived in Section 1.1 describes the ideal spectral distribution of radiant energy, which is only related to temperature and wavelength. What is the difference, then, between a real object or surface and a blackbody, pertinent to radiant energy distribution and radiation characteristics?

1.2.1   Description of Radiant Energy [3,4]

Let's first build an accurate description of "radiant energy." Radiation exists in the form of photons, having wave-particle dualism, thus it can be described using electromagnetic wave theory. An electromagnetic wave obeys the Maxwell equation as following:

$$\begin{cases} \nabla \times E = -\dfrac{\partial B}{\partial t} \\[2mm] \nabla \times H = -\dfrac{\partial D}{\partial t} + J \\[2mm] \nabla \cdot D = \rho \\[2mm] \nabla \cdot B = 0 \end{cases} \tag{1.13}$$

where E is the electric field intensity in position X at time t, denoted as $E(X, t)$. Here, we use simplified representation. H is magnetic field intensity; $D = \varepsilon E$ is an auxiliary variable, ε is the dielectric constant, B is the magnetic induction intensity, J is the current density, and ρ is the charge density.

The energy flow density of an electromagnetic wave can be expressed by Poynting vector S as follows:

$$S = E \times H \tag{1.14}$$

For linear media, $B = \mu H$, where μ is a constant. From Eq. (1.13), the following wave equation can be obtained:

$$\begin{cases} \nabla^2 E - \dfrac{1}{c^2}\dfrac{\partial^2 E}{\partial t^2} = 0 \\[2mm] \nabla^2 B - \dfrac{1}{c^2}\dfrac{\partial^2 B}{\partial t^2} = 0 \end{cases} \tag{1.15}$$

For a specific frequency (monochrome, in this case), the formal solution to Eq. (1.15) is:

$$\begin{cases} E(X,t) = E(X)e^{-i\omega t} \\[2mm] B(X,t) = B(X)e^{-i\omega t} \end{cases} \tag{1.16}$$

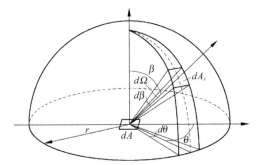

FIGURE 1.3 Definition of space of the solid angle.

where ω is the angular frequency of the electromagnetic field, and i is the imaginary number, $\sqrt{-1} = i$. For frequencies other than monochrome, the formal solution can be obtained similarly to Eq. (1.16) by Fourier analysis.

Combining Eqs. (1.14) and (1.15) results in the following:

$$S = S(X, \omega) \tag{1.17}$$

Radiant energy is a function of position X and frequency ω. This is the start point of the calculation and analysis of radiant energy, and is a vital component of thermal radiation theory.

Physical variables are introduced below in order to describe the spiral distribution of radiant energy. As shown in Fig. 1.3, there is a differential area dA on the surface of a hemisphere with radius r, thus the space of the solid angle from differential area dA to sphere center O is defined as:

$$d\Omega = \frac{dA_s}{r^2} \tag{1.18}$$

The unit of solid angle Ω is the spherical degree (sr). Of course, the solid angle from the hemisphere to the sphere's center is 2π, and the solid angle from the entire sphere to its center is 4π. The direction angle is also required to describe the space properties – because this is basic information provided in any geometry course, it is not discussed in detail here.

Once direction angle and solid angle are defined, the concepts of radiation intensity and emissive power can be established.

1. Radiation intensity. As shown in Fig. 1.4, this is the radiant energy leaving a surface per unit area normal to the pencil of rays in unit time into the unit solid angle in a wavelength range from 0~∞, denoted by I. The unit is: $W/(m^2 \cdot sr)$.

In Fig. 1.4, dA is a differential area in space, n is the normal line of dA, s is the radiation direction, dA_r is the projected area of dA normal to s, and β is

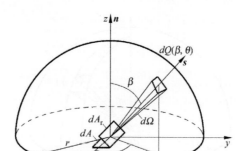

FIGURE 1.4   Definition of radiation intensity and radiation power.

the angle between s and n, namely, the direction angle in the radiation direction. Thus, $dA_r = \cos\beta dA$, where $d\Omega$ is the solid angle of any differential area in the direction s. According to these definitions, when the radiant energy in direction s is dQ, the following is obtained:

$$I\beta = \frac{dQ}{dA_r d\Omega} = \frac{dQ}{\cos\beta dA \, d\Omega} \qquad (1.19)$$

where the subscript β in I_β denotes the direction.

2. Emissive power. As shown in Fig. 1.4, this is the energy leaving a surface per unit area in unit time into the unit solid angle described by β and θ, in a wavelength range from $0 \sim \infty$, denoted by $E_{\beta,\theta}$. The unit is $W/(m^2 \cdot sr)$. Typically, the surface radiation angle is irrelevant to θ, thus E_β represents emissive power, expressed as follows:

$$E_\beta = \frac{dQ}{dA d\Omega} \qquad (1.20)$$

The relation between radiation intensity and emissive power is obtained as follows, by comparing Eq. (1.19) with Eq. (1.20):

$$E_\beta = I_\beta \cos\beta \qquad (1.21)$$

Especially when considering the total radiation power from a surface to its hemispherical space, the hemispherical radiance E $W/(m^2)$ is appropriate. The relation between E and E_β is:

$$E = \int_0^{2\pi} E_\beta d\Omega \qquad (1.22)$$

1.2.2 Physical Radiation Characteristics

The above section disclosed the characteristics that radiant energy is distributed according to space and frequencies; however, in engineering applications, we are more concerned about the radiation heat transfer between objects, such as between the hot flue gas and the water wall in a furnace, or the furnace wall and the work piece in a heating furnace. Therefore, it's still necessary to study the radiation characteristics of solid, liquid, and gas.

Objects have the ability to absorb, reflect, and penetrate external radiation, thus, absorptivity, reflectivity, and transmissivity are introduced to build a quantitative description of this ability. As shown in Fig. 1.5a, when an object receives external irradiation Q_I, some is absorbed, denoted as Q_α, some is reflected, denoted as Q_ρ, and some penetrates the object, denoted as Q_τ. See the following:

$$\text{Absorptivity } \alpha = Q_\alpha / Q_I \tag{1.23}$$

$$\text{Reflectivity } \rho = Q_\rho / Q_I \tag{1.24}$$

$$\text{Transmissivity } \tau = Q_\tau / Q_I \tag{1.25}$$

From the energy conservation relation shown in Fig. 1.5b, the following is obtained:

$$Q_I = Q_\alpha + Q_\rho + Q_\tau$$

Combining the above four equations results in:

$$\alpha + \rho + \tau = 1 \tag{1.26}$$

The above definitions suggest that absorptivity, reflectivity, and transmissivity are actually fractions of irradiation that has been absorbed, reflected, and penetrated. What are their properties, then?

Fig. 1.5 and Eqs. (1.23)–(1.25) demonstrate that the absorption, reflection, and transmission of an object relating to external projected energy are related to the characteristics of the object and the projected radiation. Generally, object properties, including physical properties, geometric structure (such as

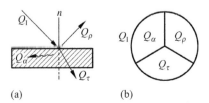

(a)　　　　　(b)

FIGURE 1.5 **Absorption, reflection, and transmission of radiant energy.** (a) Absorption, reflection, and penetration for radiation. (b) Energy conservation relationship.

species and the arrangement of atoms or molecules), temperature, and surface roughness all directly affect the absorption, reflection, and transmission of the irradiation, and therefore the frequency of irradiation. Basically, one object has different abilities to absorb, reflect, and penetrate radiation if the frequencies of projected energy are not the same.

Absorption: In quantum mechanics, the absorption of radiation is actually the process of energy level transition of particles that constitute the object (usually atoms or molecules) due to the absorption of photons under certain conditions. Because the micro states of particles with different species, temperatures, and atomic or molecular structures vary, the photon frequency/photon energy they absorb to satisfy energy level differences also vary, thus the absorption characteristics differ, and vice versa.

Reflection: Reflection is mainly dependent on the surface roughness of the object, and obeys reflection law.

Transmission: Transmission mainly occurs in gas media and is dependent on the absorption and reflection ability of the medium, which is further discussed in chapter: Emission and Absorption of Thermal Radiation.

According to the third law of thermodynamics, the temperatures of all objects are higher than 0 K, thus they all maintain thermal motion and emit thermal radiation. In order to describe the radiant energy emitted from different objects at different temperatures, the concept of emissivity is established with the blackbody as a criterion.

A blackbody is a hypothetical body that completely absorbs the radiant energy of all wavelengths from every direction. The absorptivity of the blackbody is 1 to all irradiation. The concept is the same as that of an ideal gas in thermodynamics—a blackbody is an ideal concept that does not exist in nature.

Object emissivity (also called "radiation rate") is defined as the ratio between the power emitted from the object and the power emitted from a blackbody with the same temperature as the object, denoted by ε:

$$\varepsilon = E/E_b \tag{1.27}$$

For a real object, $0 < \varepsilon < 1$. E_b is the emissive power of the blackbody, the calculation equation of which can be derived from Planck's law (see Section 1.3). It is important to note that though the blackbody appears black, an object with large emissivity is not always black. This happens because the human eye is only sensitive to visible light, while thermal radiation rays cover a broad wavelength including infrared. For example, the emissivity of black carbon at 52°C is 0.95~0.99 and this looks very black, but the emissivity of water at 0°C is 0.96~0.98 and water does not look black at all.

Other notable ideal concepts include transparent bodies, which have a transmission value of 1 for radiation from all wavelengths from any direction. Also mirror bodies, which have a reflection rate of 1 for radiation from all wavelengths from any direction. These comply with mirror reflection, as opposed to similar bodies that comply with diffuse reflection, called white bodies. In the allowable error range of engineering calculation, real bodies are abstracted into ideal objects (or the ideal objects serve as reference points) to obtain the macroscopic properties of the real surfaces. Such treatment lends a considerable amount of convenience to engineering applications, and is a common method of transmitting physics theory to engineering. For example, in engineering application, a single atom or diatom gas (He, H_2, N_2, and O_2) is treated as a transparent body as it hardly absorbs or emits radiation; conversely, a closed cavity with only one hole is treated as a blackbody.

1.2.3 Monochromatic and Directional Radiation

Radiant energy is distributed according to spatial position and wavelength, thus, the exchange process of radiant energy (ie, the emission and absorption of radiation) is also related to spatial position and wavelength. To describe the relations between emission, absorption, and wavelength, the directional forms of emissivity and absorptivity are introduced below and relevant expressions are obtained.

Monochromatic directional emissivity ($\varepsilon_{\lambda,\beta}$) is the ratio between the monochromatic radiation intensity of a real surface in direction β and that of a blackbody with the same temperature, expressed as follows:

$$\varepsilon_{\lambda\beta} = I_{\lambda\beta}/I_{b\lambda} \tag{1.28}$$

Monochromatic emissivity (ε_λ) is the ratio between the monochromatic emissive power of a real surface in wavelength λ and that of a blackbody with the same temperature and wavelength:

$$\varepsilon_\lambda = E_\lambda/E_{b\lambda} \tag{1.29}$$

Directional emissivity (ε_β) is the ratio between the monochromatic radiation intensity of a real surface in direction β and that of a blackbody:

$$\varepsilon_\beta = I_\beta/I_b \tag{1.30}$$

Monochromatic directional absorptivity ($\alpha_{\lambda\beta}$) is the ratio between the monochromatic radiation intensity absorbed in wavelength λ in direction β and the irradiation intensity with the same wavelength and direction:

$$\alpha_{\lambda\beta} = I_{\alpha\lambda\beta}/I_{1\lambda\beta} \tag{1.31}$$

Monochromatic absorptivity (α_λ) is the ratio between the absorbed monochromatic emissive power and the projected emissive power with the same wavelength:

$$\alpha_\lambda = E_{\alpha\lambda}/E_{1\lambda} \tag{1.32}$$

Directional absorptivity (α_β) is the ratio between the absorbed radiation intensity in a direction and the irradiation intensity with the same wavelength:

$$\alpha_\beta = I_{\alpha\beta}/I_{1\beta} \tag{1.33}$$

1.3   BASIC LAWS OF THERMAL RADIATION [4–10]

As discussed in Sections 1.1 and 1.2, radiant energy distribution depends on both wavelength and direction. Planck's law asserts that the radiant energy of a blackbody is distributed according to its wavelength, which is the basis of thermal radiation theory. Knowledge of this characteristic alone is not sufficient— the spatial distribution of radiant energy is also needed, which can be described under Lambert's law. Information regarding the ability to emit or absorb radiation is also necessary, described by Kirchhoff's law, in order to calculate and analyze heat transfer. This chapter discusses the engineering application form of Planck's law and its corollaries, Lambert's law and Kirchhoff's law.

1.3.1   Planck's Law and Corollaries

In Section 1.1, the monochrome radiation intensity (also called "spectral radiation intensity") of a blackbody was derived as:

$$u_\lambda = \frac{8\pi hc}{\lambda^5} \frac{1}{e^{\frac{hc}{\lambda kT}} - 1} \tag{1.12}$$

For the convenience of engineering application, the above equation is rewritten using the monochrome emissive power of a blackbody $E_{b\lambda}$. According to thermodynamic method, the following is obtained:

$$E_{b\lambda} = \frac{c}{4} u_\lambda \tag{1.34}$$

The equation is derived based on the definitions of $E_{b\lambda}$ and u_λ. Substituting Eq. (1.12) into Eq. (1.34) provides:

$$E_{b\lambda} = \frac{2\pi hc^2}{\lambda^5(e^{\frac{hc}{\lambda kT}} - 1)} \tag{1.35}$$

For simplicity, the first Planck's constant $c_1 = 2\pi hc^2 = 5.9553 \times 10^{-16} \text{W} \cdot \text{m}^2$ and the second Planck's constant $c_2 = hc/k = 1.4388 \times 10^{-2} \text{m} \cdot \text{K}$ are introduced. Eq. (1.35) then becomes:

$$E_{b\lambda} = \frac{c_1}{\lambda^5(e^{\frac{c_2}{\lambda T}} - 1)} \tag{1.36}$$

where the units of wavelength λ and temperature T are m and K, respectively.

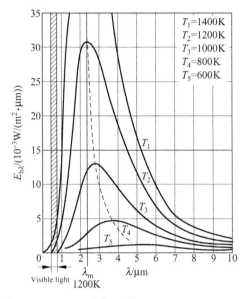

FIGURE 1.6   Relation between $E_{b\lambda}$ and λ at different temperatures.

Fig. 1.6 shows the relation between $E_{b\lambda}$ and λ at different temperatures. At the same temperature, as λ increases, $E_{b\lambda}$ increases first to its maximum value and then decreases. This phenomenon reflects Wien's displacement law, which will be discussed later.

Planck's law describes the radiation of a blackbody in a vacuum; for radiation in nonvacuum media, the value of light speed in a vacuum in Eqs. (1.35) and (1.36) can simply be replaced with the speed in the medium.

Certain corollaries can be obtained through Planck's law. When temperature is high and wavelength is long ($hc/\lambda kT \ll 1$), the term $e^{hc/\lambda kT}$ in Eq. (1.35) can be expanded as follows:

$$e^{\frac{hc}{\lambda kT}} = 1 + \frac{hc}{\lambda kT} + \frac{1}{21}\left(\frac{hc}{\lambda kT}\right)^2 + \cdots$$

Because $hc/\lambda kT \ll 1$, high-order terms can be neglected. $E_{b\lambda}$ can then be calculated as:

$$E_{b\lambda} = 2\pi kc\frac{T}{\lambda^4} \tag{1.37}$$

This is the Rayleigh–Jeans law, which is mainly applicable to infrared.

When temperature is low and wavelength is short ($hc/\lambda kT \gg 1$), $e^{hc/\lambda kT} - 1 \approx e^{hc/\lambda kT}$ in Eq. (1.35), thus:

$$E_{b\lambda} = \frac{2\pi hc^2}{\lambda^5 e^{\frac{hc}{\lambda kT}}} \tag{1.38}$$

This is the Wien equation derived by the semiempirical method, which is applicable to ultraviolet.

Fig. 1.6 shows that at a certain temperature, energy density reaches its maximum corresponding to a certain wavelength. This wavelength is called "the most probable wavelength," denoted by λ_m. According to the extreme conditions of Eq. (1.36):

$$\frac{\partial E_{b\lambda}(\lambda, T)}{\partial \lambda} = 0$$

λ_m at different temperatures can be obtained as follows:

$$\frac{\partial E_{b\lambda}(\lambda, T)}{\partial \lambda} = -\frac{5}{\lambda^6 (e^{\frac{hc}{\lambda kT}} - 1)} + \frac{hc}{\lambda^7 kT} \frac{e^{\frac{hc}{\lambda kT}}}{(e^{\frac{hc}{\lambda kT}} - 1)^2} = 0$$

By defining the dimensionless variable $x = \dfrac{hc}{\lambda kT}$, the above equation can be rewritten:

$$5e^{-x} + x = 5$$

The numerical solution to the equation is $x = 4.965$, thus:

$$\lambda_m T = \frac{hc}{xk} = 2.8978 \times 10^{-3} \, (\text{m} \cdot \text{K}) \tag{1.39}$$

This is Wien's displacement law, which states that there is an inverse relationship between the most probable wavelength λ_m and temperature T; that is, λ_m decreases as T increases, so the higher the temperature, the more likely the energy is distributed in a high-frequency range.

To illustrate the relation between the blackbody's emissive power and temperature, Eq. (1.35) is integrated from $\lambda = 0$ to $\lambda = \infty$ as follows:

$$E_b = \int_0^\infty E_{b\lambda} \, d\lambda = 2\pi hc^2 \int_0^\infty \frac{1}{\lambda^5 (e^{\frac{hc}{\lambda kT}} - 1)} \, d\lambda$$

By defining $x = \dfrac{hc}{\lambda kT}$ and using definite integral relation $\int_0^\infty \dfrac{x^3}{e^x - 1} \, dx = \dfrac{\pi^4}{15}$ in the above equation:

$$E_b = \frac{2\pi^5 k^4}{15 c^2 h^3} T^4 \tag{1.40}$$

For simplicity, the radiation constant is defined as $\sigma = \dfrac{2\pi^5 k^4}{15c^2 h^3}$ and all known values are plugged in to obtain $\sigma = 5.6693 \times 10^{-8} \text{W/(m}^2 \cdot \text{K}^4)$ so the above equation becomes:

$$E_b = \sigma T^4 \tag{1.41}$$

This is the Stefan–Boltzmann law, also known as the fourth-power law—a very practical application commonly applied to radiation heat transfer.

1.3.2   Lambert's Law

To describe the distribution property of radiant energy in different directions, the concepts of directional emissive power and directional radiation intensity are introduced below. Similarly to the definition of the spatial distribution property of the directional emissive power of a blackbody, here, we introduce Lambert's law, which states that the directional radiation intensity of hemispheric space is the same in any direction:

$$I_{\beta_1} = I_{\beta_2} = I_{\beta_3} = \cdots = I_{\beta_i} \tag{1.42}$$

where β_i denotes any direction within hemispheric space. Blackbody radiation obeys Lambert's law completely. The equation can be rewritten as follows to apply to a blackbody:

$$I_{b\beta_1} = I_{b\beta_2} = I_{b\beta_3} = \cdots = I_{b\beta_i} = I_b \tag{1.43}$$

Considering the relation between directional emissive power and directional radiation intensity [Eq. (1.21)], Eq. (1.43) becomes:

$$E_{b\beta} = I_b \cos\beta \tag{1.44}$$

For a blackbody, Eq. (1.44) is also applicable in monochromatic conditions:

$$E_{b\lambda\beta} = I_b \cos\beta \tag{1.45}$$

Due to the specific form that Eqs. (1.44) and (1.45) take, Lambert's law is also called "cosine law."

A surface which obeys Lambert's law is called a diffuse surface, whether or not it is a blackbody. Eq. (1.44) can also be written as follows:

$$E_\beta = I \cos\beta \tag{1.46}$$

Upon integrating the above equation in hemispheric space, we obtain:

$$E = \int_{2\pi} E_\beta \, d\Omega = I \int_{2\pi} \cos\beta \, d\Omega$$

where $d\Omega = dA/r^2 = \sin\beta \, d\beta \, d\varphi$, thus:

$$E = I \int_{\varphi=0}^{2\pi} d\phi \int_{\beta=0}^{\frac{\pi}{2}} \sin\beta \cos\beta \, d\beta = \pi I \tag{1.47}$$

This is the general expression of Lambert's law for a diffuse surface. It is important to note that Eq. (1.47) is not applicable under all possible conditions, as it is derived under the premise that the solid angle is defined by hemispheric space.

1.3.3   Kirchhoff's Law

The relation between the emission and the absorption of a real surface is an important question in thermal radiation theory, which is quite adequately described by Kirchhoff's law.

Consider the radiation heat transfer of an isolated system consisting of two infinite parallel walls, as shown in Fig. 1.7. Surface 1 is a blackbody, with temperature of T_b, emissive power of E_b, and absorptivity $\alpha_b = 1$. Surface 2 is an arbitrary surface with temperature of T, emissive power of E, and absorptivity of α. The emissive power emitted from Surface 1 is E_b, and after being absorbed by Surface 2, the remaining $(1 - \alpha)E_b$ is reflected back to Surface 1 until finally completely absorbed by Surface 1. Similarly, the emissive power E from Surface 2 to Surface 1 is absorbed completely by the blackbody. Thus, the radiation exchange heat is expressed as:

$$q = \alpha E_b - E \tag{1.48}$$

When $T_1 = T_2$, that is, when the system is in thermal radiation equilibrium, the radiation exchange heat is zero, thus Eq. (1.48) becomes:

$$\frac{E}{\alpha} = E_b \tag{1.49}$$

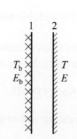

FIGURE 1.7   Radiation heat transfer of an isolated system consisting of two infinite parallel walls.

This is Kirchhoff's law. Note that E and E_b denote the emissive power of a real surface and blackbody at the same temperature. The following relations between emission and absorption can be obtained on the basis of this law:

1. The ratio between the emissive power of any object and its absorptivity to a blackbody equals the blackbody's emissive power at the same temperature, thus the emissive power of any object is only related to temperature.
2. The larger the emissive power of an object is, the greater its absorptivity is—objects with strong emission ability also have strong absorption ability.
3. At the same temperature, the emissive power of a blackbody is the strongest, because the absorptivity of real surfaces is always smaller than 1.

There is another form of Kirchhoff's law—after comparing Eq. (1.49) with the definition of emissivity:

$$\varepsilon = \frac{E}{E_b}$$

The following can be obtained:

$$\varepsilon = \alpha \tag{1.50}$$

Eq. (1.50) is only true at thermal radiation equilibrium. Thus, Kirchhoff's law can also be described as an arbitrary body emitting and absorbing thermal radiation in thermodynamic equilibrium, where emissivity equals absorptivity. This expression states relation (2) more directly.

Kirchhoff's law is also applicable to monochromatic conditions, expressed as follows:

$$\varepsilon_\lambda = \alpha_\lambda \tag{1.51}$$

Kirchhoff's law is not experimental. It is derived strictly based on thermo-dynamic laws.

1.4   RADIATIVITY OF SOLID SURFACES

When it comes to general technical problems in engineering, the majority of radiative heat exchange involves solid surfaces; to this effect, mastering the radiation properties of a solid surface is necessary in order to calculate and analyze radiation heat transfer.

1.4.1   Difference Between Real Surfaces and Blackbody Surfaces

A blackbody is an ideal object, defined by researchers, which perfectly obeys the thermal radiation laws discussed in Section 1.3. Naturally, the emission and absorptivity of real surfaces are different. Fig. 1.8 shows a basic comparison between the surface emission and absorptivity of real bodies and blackbodies.

1. Radiativity

The radiation spectrum of a blackbody's surface obeys Planck's law completely, thus its monochromatic emissivity is always 1. In addition, blackbody radiation distribution in each direction is uniform, completely obeying Lambert's law.

The surface radiation spectrum energy of a real body is less than that of a blackbody at the same temperature, thus its monochromatic emissivity is less than 1. The wavelength distribution of the spectrum has selectivity relative to the structure of molecules and atoms which comprise the object's surface, as well as temperature. Spatial distribution of the surface radiation of a real body is not uniform, and its radiation intensity is a function of direction angle, so the surface fails to obey Lambert's law—its properties are related to the geometric structure of the surface.

Fig. 1.8a illustrates the difference described earlier, where the blackbody value is denoted by 1 and the real body value is represented by 2. The spatial radiation intensity of the blackbody is uniform, as opposed to the real body.

2. Absorptivity

A blackbody surface completely absorbs external irradiation, regardless of direction and wavelength distribution. Its surface absorptivity is always 1, so $\alpha_\lambda = \alpha = 1$. The reflectivity of a blackbody is always zero. The monochromatic absorptivity of a real body surface is less than 1 for external irradiation, thus its reflection ratio is not zero relative to direction and wavelength.

Fig. 1.8b illustrates the relation between absorptivity and wavelength of a blackbody and a real surface. By comparison, the radiation and absorption of real bodies differ from those of blackbodies, but all basic laws of thermal radiation are based on the blackbody—thus, in order to calculate and analyze the radiation heat transfer of real bodies, certain secondary factors must be neglected accordingly. If the calculation error is kept in an allowable range, it is possible to refer to blackbody radiation law; the following sections introduce relevant concepts of "graybodies," "diffuse surfaces," and other ideal concepts, as they utilize blackbody laws to achieve accurate results for real surfaces. This method is commonly used in engineering calculations.

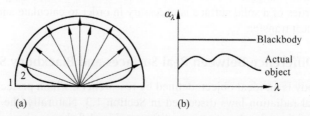

(a) (b)

FIGURE 1.8 Surface property differences between a real body and a blackbody. (a) The difference of surface radiation intensity in spatial distribution between blackbody and actual object. (b) The relationship between surface adsorption and wavelength of blackbody and actual object.

1.4.2   Graybody

According to the properties of real surfaces, the thermal radiation spectrums of which do not obey Planck's law and the wavelengths of which have irregular distribution, an ideal simplified concept, the graybody, is introduced. A graybody is defined as an object that lacks radiation selectivity, thus its emissive power is proportional to that of a blackbody at the same temperature. This proportional coefficient is simply the emissivity of the object. The emissive power of a graybody equals that of a real surface, as demonstrated by the following equations:

$$E_{G\lambda} = \varepsilon E_{b\lambda} \tag{1.52}$$

$$E_G = \varepsilon E_b \tag{1.53}$$

$$E_G = E_\varepsilon \int_0^\infty E_{b\lambda}(\lambda, T)d\lambda \tag{1.54}$$

The monochromatic emissivity of a graybody is independent of wavelength and equals hemispheric emissivity:

$$\varepsilon_\lambda = \varepsilon_G \tag{1.55}$$

According to Kirchhoff's law,

$$\alpha_\lambda = \varepsilon_\lambda$$

thus:

$$\alpha_G = \varepsilon_G \tag{1.56}$$

The absorptivity of a graybody therefore equals its emissivity, and shows no selectivity in wavelength. It exists independent of the temperature of the radiation source.

The relation between emissive power and wavelength of a blackbody, a graybody, and a real body is illustrated in Fig. 1.9. Fig. 1.10 shows the relation between the emissivity and wavelength of a graybody.

For metal and nonmetal materials commonly used in engineering, selectivity in the long-wave range (infrared region) is weak. The temperature of boiler furnace radiation is generally not higher than 2000 K, so most common engineering materials can be considered graybodies.

1.4.3   Diffuse Surfaces

Radiant energy is energy distributed according to wavelength and spatial position. The introduction of graybodies makes the radiation of a real body similiar to that of a blackbody in wavelength distribution, but radiation intensity is uniformly decreased by a factor of emissivity. The spatial distribution

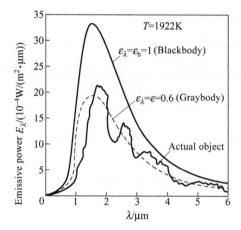

FIGURE 1.9 Relation between emissive power and wavelength of a blackbody, graybody, and real body.

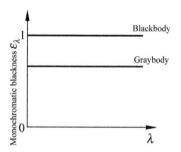

FIGURE 1.10 Spectral distribution and emissivity of a graybody.

characteristics of radiant energy mainly refer to the fact that heat rays are related to surface direction. The concept of diffused reflection is introduced below to simplify the calculation difficulties inherent to nonuniform spatial distribution of emission and reflected thermal radiation.

Diffused radiation describes the emission of thermal radiative rays according to Lambert's law, where radiant intensity is distributed uniformly in the hemisphere. Similarly, diffused reflection is the reflection of irradiation according to Lambert's law, where the radiant intensity of reflection radiation is distributed uniformly in the hemisphere. Diffuse surfaces that comply with graybody properties are called diffuse gray surfaces.

For simplicity, solid surfaces are treated as diffuse gray surfaces. This assumption is unsuitable in engineering applications only when calculation error is not tolerable. In the majority of furnaces, which are very rough due to the existence of fouling layers, the use of this diffuse gray assumption is allowed.

According to Kirchhoff's law, a diffused radiation surface must show diffused absorption but not necessarily diffused reflection, and vice versa.

1.5 THERMAL RADIATION ENERGY

1.5.1 Thermal Radiation Forms

Any object, be it solid, liquid, or gas, is known to constantly radiate heat according to Planck's law, as its temperature cannot reach absolute zero (0 K), though radiation projected on its surface is absorbed, reflected, and penetrated through energy exchange. In order to further describe these processes and behaviors, relevant descriptions of various forms of thermal radiation are provided in this section. These are defined according to the engineering field, specifically, not just in terms of physics—the treatment is a transition from fundamental physical laws and strict physical methods into engineering practice, as a combination of scientific theory and engineering application.

Self-radiation: Self-radiation refers to a body with temperature above 0 K, which radiates heat constantly, where emission is caused by temperature. Self-radiation energy and power are represented by Q_s and E_s, respectively. Given the emissivity of an object, ε, self-radiation can be expressed as:

$$E_s = \varepsilon E_b \tag{1.57}$$

where E_b is a blackbody's emissive power with the same temperature as the object, which can be calculated by the Stefan–Boltzmann law: $E_b = \sigma T^4$.

Irradiation: Irradiation is the radiation projected onto an object by its specific external environment. The irradiation energy and power are represented by Q_I and E_I, respectively.

Absorbed radiation: A body with absorptivity that is not zero absorbs the radiant energy projected onto it (represented by Q_I); the absorbed part is called absorbed radiation. Absorbed radiation energy and power are represented by Q_α and E_α, respectively. Given the absorptivity of α, then:

$$Q_\alpha = \alpha Q_I \tag{1.58a}$$

$$E_\alpha = \alpha E_I \tag{1.58b}$$

Reflected radiation: A body with zero transmissivity reflects radiant energy that is projected onto it; the reflected part is called reflected radiation. Reflected radiation energy and power are represented by Q_ρ and E_ρ, respectively. Given the reflectivity of ρ, then:

$$Q_\rho = \rho Q_I \tag{1.59a}$$

$$E_\rho = \rho E_I \tag{1.59b}$$

$$E_\rho = E_I - E_\alpha = (1-\alpha)E_I$$

Transmissive radiation: For a body with transmissivity that is not zero, some of the radiant energy projected onto it will penetrate into the object. The

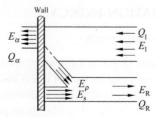

FIGURE 1.11   Flows of radiation heat.

transmission part is called transmissive radiation. Transmissive radiation energy and power are represented by Q_τ and E_τ, respectively. Given the transmissivity of τ, then:

$$Q_\tau = \tau Q_I \tag{1.60a}$$

$$E_\tau = \tau E_I \tag{1.60b}$$

Radiosity (effective radiation): It is difficult to distinguish the self-radiation and reflected radiation of a real body, but there is no need to strictly distinguish them due to their macroscopic effects in engineering. For simplicity, the concept of radiosity, which is the sum of both, is introduced. As shown in Fig. 1.11, radiosity energy and power are represented by Q_R and E_R. See the following:

$$Q_R = Q_s + Q_\rho \tag{1.61}$$

$$E_R = E_s + E_\rho \tag{1.62}$$

The sum of absorptivity, reflectivity, and transmissivity of an object equals 1: $\alpha + \tau + \rho = 1$. Thus, the sum of absorbed radiation, reflected radiation, and transmissive radiation is equal to irradiation:

$$Q_\alpha + Q_\tau + Q_\rho = Q_I \tag{1.63}$$

Radiation heat transfer: By definition, radiation heat transfer is the net radiation energy an object acquires or loses. Generally, radiation heat transfer is positive when an object is heated, and vice versa. Radiative heat flux can be directly calculated according to irradiation and radiosity ($q > 0$ when adding heat to the object):

$$q = E_I - E_R \tag{1.64}$$

1.5.2 Radiosity

The radiation heat transfer between graybody surfaces is more complicated than between blackbodies, because graybodies cannot completely absorb radiant energy projected onto them—instead some of the energy is reflected, generating multiple reflection and absorption values between the surfaces. The introduction of radiosity greatly simplifies the calculation of this phenomenon.

The total radiative energy emitted from a unit area of a solid surface in unit time is called effective emissive power (or "radiosity," for short). Similarly, the total radiant energy projected onto a unit area of an object based on its external environment in unit time is called projected emissive power, or "irradiation." Radiation heat transfer can thus be calculated in a concise manner, according to Eq. (1.64).

Projected emissive power is difficult to measure, thus the following provides a simplified method of calculating radiation heat transfer, built by combining radiosity with the object's properties and state.

See the following:

$$E_R = E_s + E_\rho$$

Because $E_s = \varepsilon E_b, E_\rho = (1-\alpha)E_I$:

$$E_R = \varepsilon E_b + (1-\alpha)E_I \tag{1.65}$$

For a graybody, $\varepsilon = \alpha$, thus:

$$E_R = \varepsilon E_b + (1-\varepsilon)E_I \tag{1.66}$$

This is an expression of radiosity derived from its definition.

Considering the thermal equilibrium between the object and its external environment, however:

$$q = E_I - E_R, E_R = E_I - q$$

Considering the thermal equilibrium of the inner object:

$$q = E_\alpha - E_s, E_\alpha = E_s + q$$

In addition, $E_I = \dfrac{1}{\alpha}E_\alpha, E_s = \varepsilon E_b$, so:

$$E_R = \frac{\varepsilon}{\alpha}E_b + \left(\frac{1}{\alpha}-1\right)q \tag{1.67}$$

For a graybody, this can be rewritten:

$$E_R = E_b + \left(\frac{1}{\varepsilon}-1\right)q \tag{1.68}$$

This is an expression derived from radiation heat transfer, which associates radiosity E_R, radiation properties (ε, α), temperature (T, E_b), and heat transfer flux q with each other to simplify calculation.

A remarkable conclusion can be drawn from Eq. (1.68): if an object has no radiant energy exchange with its external environment ($q = 0$), then $E_R = E_b$ whether or not the object is a graybody and whether or not its radiation properties are uniform. The radiosity of the object equals that of a blackbody at the same temperature. A heating furnace which heats a workpiece or steel slab, a kiln which fires refractory materials, or a boiler furnace with no heating surface all fall into this category. Additionally, although this type of object's radiosity equals that of a blackbody and it can technically be considered a blackbody, the emissivity of the object is not 1, because its self-radiation does not conform to the blackbody radiation spectrum.

The concept of radiosity is very useful for calculating radiation heat transfer, and will be further described in chapter: Radiative Heat Exchange Between Isothermal Surfaces.

1.6   RADIATIVE GEOMETRIC CONFIGURATION FACTORS [20]

Configuration factors reflect the effects of geometric relationships of surfaces to radiation heat transfer; they describe the proportion of the radiation which leaves one surface and projects onto another surface. A configuration factor is also called a "shape coefficient" or "shape factor," and is an essential parameter of radiation heat transfer.

1.6.1   Definition of the Configuration Factor

Consider two arbitrarily placed blackbody surfaces, dA_1 and dA_2, with temperatures of T_1 and T_2 (Fig. 1.12), between which is a vacuum or transparent medium. The distance between the two surfaces is r, the angles between the two surfaces' attachment lines and each surface's normal line are β_1 and β_2. The configuration factor φ_{d_1,d_2} is defined as the proportion of the radiation which leaves surface dA_1 and reaches surface dA_2. Thus, the configuration factor from dA_1 to dA_2 is:

$$\phi_{d_1,d_2} = \frac{Q_{d_1,d_2}}{Q_{d_1}} \tag{1.69}$$

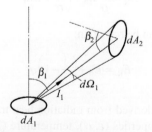

FIGURE 1.12   Angular coefficient between cell surfaces.

where Q_{d_1,d_2} is the radiant energy leaving surface dA_1 for surface dA_2 in unit time. According to the definition of radiation intensity, Q_{d_1,d_2} can be written as:

$$Q_{d_1,d_2} = I_{b_1} dA_1 \cos\beta_1 d\Omega_1$$

where $d\Omega_1$ is the solid angle from dA_1 to dA_2 which can be expressed as:

$$d\Omega_1 = \frac{\cos\beta_2 dA_2}{r^2}$$

Thus,

$$Q_{d_1,d_2} = \frac{I_{b_1} \cos\beta_1 \cos\beta_2 dA_1 dA_2}{r^2}$$

The denominator in Eq. (1.69) is:

$$Q_{d_1} = E_{b_1} dA_1 = \pi I_{b_1} dA_1$$

Upon substituting the above equations into Eq. (1.69), the following is obtained:

$$\phi_{d_1,d_2} = \frac{\cos\beta_1 \cos\beta_2}{\pi r^2} dA_2 \tag{1.70}$$

Similarly, the configuration factor from dA_2 to dA_1 is:

$$\phi_{d_2,d_1} = \frac{\cos\beta_2 \cos\beta_1}{\pi r^2} dA_1 \tag{1.71}$$

Once Eqs. (1.70) and (1.71) are obtained, the configuration factors between a differential surface and finite surface, or between finite surfaces, can be calculated effectively.

As shown in Fig. 1.13, both the differential surface dA_1 and the finite surface dA_2 are blackbody surfaces. To calculate the configuration factor between dA_1 and A_2, consider a differential surface dA_2 on surface A_2.

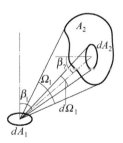

FIGURE 1.13   Configuration factors between a differential surface and finite surface.

First, calculate the radiant energy projected from dA_1 to A_2 as follows:

$$Q_{d_1,2} = \int_{A_2} \frac{I_{b_1}\cos\beta_1\cos\beta_2}{r^2}dA_1dA_2$$

Then, calculate the total energy emitted from dA_1:

$$Q_{d_1} = E_{b_1}dA_1 = \pi I_{b_1}dA_1$$

Substituting the above two equations into Eq. (1.69) gives:

$$\varphi_{d_1,2} = \frac{Q_{d_1,2}}{Q_{d_1}} = \int_{A_2} \frac{\cos\beta_1\cos\beta_2}{\pi r^2}dA_2 \tag{1.72}$$

Eq. (1.72) is the result of Eq. (1.70) after integrating over A_2. In fact, this is an inevitable result of defining the configuration factor as energy distribution.

Similarly, the configuration factor from finite surface A_2 to differential surface dA_1 can be calculated as follows:

$$Q_{2,d_1} = \int_{A_2} Q_{d_2,d_1} = \int_{A_2} \frac{I_{b_1}\cos\beta_1\cos\beta_2}{r^2}dA_1dA_2$$

$$Q_2 = E_{b_2}A_2 = \pi I_{b_2}A_2$$

$$\varphi_{2,d_1} = \frac{Q_{2,d_1}}{Q_2} = \frac{1}{A_2}\int_{A_2} \frac{\cos\beta_1\cos\beta_2}{\pi r^2}dA_2dA_1 \tag{1.73}$$

For blackbody surfaces, the configuration factor between finite surfaces is completely similar to the above circumstance and need not be derived again. Results are provided below.

The configuration factor from finite surface A_1 to A_2 is:

$$\phi_{12} = \frac{1}{A_1}\int_{A_1}\int_{A_2} \frac{\cos\beta_1\cos\beta_2}{\pi r^2}dA_1dA_2 \tag{1.74a}$$

The configuration factor from finite surface A_2 to A_1 is:

$$\varphi_{21} = \frac{1}{A_2}\int_{A_1}\int_{A_2} \frac{\cos\beta_1\cos\beta_2}{\pi r^2}dA_1dA_2 \tag{1.74b}$$

In the above derivations, surfaces are assumed to be blackbodies whether or not they are differential or finite surfaces. As long as a surface is uniform in temperature and radiation, and is a diffuse gray surface, the result is the same as above. Basically, as long as the above conditions are met, the properties and calculation methods described as follows are suitable even for nonblackbody surfaces.

1.6.2 Configuration Factor Properties

The configuration factor of a surface only reflects the effects of the surface's physical geometric characteristics such as its shape, dimensions, and relative positions on the radiation heat transfer. The configuration factor has nothing to do with the temperature or radiation properties (such as emissivity) of the surface under the premise of a diffuse gray surface, uniform temperature, and uniform radiation.

Configuration factor properties derived from its definition and integral equation are as follows.

1. **Reciprocity**

    For any two surfaces i and j with areas A_i and A_j, the configuration factor from surface i to surface j is φ_{ij} and the configuration factor from surface j to surface i is φ_{ji}. The following is then obtained based on Eqs. (1.74a) and (1.74b):

$$\phi_{ij}A_i = \phi_{ji}A_j \tag{1.75}$$

    where there is no special requirement for surface shape, for example, the surface need not be concave. The equation is applicable to surfaces that satisfy the simplified engineering conditions (Section 1.7) of diffuse gray surface, uniform temperature, and uniform radiation.

2. **Additivity**

    The very definition of configuration factor implies additivity. As shown in Fig. 1.14, surface A_2 is composed of the surfaces A_{2A} and A_{2B}, that is, $A_2 = A_{2A} + A_{2B}$.

    The projected radiant energy from A_1 to A_2 is:

$$Q_{12} = Q_{1,2A} + A_{1,2B}$$

    Thus, the configuration factor from A_1 to A_2 is:

$$\varphi_{12} = \frac{Q_{12}}{Q_1} = \frac{Q_{1,2A}}{Q_1} + \frac{Q_{1,2B}}{Q_1} = \varphi_{1,2A} + \varphi_{1,2B} \tag{1.76}$$

    The additivity of the configuration factor simplifies the calculation when determining the configuration factor from a surface to another complicated sur-

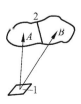

FIGURE 1.14 Additivity of the configuration factor.

face. The complicated surface can be divided into several simple surfaces to calculate each configuration factor before adding the results according to additivity.

3. Integrity

All energy emitted from a nonconcave surface A_1 is Q_1, and the energy projected onto the surfaces surrounding it is represented by A_k $(k = 1,2,...,n)$. According to energy conservation:

$$Q_1 = \sum_{k=1}^{n} Q_{1k}$$

Then,

$$\sum_{k=1}^{n} \phi_{1k} = 1 \tag{1.77}$$

This equation states that the sum of the configuration factors from A_1 to A_k equals 1. In fact, this conclusion can be drawn directly from the definition of "configuration factor." The integrity of the configuration factor is very important when calculating using algebraic methods.

When surface A_1 is concave, some of the radiant energy of A_1 is reflected onto itself, so Eq. (1.77) must add another term, φ_{11}:

$$\phi_{11} + \sum_{k=1}^{n} \phi_{1k} = 1 \tag{1.78}$$

4. Equivalence

As shown in Fig. 1.15a, surfaces A_1 and A_2 centered in differential dA_0 have the same solid angle Ω:

$$\Omega = \frac{A_1 \cos\theta_1}{r_1^2} = \frac{A_2 \cos\theta_2}{r_2^2}$$

The energy radiated from dA_0 to A_1 is:

$$dQ_{01} = \varphi_{d_0,1} E dA_0$$

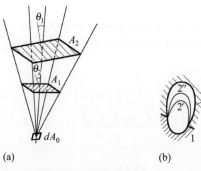

(a)   (b)

FIGURE 1.15 **Equivalence of the configuration factor.** (a) Schematic diagram of derivation of equivalence. (b) Schematic diagram of equivalence.

The energy radiated from dA_0 to A_2 is:

$$dQ_{02} = \varphi_{d_0,2} E dA_0$$

By definition, $dQ_{01} = dQ_{02}$ when the solid angle is the same. Thus:

$$\varphi_{d_0,1} = \varphi_{d_0,2} \tag{1.79}$$

This is the equivalence of configuration factor, which reveals that as long as the solid angles of objects from dA_0 are the same, the configuration factors from dA_0 to the objects are also the same, regardless of distance and shape. As shown in Fig. 1.15b, $\varphi_{12} = \varphi_{12'} = \varphi_{12''}$. This property can be used to calculate the configuration factor of curved surfaces or irregular surfaces, because the surface can be substituted by its projection plane.

1.6.3 Configuration Factor Calculation

When calculating radiation heat transfer, analysis and calculation of configuration factor is a crucial consideration, or even the primary consideration. Generally, in engineering and scientific calculations, calculating configuration factors is highly complicated—sometimes analytical methods are not sufficient, and graphical or numerical methods are necessary. The graphic method and numerical method based on Monte-Carlo's method are not discussed here, but details are available in previous research [18,20].

The analytical method for calculating the configuration factor includes the integral method, differential method, and algebraic method. The integral method can be further divided into the direct integral method and loop integral method. The direct integral method calculates according to the definition of the configuration factor, that is, by substituting geometric parameters into Eqs. (1.69) and (1.74a) to obtain the expression of the configuration factor after multiple integrations. This is the most commonly used method, but also the most complicated.

1. Integral method
   a. Configuration factor of two differential surfaces

Example 1.1
As shown in Fig. 1.16, find the configuration factor between two differential surfaces on two narrow strips with parallel normal characteristics.

Solution
According to the equation derived above, the configuration factor of two differential surfaces is:

$$\varphi_{d_1,d_2} = \frac{\cos\beta_1 \cos\beta_2 dA_2}{\pi l^2} = \cos\beta_1 d\Omega_1$$

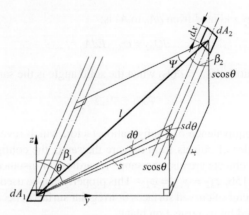

FIGURE 1.16   **Calculation of the configuration factor between two differential surfaces on two parallel narrow strips.**

As Fig. 1.16 suggests:

$$l^2 = s^2 + x^2$$

and,

$$\cos\beta_1 = \frac{s\cos\theta}{l} = \frac{s\cos\theta}{(s^2+x^2)^{1/2}}$$

Consider the solid angle of dA_2 from dA_1:

$$d\Omega_1 = \frac{\text{projected area of } dA_2}{l^2} = \frac{\text{projected width of } dA_2 \times \text{projected length of } dA_2}{l^2}$$

$$= \frac{(sd\theta)(dx\cos\psi)}{l^2} = \frac{(sd\theta dx\cos\psi)}{l^2}$$

Because

$$\cos\psi = \frac{s}{l}$$

$$d\Omega_1 = \frac{s^2 d\theta dx}{l^3} = \frac{s^2 d\theta dx}{(s^2+x^2)^{3/2}}$$

then,

$$\varphi_{d_1,d_2} = \frac{s\cos\theta}{(s^2+x^2)^{1/2}} \frac{1}{\pi} \frac{s^2 d\theta dx}{(s^2+x^2)^{3/2}}$$

and finally,

$$\varphi_{d_1,d_2} = \frac{s^3\cos\theta d\theta dx}{\pi(s^2+x^2)^2} \tag{1.80}$$

This is the equation for the configuration factor from dA_1 to dA_2.

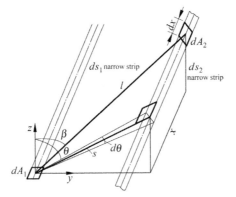

FIGURE 1.17 **The configuration factor from differential surface dA_1 to narrow infinite strip ds_2.**

If dA_2 is a narrow infinite strip, integrate x from $-\infty$ to $+\infty$, then the configuration factor from the differential surface dA_1 to the narrow infinite strip ds_2 can be calculated. Its geometric shape and relative position are shown in Fig. 1.17.

$$\varphi_{d_1,d_2} = \frac{s^3\cos\theta d\theta}{\pi} \int_{-\infty}^{+\infty} \frac{dx}{(s^2+x^2)^2}$$

$$= \frac{s^3\cos\theta d\theta}{\pi} \left[\frac{x}{2s^2(s^2+x^2)} + \frac{l}{2s^2}\arctan\frac{x}{s} \right]_{-\infty}^{+\infty}$$

$$= \frac{\cos\theta d\theta}{2}$$

$$= \frac{1}{2}d(\sin\theta) \tag{1.81}$$

where θ is on plane yz. Regardless of the way the position of dA_1 varies on narrow strip ds_1, Eq. (1.81) can be used to calculate the coefficient. So the configuration factor between two infinite narrow strips with parallel normal is also $\frac{1}{2}d(\sin\theta)$, and:

$$\varphi_{d_1,d_2} = \frac{1}{2}d(\sin\theta)$$

b. Configuration factor between a differential surface and a finite surface
   The configuration factor between a differential surface and a finite surface can be derived according to the definition of the configuration factor and the configuration factor equation of differential surfaces. Consider the radiation heat transfer between isothermal differential surface dA_1 and isothermal blackbody surface A_2, as shown in Fig. 1.18.

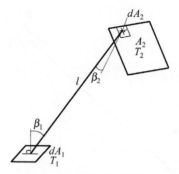

FIGURE 1.18   Radiation heat transfer between dA_1 and A_2.

Total energy emitted from dA_1 is:

$$dQ_1 = \sigma T_1^4 dA_1$$

The energy radiated from dA_1 on dA_2 is:

$$dQ_{d_1,d_2} = \sigma T_1^4 \frac{\cos\beta_1 \cos\beta_2}{\pi l^2} dA_1 dA_2$$

Integrating the finite surface A_2 provides the energy from dA_1 to A_2:

$$dQ_{d_1,2} = \int_{A2} \sigma T_1^4 \frac{\cos\beta_1 \cos\beta_2}{\pi l^2} dA_1 dA_2$$

According to the definition of the configuration factor:

$$\varphi_{d_1,2} = \frac{dQ_{d_1,2}}{dQ_1}$$

$$= \int_{A_2} \frac{\sigma T_1^4 \dfrac{\cos\beta_1 \cos\beta_2}{\pi l^2} dA_1 dA_2}{\sigma T_1^4 dQ A_1}$$

$$= \int_{A_2} \frac{\cos\beta_1 \cos\beta_2}{\pi l^2} dA_2 \qquad (1.82)$$

The expression in integral brackets is simply the configuration factor between dA_1 and dA_2, that is, φ_{d_1,d_2}, thus Eq. (1.82) can be rewritten as follows:

$$\varphi_{d_1,2} = \int_{A_2} \varphi_{d_1,d_2} \qquad (1.83)$$

Example 1.2

As shown in Fig. 1.19a, differential surface dA_1 is perpendicular to the disc A_2 of radius r_0. Given the expression of the configuration factor from dA_1 to dA_2 using parameters h, l, and r_0.

Solution

First, substitute the known parameters into Eq. (1.82) and then simplify it.

To determine $\cos\beta_1$, $\cos\beta_2$, and l, two auxiliary charts are plotted as shown in Fig. 1.19b and c.

The area of differential surface dA_2 can be calculated by the radius r and angle θ of the disc:

$$dA_2 = rdrd\theta$$

Using the auxiliary chart in Fig 1.19b:

$$\cos\beta_1 = \frac{s + r\cos\theta}{l}$$

$$\cos\beta_2 = \frac{h}{l}$$

Using the auxiliary chart in Fig. 1.19c:

$$l^2 = h^2 + b^2$$

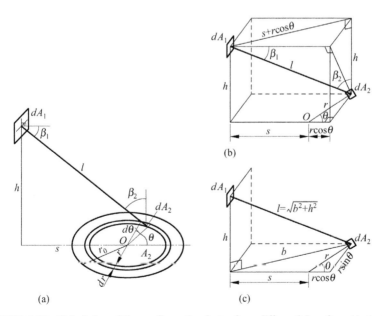

(b)

(a)

(c)

FIGURE 1.19 **Calculation of the configuration factor from differential surface dA_1 to disc A_2.** (a) Geometrical relationship between differential surface and disk. (b) Auxiliary chart for determining $\cos\beta_1$, $\cos\beta_2$. (c) Auxiliary chart for determining l.

where

$$b^2 = (s + r\cos\theta)^2 + (r\cos\theta)^2$$
$$= s^2 + r^2 + 2sr\cos\theta$$

Substituting the above equations into Eq. (1.82) yields:

$$\varphi_{d_1,2} = \int_{A_2} \frac{\cos\beta_1\cos\beta_2}{\pi l^2}dA_2 = \int_{A_2} \frac{h(s + r\cos\theta)}{\pi l^4}rdrd\theta$$

$$= \frac{h}{\pi} \int_{r=0}^{r_0} \int_{\theta=0}^{2\pi} \frac{r(s + r\cos\theta)}{(h^2 + s^2 + r^2 + 2sr\cos\theta)^2}drd\theta$$

According to shape symmetry, dimensionless variables are defined as follows:

$$H = h/s, R = r_0/r, P = r/s$$

Divide the numerator and denominator by s^4, and finally:

$$\varphi_{d_1,2} = \frac{2h}{\pi} \int_{r=0}^{r_0} \int_{\theta=0}^{\pi} \frac{r(s + r\cos\theta)}{(h^2 + r^2 + s^2 + 2rs\cos\theta)^2}d\theta dr$$

$$= \frac{2H}{\pi} \int_{P=0}^{R} \int_{\theta=0}^{\pi} \frac{P(1 + P\cos\theta)}{(H^2 + P^2 + 1 + 2P\cos\theta)^2}d\theta dP \qquad (1.84)$$

$$= \frac{H}{2}\left\{\frac{H^2 + R^2 + 1}{\left[(H^2 + R^2 + 1)^2 - 4R^2\right]^{1/2}} - 1\right\}$$

---

c. Configuration factor between two finite surfaces
As shown in Fig. 1.20, there are two blackbody surfaces A_1 and A_2, and dA_1, dA_2 are the differential surfaces on A_1 and A_2, respectively.

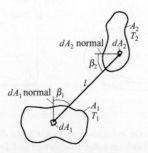

FIGURE 1.20 Radiation heat transfer between surfaces A_1 and A_2.

The energy radiated to dA_2 from dA_1 is:

$$d^2Q_{d_1,d_2} = \sigma T^4 \frac{\cos\beta_1\cos\beta_2}{\pi l^2}dA_1dA_2$$

Then, integrating over the finite area for A_1 and A_2 yields:

$$Q_{12} = \int\limits_{A_1}\int\limits_{A_2} \sigma T^4 \frac{\cos\beta_1\cos\beta_2}{\pi l^2}dA_1dA_2$$

The total energy leaving A_1 is $\sigma A_1 T^4$, so the configuration factor from A_1 to A_2 is:

$$\varphi_{12} = \frac{\displaystyle\int\limits_{A_1}\int\limits_{A_2}\sigma T^4 \frac{\cos\beta_1\cos\beta_2}{\pi l^2}dA_1dA_2}{\sigma A_1 T^4} = \frac{1}{A_1}\int\limits_{A_1}\int\limits_{A_2}\frac{\cos\beta_1\cos\beta_2}{\pi l^2}dA_1dA_2$$

Because

$$\varphi_{d_1,d_2} = \frac{\cos\beta_1\cos\beta_2}{\pi l^2}dA_2$$

then,

$$\varphi_{12} = \frac{1}{A_1}\int\limits_{A_1}\int\limits_{A_2}\varphi_{d_1d_2}dA_1$$

also,

$$\varphi_{d_1,2} = \int\limits_{A_2}\varphi_{d_1,d_2}$$

so:

$$\varphi_{12} = \frac{1}{A_1}\int\limits_{A_1}\varphi_{d_1,2}dA_1 \tag{1.85}$$

Similarly, the configuration factor from A_1 to A_2 is:

$$\varphi_{21} = \frac{1}{A_2}\int\limits_{A_2}\varphi_{d_2,1}dA_2 \tag{1.86}$$

According to the interchangeability of the configuration factor:

$$A_1\varphi_{12} = A_2\varphi_{21}$$

Then,

$$\varphi_{21} = \frac{A_1}{A_2}\varphi_{12} = \frac{A_1}{A_2}\frac{1}{A_1}\int\limits_{A_1}\varphi_{d_1,2}dA_1$$

$$= \frac{1}{A_2}\int\limits_{A_2}A_2\varphi_{2,d_1} = \int\limits_{A_1}\varphi_{2,d_1} \tag{1.87}$$

Example 1.3

There are two flats with finite width, infinite length, and intersected angle of α, as shown in Fig. 1.21. The configuration factor of these two flats is identified as follows.

Solution

The solution can be divided into two steps: first, calculating the configuration factor from the narrow strip dx to the surface with the width of s, $\varphi_{dx,s}$; then obtaining $\varphi_{s',s}$ by integrating x.

Draw the auxiliary chart as shown in Fig. 1.21b, then, according to Eq. (1.81):

$$\varphi_{dx,dz} = \frac{1}{2}d(\sin\theta)$$

where angle θ is in the normal plane of the two differential narrow strips, which is positive in the clockwise direction of the normal line of dx.

The configuration factor from dx to the surface with the width of s is:

$$\varphi_{dx,s} = \int_{z=0}^{s} \varphi_{dx,dz} = \int_{\theta=-\frac{\pi}{2}}^{0} \frac{1}{2}d(\sin\theta) + \int_{0}^{\theta'} \frac{1}{2}d(\sin\theta)$$

$$= \frac{\sin\theta}{2}\Big|_{\theta=-\frac{\pi}{2}}^{0} + \frac{\sin\theta}{2}\Big|_{\theta=0}^{\theta'} = \frac{1}{2} + \frac{1}{2}\sin\theta'$$

The expression of function $\sin\theta'$ from Fig. 1.21b is identified as:

$$\sin\theta' = \frac{b}{c} = \frac{s\cos\alpha - x}{(x^2 + s^2 - 2xs\cos\alpha)^{1/2}}$$

Thus,

$$\varphi_{dx,s} = \frac{1}{2} + \frac{s\cos\alpha - x}{2(x^2 + s^2 - 2xs\cos\alpha)^{1/2}}$$

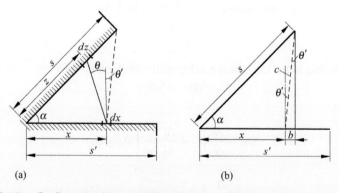

(a)   (b)

FIGURE 1.21   Configuration factor calculation between two flats. (a) Cross-section view of two flats with intersected angle of α. (b) Auxiliary chart for determining $\sin\theta'$.

According to interchangeability, another configuration factor is:

$$\varphi_{s,dx} = \frac{dx}{s}\varphi_{dx,s} = \left[\frac{1}{2s} + \frac{\cos\alpha - \dfrac{x}{s}}{2(x^2 + s^2 - 2xs\cos\alpha)^{1/2}}\right]dx \qquad (1.88)$$

From Eq. (1.87):

$$\varphi_{s,s'} = \int\limits_{x=0}^{s'} \varphi_{s,dx} = \int\limits_{0}^{s'}\left[\frac{1}{2s} + \frac{\cos\alpha - \dfrac{x}{s}}{2(x^2 + s^2 - 2xs\cos\alpha)^{1/2}}\right]dx$$

According to subject meaning, $s = s'$. Define a dimensionless variable $\dfrac{x}{s} = X$, then the above equation can be rewritten as follows:

$$\varphi_{s,s'} = \int\limits_{0}^{1}\left[\frac{1}{2} + \frac{\cos\alpha - X}{2(X^2 + 1 - 2X\cos\alpha)^{1/2}}\right]dX$$

Integrate the above equation, then:

$$\varphi_{s,s'} = 1 - \left(\frac{1 - \cos\alpha}{2}\right)^{\frac{1}{2}} = 1 - \sin\frac{\alpha}{2} \qquad (1.89)$$

Eq. (1.89) suggests that the configuration factor of flats of the same width is only related to the intersected angle. Because the areas of two flats are the same, according to interchangeability:

$$\varphi_{s,s'} = \varphi_{s',s}$$

The above derivations exist under the premise that the surfaces are isothermal blackbody surfaces. Their applicability will be discussed in detail in Section 1.7.

---

2. Algebraic analysis method

The integral method determines the configuration factor through integral operation, thus it is rather difficult to calculate for complicated geometries. Determining the geometric configuration factor with some known configuration factor expressions can be realized by algebraic analysis, which is quite simple and successfully avoids integral operation. The algebraic analysis method is based on the energy conservation law and the interchangeability of the configuration factor. For example, as shown in Fig. 1.22, if the configuration factor φ_{12} from surface A_1 to surface A_2 is known, and the area of A_2 is the sum of A_3 and A_4, according to the energy conservation law:

$$\varphi_{12} = \varphi_{1,(3+4)} = \varphi_{13} + \varphi_{14}$$

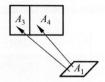

FIGURE 1.22 Example of configuration factor calculation with the algebraic analysis method.

where $A_3 + A_4 = A_2$. If φ_{12} and φ_{14} are known, the calculation of φ_{14} or φ_{41} becomes very easy. φ_{14} can be calculated as follows:

$$\varphi_{14} = \varphi_{12} - \varphi_{13}$$

Then, according to the interchangeability of the configuration factor, φ_{41} can be calculated as:

$$\varphi_{41} = \frac{A_1}{A_4}\varphi_{14} = \frac{A_1}{A_4}(\varphi_{12} - \varphi_{13}) \tag{1.90}$$

Example 1.4
Determine the radiation configuration factor φ_{14} in Fig. 1.23.

Solution
Draw an auxiliary line from the surfaces of area A_2 and area A_3 to obtain:

$$A_{12} = A_1 + A_2, A_{34} = A_3 + A_4$$

The radiation configuration factor must be calculated with known results. According to the energy conservation law:

$$A_{12}\varphi_{12,34} = A_{12}\varphi_{12,3} + A_{12}\varphi_{12,4}$$

$$A_{12}\varphi_{12,4} = A_1\varphi_{14} + A_2\varphi_{24}$$

$$A_2\varphi_{2,34} = A_2\varphi_{23} + A_2\varphi_{24}$$

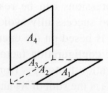

FIGURE 1.23 The configuration factor between two nonadjacent rectangular surfaces.

From the relations above, the following can be derived:

$$A_1\varphi_{14} = A_{12}\varphi_{12,4} - A_2\varphi_{2,24}$$

$$= (A_{12}\varphi_{12,34} - A_{12}\varphi_{12,3}) - (A_2\varphi_{2,34} - A_2\varphi_{23})$$

$$\varphi_{14} = \frac{1}{A_1}(A_{12}\varphi_{12,34} - A_{12}\varphi_{12,3} - A_2\varphi_{2,34} + A_2\varphi_{23}) \tag{1.91}$$

where $\varphi_{12,34}, \varphi_{12,3}, \varphi_{2,34},$ and φ_{23} can be calculated by the known equations, or determined from charts and tables.

Example 1.5

Determine the configuration factor from differential surface dA_1 to coaxial parallel ring A_1 (Fig. 1.24). The inner radius and outer radius of the ring are R_1 and R_2, and the distance between the two surfaces is D.

Solution

By checking the chart or integration, the configuration factor from the differential surface to the coaxial parallel rings is:

$$\varphi_{dA_1,A_{R_1}} = \frac{R_1^2}{R_1^2 + D^2}$$

$$\varphi_{dA_1,A_{R_2}} = \frac{R_2^2}{R_2^2 + D^2}$$

The area of the ring A_2 is calculated as $A_2 = A_{R2} - A_{R1}$, thus:

$$\varphi_{dA_1,A_2} = \varphi_{dA_1,A_{R_2}} - \varphi_{dA_1,A_{R_2}}$$

$$= \frac{R_2^2}{R_2^2 + D^2} - \frac{R_1^2}{R_1^2 + D^2} \tag{1.92}$$

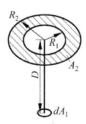

FIGURE 1.24 **The configuration factor from differential surface dA_1 to coaxial parallel ring A_1.**

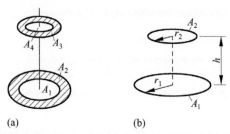

(a)   (b)

FIGURE 1.25   The configuration factor between two parallel coaxial rings A_2 and A_3. (a) Configuration factor between two parallel coaxial rings A_2 and A_3. (b) Configuration factor between two parallel coaxial disks.

Example 1.6

Determine the configuration factor between two parallel coaxial rings A_2 and A_3. Relevant geometric sizes are shown in Fig. 1.25a.

Solution

By checking the chart or integration, the configuration factor between two parallel coaxial discs is:

$$\begin{cases} R_1 = \dfrac{r_1}{h} \\[2mm] R_2 = \dfrac{r_2}{h} \\[2mm] X = 1 + \dfrac{1+R_2^2}{R_1^2} \\[2mm] \varphi_{12} = \dfrac{1}{2}\left[X - \sqrt{X^2 - 4\left(\dfrac{R_2}{R_1}\right)^2} \right] \end{cases} \tag{1.93}$$

This question can be solved by the known configuration factor in Eq. (1.93):

$$\varphi_{23} = \varphi_{2,(3+4)} - \varphi_{24} \tag{1.94a}$$

Because

$$A_2\varphi_{2,(3+4)} = (A_3 + A_4)(\varphi_{(3+4),(1+2)} - \varphi_{(3+4),1})$$
$$= (A_3 + A_4)\varphi_{(3+4),(1+2)} - (A_3 + A_4)\varphi_{(3+4),1}$$

thus,

$$A_2\varphi_{2,(3+4)} = (A_3 + A_4)\varphi_{(3+4),(1+2)} - A_1\varphi_{1,(3+4)}$$

and

$$\varphi_{24} = \frac{A_4}{A_2}\varphi_{42} = \frac{A_4}{A_2}(\varphi_{4,(1+2)} - \varphi_{41})$$

$$= \frac{1}{A_2}\left[(A_1 + A_2)\varphi_{(1+2),4} - A_1\varphi_{14}\right]$$

By substituting φ_{24} and $\varphi_{2,(3+4)}$ into Eq. (1.94a), the following is finally obtained:

$$\varphi_{23} = \frac{A_1 + A_2}{A_2}\left[\varphi_{(1+2),(3+4)} - \varphi_{(1+2),4}\right] - \frac{A_1}{A_2}\left[\varphi_{1,(3+4)} - \varphi_{14}\right] \tag{1.94b}$$

where $\varphi_{(1+2),(3+4)}$, $\varphi_{(1+2),4}$, $\varphi_{1,(3+4)}$, and φ_{14} all are the radiation configuration factors between two parallel coaxial discs.

---

3. Common calculation formulas of configuration factor (See Appendix B.).

1.7 SIMPLIFIED TREATMENT OF RADIATIVE HEAT EXCHANGE IN ENGINEERING CALCULATIONS [12]

In order to successfully apply fundamental thermal radiation laws to engineering application, it remains necessary to simplify the actual problem by way of a few reasonable assumptions.

The primary tenets of simplification are the applicability of relevant laws, and the required accuracy of solutions. Essentially, the goal of simplification is to apply laws simply and effectively to engineering practice with acceptable accuracy. Restrictions are set based on mastery of engineering data, for example, determining whether there are sufficient experimental data to validate the simplification method, or whether there are sufficient experimental data to provide empirical equations for simplification. The following will further discuss the simplification treatment commonly used in engineering.

1.7.1 Simplification Treatment of Radiation Heat Transfer in Common Engineering Calculations

Based on the fundamental concepts discussed earlier, the following simplified conditions are typically utilized in heat and power engineering fields:

1. Isothermal surface
2. Diffuse gray surface
3. Homogeneous surface radiation properties
4. Homogeneous surface radiosity

1.7.2 Discussion on Simplified Conditions

Unless otherwise noted, the above conditions are assumed to be true by default in this book; they likewise hold true in common heat transfer calculations for furnaces. The following provides explanations on four relevant simplification conditions.

1. The isothermal surface assumption is not only applicable to surfaces, but also to media such as flue gas. If the entire body's isothermal assumption introduces significant error, the surface can be divided into several partitions and be calculated separately. The division method depends on the requirement for calculation accuracy. If the number of partitions is large, numerical calculation is required.

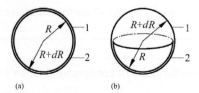

FIGURE 1.26 Surfaces satisfying the first three conditions. (a) Satisfying conditions (4).
(b) Not satisfying condition (4).

2. Diffuse surfaces are only used to simplify surfaces, while graybodies are applicable to both surfaces and media. If the diffuse gray assumption introduces significant error, the surface can be divided into several partitions which can be considered diffuse gray and calculated separately. The division method is usually the same as that used in the isothermal surface assumption.

3. When the assumption of homogeneous surface radiation properties or homogeneous surface radiosity is not satisfied, the above division method can also be applied. (Please note that special calculation is required if any of these conditions is not satisfied.) The treatment principles for these four conditions are the same, but the specific results (such as the number of partitions or partition size) may vary due to different influencing factors. Theoretically, the mathematical union method should be applicable, in which division satisfies the most stringent requirements; however, in engineering practice, simplified calculation usually seizes on major contradictions and neglects minor ones.

4. A surface that satisfies the first three conditions does not always satisfy the fourth condition. For example, the surface in Fig. 1.26a satisfies the first three conditions as well as the fourth, but the surface in Fig 1.26b does not.

The difference between the two can be explained through the definitions and physical implications of related concepts. If surface radiosity is to be uniformly distributed, the reflected radiation should be uniformly distributed; if surface reflected radiation is to be uniformly distributed, irradiation should be uniformly distributed; if surface irradiation is to be uniformly distributed, the radiosity of all radiation source surfaces must be uniformly distributed, and the radiant energy projected uniformly on all parts of the surface. According to the definition of the configuration factor, the fraction of energy projected onto each part of the surface is the same, so the configuration factors are the same. Assume that surface i and surface j are two surfaces which form a closed system—the configuration factor from differential element di on surface i to surface j then equals the configuration factor between the two surfaces:

$$\varphi_{d_i,j} = \varphi_{i,j}$$

This is the mathematical expression of the fourth condition described earlier.

Notably, the characteristic of radiation heat transfer and calculation equations of the configuration factor described earlier are only true when these simplified conditions are met. Otherwise, the configuration factor is unable to represent the energy allocation relation with a purely geometric relation.

In general, if the fourth condition is not satisfied, the above simplified algorithms can still be used because the resultant error is negligible; thus, the fourth condition is usually neglected. Example 3.1 in Chapter: Radiation Heat Exchange Between Isothermal Surfaces analyzes this phenomenon in detail.

Chapter 2

Emission and Absorption of Thermal Radiation

Chapter Outline

Generally speaking, a medium can exist in any of the four states of matter: solid, liquid, gas, or plasma. As for thermal radiation, it is important to note that most solids and liquids are strongly absorbing media (that is, they very readily absorb thermal radiation). The thickness of a solid metal thermal absorption layer is only a few hundred nanometers, and is usually only a fraction of a millimeter for insulating materials. The thermal radiation characteristics of plasma, which consists of a large number of charged particles, are quite atypical and not involved in furnaces. Solid, liquid, and plasma media are not further discussed in this book. By contrast, the thermal radiation absorption capacity of gas, which is quite weak, is the main topic of this chapter.

The main components of air are O_2 and N_2, both of which are transparent to thermal radiation. The thermal radiation medium in a furnace is usually a combination of fuel combustion products, such as CO_2, H_2O, CO, SO_2, and NO_x in the flue gas, and, of course, N_2 and O_2 as well. The components of the media in the furnace and the surrounding air are different, so their radiation characteristics are also different. The flue gas in the combustion chamber also contains fly ash, carbon black, coke, and other solid particles in a solid-fuel-fired boiler,

Theory and Calculation of Heat Transfer in Furnaces. http://dx.doi.org/10.1016/B978-0-12-800966-6.00002-8

or coke particles in a liquid- or gas-fuel-fired boiler. The flue gas in a boiler designed to combust municipal solid waste and industrial waste, may contain HCl, Cl_2, HF, or other components. (The contents of a fluidized bed which combusts solid fuel are so unique that they are discussed separately, in chapter: Heat Transfer in Fluidized Beds.)

This chapter will attempt to answer several important questions: Which components in the medium are most closely related to thermal radiation in the system? How do they emit and absorb thermal radiative rays, and how does an engineer quantify their characteristics?

2.1 EMISSION AND ABSORPTION MECHANISMS

2.1.1 Molecular Spectrum Characteristics

The atomic spectrum of a chemical element is induced by the energy state transition of its electrons moving around its nucleus. The spectrum is lineal, and each line has a corresponding frequency. Spectral line distributions are relatively rare on the whole, and only become dense on the end. The molecular spectrum differs significantly from the atomic spectrum, which is much more complex, mainly due to the general complexity of molecular structures and the diversity of internal molecular motion.

In addition to the motion of electrons, "molecular motion" also refers to translation, rotation, and vibration, each corresponding to a certain energy. Translational energy is caused by the thermal motion of molecules and is a function of temperature. The translation of molecules does not alter molecular dipole moment; as such, it does not include photon emission or absorption. The energy related to thermal radiation includes extranuclear electron energy, vibrational atom energy, and molecular rotation energy. Changes in energy cause the emission or absorption of thermal radiative rays.

Nonpolar molecules in the air or flue gas, such as O_2, N_2, and H_2, are symmetrical diatomic molecules. Their positive and negative charge coincides with the center of the molecule and their dipole moments are zero, so their electronics are distributed symmetrically. Rotation of the molecules and vibration of the internal nuclei do not cause any changes in the electric field surrounding the molecules or any energy state transition, thus no photon emission or absorption occurs. Nonpolar molecules can therefore be considered transparent to thermal radiation.

As for polar molecules such as CO_2, although they have symmetrical centers and zero permanent dipole moments, the center of their positive and negative charges may be relatively offset by internal molecular vibration, resulting in instantaneous changes in dipole moment. Therefore, their energy state transitions lead to emission or absorption of thermal radiative rays.

Polar molecules in certain media, such as H_2O, CO_2, CO, and HCl, have incongruous positive and negative charge centers in the molecule, thus the dipole moments of the molecules are nonzero. When the vibrational and rotational energy states of the polar molecules change, there is emission and absorption of radiant energy. To this effect, polar molecules in the medium are opaque—that is, they have radiation and absorption capacity.

2.1.2 Absorption and Radiation of Media

Studying radiation heat transfer usually involves referring to solely gas media. Liquid media, due to their relatively large density and very small mean beam length, show negligible radiation heat transfer compared with heat transfer through convection and conduction. Of course, a gas medium may contain solid particles (such as coke or ash) but in fairly low concentrations – generally between a few hundred milligrams and a few grams per cubic meter in the furnace of a pulverized coal boiler or grate boiler, and up to several hundred grams to several kilograms per cubic meter in the dilute zone of a circulating fluidized bed boiler. The radiation, absorption, and scattering phenomena of gas media are the focus of this book. To illustrate the mechanisms of media absorption and radiation in the industry, the simplest case (ie, a medium of purely flue gas) is discussed first; specifically, we will explain why some molecules and atoms in a flue gas have radiation and absorption capacity, while others do not.

Quantum mechanics tells us that thermal radiation involves photons within infrared and visible light frequencies, called thermal radiative rays, which are the basic unit of radiant energy. When a medium emits photons (thermal radiative rays), microscopically, the atoms (or molecules) of the medium emit photons due to energy state transition in the atoms (or molecules). Fig. 2.1 shows the three types of absorption and radiation that form when a medium absorbs or emits photons (ie, bound state-bound state, bound state-free state, and free state-free state). E_1 represents zero energy level, which is the ground state of the medium's particles (atoms or molecules) and is also the lowest level of bound states. E_I is the ionization potential of the particles, and energy levels between E_1 and E_I (E_2, E_3, etc.) are bound state levels. Ionization potential energy E_I is the minimum energy that a particle requires for ionization from the ground state.

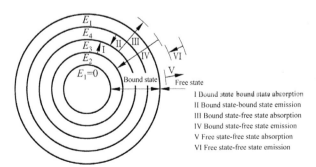

FIGURE 2.1 Schematic of energy-level transition due to emission and absorption.

Particle energy levels in the bound state are in a discrete, steady state, that is, the energy difference between the different levels is fixed. Thus, the generated or absorbed energy due to bound state- bound state transition is also fixed, and is proportional to photon frequency. See the following:

$$\varepsilon = hv \tag{2.1}$$

where h is Planck's constant (6.63×10^{-34} Js) and v is the photon frequency.

In the case of photon emission due to energy level transition, such as when a photon is emitted due to energy level transition from E_4 to E_3 (as shown in Fig. 2.1), the energy transition can be expressed as follows:

$$\Delta E = E_4 - E_3 \tag{2.2}$$

The photon frequency can be obtained based on Eq. (2.1) as follows:

$$v = \frac{\Delta E}{h} = \frac{E_4 - E_3}{h} \tag{2.3}$$

The frequencies of the photons emitted from bound state-bound state transitions are constant, thus the transitions between different energy levels of bound states generate a series of discrete spectral lines at different frequencies. Conversely, when a molecule or atom absorbs a photon, it can transit from its bound state to a higher energy level. The energy levels are discrete, so the frequency of the absorbed photon is discontinuous. Ideally, the transitions between bound states will form the same spectral line regardless of whether photons are being absorbed or emitted. In actuality, they are widened due to various causes (including natural broadening, Doppler broadening, and Stack broadening). The spectral lines have a certain width, and the conditions become complicated.

There is no ionization or complex of ions and electrons during bound state-bound state energy level transition; when an atom or a molecule absorbs or emits a photon under these conditions, it is actually the bound state that is responsible for absorbance or emission. The atom or molecule transitions from a determined bound state to another state which can be rotated, vibrated, or undergo electron or molecular motion.

The vibrational and rotational energy levels of a molecule or atom are, by definition, mutually associated. The superstition of the rotation spectrum from rotational energy level transition, plus the vibration spectrum from the vibration level transition, generates a small spectral band. If the band exists on a continuous interval on average, the vibration-rotational band can be observed. Within the temperature range (500–2000 K) commonly encountered in the industry, emission and absorption are mainly caused by vibrational and rotational transitions. Generally, the rotational transition spectral lines are within the long-wave range (8–1000 μm) and the vibration-rotational transition lines are within the infrared range (1.5–20 μm). Electronic transition is an important cause of emission and absorption, which forms at high temperatures (T > 2000 K) and has spectral lines mainly within the visible spectral range (0.4–0.7 μm) and the adjacent infrared and ultraviolet ranges.

2.2 RADIATIVITY OF ABSORBING AND SCATTERING MEDIA [24,26,30]

2.2.1 Absorbing and Scattering Characteristics of Media

2.2.1.1 Absorption

Absorption is a phenomenon through which the energy of thermal radiative rays shrinks as they pass through the medium—the medium is called an "absorbing medium" for this reason. The radiation and absorption characteristics of the absorbing medium can be summarized as follows:

1. The radiation and absorption processes are carried within the entire volume. As discussed in Section 2.1, microscopically, the absorbing medium contains a large number of molecules or particles capable of absorbing and emitting thermal radiative rays. These molecules or particles emit thermal radiative rays while also absorbing thermal radiative rays continually under certain thermodynamic conditions. Macroscopically, these molecules or particles always fill the entire volume, as they are gaseous. If the density of molecules or particles is low, the magnitude of the beam length of travel and the volume are the same or similar. Therefore, the radiation or absorption of the medium always occurs throughout its entire volume.

2. Absorption and radiation may show a significant amount of selectivity. Radiation and absorption only occur within a certain range of wavelengths in the medium called the "light band." There is neither radiation nor absorption in the medium beyond the light band, as shown in Fig. 2.2.

Theoretically, when it comes to the molecules or particles having the ability to emit and absorb thermal radiative rays, photon emission and absorption are carried out at a specific frequency (spectrum) characterized by their energy level distribution and thermodynamic state; these frequencies determine the range of the light band.

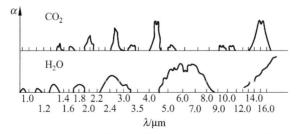

FIGURE 2.2   **Main light band of CO_2 and H_2O (gas).** Note: The main light band of CO_2 is (2.65–2.80, 4.15–4.45, 13.0–17.0) μm; the main light band of H_2O (gas) is (2.55–2.84, 5.6–7.6, 12.0–13.0) μm.

The emission and absorption spectrums of the absorbing medium are not continuous, thus, in effect, this type of medium cannot be considered a gray-body.

2.3   SCATTERING

As indicated in physics, there are four types of scattering: elastic, inelastic, isotropic, and anisotropic. Elastic scattering is the process through which the energy, frequency, and wavelength of the photon remain unchanged due to the scattering of collision, but the motion direction of the photon does change, thereby changing its momentum. Inelastic scattering is a process through which the energy and momentum of a photon both change. Isotropic scattering occurs when scattering energy distribution is equal in all directions; if unequal, the scattering is anisotropic.

When thermal radiative rays pass through a medium, elastic scattering is predominant—that is, the photon energy stays constant but its momentum changes. Scattering, like absorption and radiation, occurs throughout the entire volume of the medium and has additional selectivity.

2.4   ABSORPTION AND SCATTERING OF FLUE GAS

Generally, the flue gases from boilers and industrial furnaces burning fossil fuels contain diatomic gases (such as N_2, O_2, and CO), triatomic gases (CO_2, H_2O, and SO_2), and suspended particles (carbon black, ash, and coke). As mentioned earlier, the radiation and absorption capacities of N_2 and O_2 are very weak, so they are considered transparent to radiation, and the volume concentration of CO is generally low (eg, $50 \times 10^{-6} \sim 300 \times 10^{-6}$), so it can be neglected. The concentration of triatomic gases is high, up to 5–30% or more, and the concentration of suspended particles can reach the level of a few grams to several hundred milligrams per cubic meter of flue gas. Therefore, when it comes to heat transfers, triatomic gases, and suspended particles are the main radiation and absorption components in the flue gas.

The scattering capacity of gases is weak enough to be neglected; however, gases containing suspended particles are considered absorbing-scattering media due to the effects of particle scattering (including reflection, refraction, diffraction, and so on).

2.4.1   Radiation Intensity Characteristics

As defined in chapter: Theoretical Foundation and Basic Properties of Thermal Radiation, radiation intensity is the amount of energy emitted per unit time from the unit surface area normal to the radiation direction through the unit solid angle. [The unit is expressed as $W/(m^2 \cdot sr)$.] Radiation intensity can thus be defined for a solid surface.

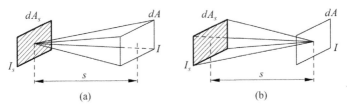

FIGURE 2.3 Schematic of radiation intensity. (a) Heat determined by radiation intensity of emitter, (b) heat determined by radiation intensity of receiver.

Let's assume that there is a surface in the absorbing medium, thus the radiation intensity through the surface can be defined as the energy emitted per unit time per the unit solid angle or per the unit projection area normal to the radiation direction—these two definitions are exactly the same.

An important characteristic of radiation intensity is that it remains constant in the transparent medium in any given direction; the following provides simple proof of this phenomenon.

As shown in Fig. 2.3, the radiation intensity of the radiation source with an area of dA_s is I_s, the area receiving the radiation is dA, dA_s, and dA are separated by s distance, and the medium is transparent. Referring to Fig. 2.3a, heat flux d^2Q can be expressed according to radiation intensity I_s as follows:

$$d^2Q = I_s dA_s d\Omega$$

Since $d\Omega = \dfrac{dA}{s^2}$, then

$$d^2Q = \frac{I_s dA_s dA}{s^2} \tag{2.4}$$

Referring to Fig. 2.3b, the radiation heat flux d^2Q flowing through dA along the direction of s can be described as:

$$d^2Q = I dA d\Omega_s$$

Substitute $d\Omega_s = \dfrac{dA_s}{s^2}$ into the above equation to obtain:

$$d^2Q = \frac{I dA dA_s}{s^2} \tag{2.5}$$

Comparing Eq. (2.4) with Eq. (2.5) gives us: $I = I_s$. Because distance s is arbitrary, the radiation intensity I remains constant throughout the entire medium.

2.4.1.1 Transmission and Absorption of Radiant Energy

The above section discussed the invariant property of radiation intensity in a transparent medium. Does intensity I remain constant through an absorption-scattering medium?

Experimental results show that when monochromatic radiation with wavelength λ moves vertically through an absorbing medium with layer thickness of ds, the change in radiation intensity is proportional to the intensity. See the following:

$$\frac{dI_{\lambda,s}}{ds} = -K_\lambda I_{\lambda,s} \tag{2.6}$$

where K_λ is a proportionality constant, characterizing the radiation intensity attenuation per unit distance, which is called the "monochromatic extinction coefficient" (m^{-1}). K_λ is influenced by the properties, pressure, temperature, and wavelength λ in the gas. The negative sign indicates that the radiation intensity decreases as the thickness of the gas increases. It is important to note that K_λ includes both absorption and scattering effects.

Integral Eq. (2.6) and the radiation intensity at $s = 0$ are set to $I_{\lambda,0}$, that is, if $I_{\lambda,0}$ is the irradiation intensity, then:

$$I_{\lambda,s} = I_{\lambda,0} \exp\left(-\int_0^s K_\lambda \, ds\right) \tag{2.7}$$

This is an experimental law describing the gas absorption process called Bouguer's law.

In Eq. (2.7), when K_λ remains constant, then optical thickness $\delta_{\lambda,s}$ can be defined as $\delta_{\lambda,s} = K_\lambda s$.

To define monochrome transmissivity $\tau_{\lambda,s}$ and monochrome absorptivity $\alpha_{\lambda,s}$, assume K_λ to be constant, then calculate as follows:

$$\tau_{\lambda,s} = \frac{I_{\lambda,s}}{I_{\lambda,0}} = \exp(-K_\lambda s) = e^{-\delta_{\lambda,s}} \tag{2.8}$$

$$\alpha_{\lambda,s} = \frac{I_{\lambda,0} - I_{\lambda,s}}{I_{\lambda,0}} = 1 - e^{-\delta_{\lambda,s}} = 1 - \tau_{\lambda,s} \tag{2.9}$$

Fig. 2.4 shows the curves of $\tau_{\lambda,s}$ and $\alpha_{\lambda,s}$ plotted according to Eqs. (2.8) and (2.9).

The total transmissivity τ_s and total absorptivity α_s can be derived by calculating radiation intensity changes along the entire spectrum as follows:

$$\tau_s = \frac{I_s}{I_0} = \frac{\int_0^\infty I_{\lambda,0} e^{-K_\lambda s} \, d\lambda}{\int_0^\infty I_{\lambda,0} \, d\lambda} \tag{2.10}$$

$$\alpha_s = 1 - \tau_s \tag{2.11}$$

Therefore, the total transmissivity τ_s and the total absorptivity α_s are associated with medium properties (such as K_λ), the spectral distribution of irradiation, and the bulk size of the medium s.

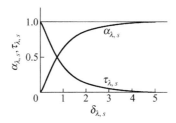

FIGURE 2.4   Curves of $\tau_{\lambda,s}$ and $\alpha_{\lambda,s}$.

By Kirchhoff's law, monochrome emissivity equals monochrome absorptivity ($\varepsilon_{\lambda,s} = \alpha_{\lambda,s}$), so total emissivity can be calculated as follows:

$$\varepsilon_s = \frac{\int_0^\infty E_{b\lambda}\left(1 - e^{-K_{\lambda s}}\right)d\lambda}{\sigma T^4} \tag{2.12}$$

2.4.1.2   Classifying the Absorbing Medium

A medium can be classified into one of the three following categories (graybody, selective, or selective graybody), according to the relation between its monochromatic extinction coefficient K_λ and wavelength λ.

1. Graybody medium

As shown in Fig. 2.5, I_0 is the original spectral distribution of the irradiation in a graybody. As the thickness of the medium crossed by a ray increases, monochromatic radiation intensity I_λ evenly decreases while the monochromatic extinction coefficient K_λ remains constant, that is, the shapes of the curves in Fig. 2.5 remain similar but their heights become smaller. Let $K_\lambda = K$, $\lambda \in [0, \infty]$, then the above fundamental relations can be simplified as follows:

$$dI = -KI_0 ds \tag{2.13}$$

$$I = I_0 e^{-Ks} \tag{2.14}$$

$$\tau = e^{-Ks} \tag{2.15}$$

$$\alpha = \varepsilon = 1 - e^{-Ks} \tag{2.16}$$

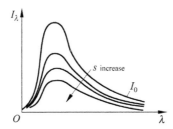

FIGURE 2.5   Attenuation characteristics of gray medium.

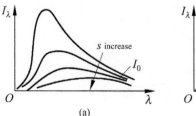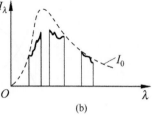

(a)   (b)

FIGURE 2.6   **Attenuation characteristics of selective media.** (a) Continuous spectrum. (b) Discontinuous spectrum.

2. Selective medium

A graybody medium has no selectivity and a continuous spectrum, making it an ideal medium. A real medium, conversely, approximates to a selective medium, which has a continuous spectrum but K_λ varies with λ (the case for flue gas containing small carbon particles), or a discontinuous spectrum, where for some light bands, K_λ varies with λ and for the remaining light bands, it becomes transparent (the case for a real gas).

Fig. 2.6a depicts a continuous spectrum—as the thickness of the medium s crossed by a ray increases, not only does monochromatic radiation intensity I_λ decrease, but monochromatic extinction coefficient K_λ also decreases, making the shape of the curves slope down. Fig. 2.6b depicts the same curves for a discontinuous spectrum.

3. Selective graybody medium

A selective graybody medium is one that has absorbing ability and a constant K_λ for certain light bands, and $K_\lambda = 0$ for other light bands. Because the medium has no absorbing ability for some of its light bands, the projection radiation from the continuous spectrum cannot be absorbed completely, even if for some of the spectral lines the medium is thick enough (even if it is infinite), thus, the absorptivity of the medium must be smaller than 1.

In engineering calculations, most media are considered graybodies – gas-fired boilers and industrial furnaces, as opposed to pulverized coal boilers (PCBs) or oil furnaces, are often considered selective graybodies due to the selectivity of flue gas.

2.4.2   Exchange and Conservation of Radiant Energy

The radiation characteristics of absorption-scattering media were discussed in Section 2.2; based on this information, the following section discusses the radiant energy transfer law of media.

2.4.2.1   Differential Volume Radiation

Fig. 2.7 depicts a differential medium dV with extinction coefficient K (which remains constant in the dV) situated in the center of a large, hollow blackbody

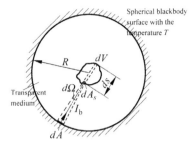

FIGURE 2.7   Differential volume radiation.

sphere, the wall temperature of which is T. The medium filled between dV and the sphere wall is transparent, and dA is a differential area of the wall of the blackbody sphere. The temperature of the entire system (including dV) is uniform.

The radiation intensity I_b from dA to dA_s is attenuated as follows when moving into dV for ds distance:

$$I = I_b e^{-Kds} \tag{2.17}$$

The change in the radiation intensity is thus:

$$dI = I - I_b = -I\left(1 - e^{-Kds}\right) \approx -I_b Kds \tag{2.18}$$

(first-order series expansion)

The radiant energy emitting from dA and absorbed by differential volume $dA_s ds$ is:

$$d^3 Q_\alpha = -dI dA_s d\Omega = I_b Kds dA_s d\Omega \tag{2.19}$$

where $d\Omega = dA/R^2$, thus the radiant energy emitting from dA and absorbed by dV is:

$$d^2 Q_\alpha = \int_{dV} d^3 Q_\alpha = I_b K \, dV \, d\Omega \tag{2.20}$$

The radiant energy projected onto dV from the entire sphere wall and absorbed by dV is:

$$dQ_\alpha = \int_{\Omega = 4\pi} d^2 Q_\alpha = I_b K \, dV \int_{4\pi} d\Omega = 4\pi K I_b \, dV = 4K E_b \, dV \tag{2.21}$$

where $E_b = \varepsilon T^4$ is blackbody emissive power.

Because the temperature of the entire system is uniform, it can be calculated based on the energy balance equation, where the radiation from dV is:

$$d^2 Q_s = d^2 Q_\alpha = 4K E_b dV \tag{2.22}$$

Without considering the gas medium's capacity to absorb its own radiation, if the radiation intensity emitting from dV in all directions is uniform, then:

$$dI_s = \frac{dQ_s}{4\pi dA_n} = \frac{4KE_b dV}{4\pi dA_n} = KI_b dS \qquad (2.23)$$

where $dA_n = \dfrac{dV}{ds}$ is the projection area of dV on the plane normal to the emitting direction, and ds is the thickness of dV parallel to the emission direction.

For a nongray medium, Eqs. (2.22) and (2.23) can be rewritten as follows:

$$d^2 Q_{s\lambda} = 4K_\lambda E_{b\lambda} dV d\lambda \qquad (2.24)$$

$$dI_{s\lambda} = K_\lambda I_{b\lambda} ds \qquad (2.25)$$

The calculation equations for radiant energy and radiation intensity of the absorbing differential medium dV as derived earlier are the foundation for establishing the radiative transfer equation.

2.4.2.2 Radiation Transfer Equation

We now know that radiant energy is attenuated when passing through an absorption-scattering medium, obeying Bouguer's law, while radiant energy is also enhanced because of the medium's own radiation. Thus, a radiative transfer equation that considers both effects must be established to account for all radiant energy transfer characteristics in the medium.

Let's first consider an example using a purely absorbing medium, the attenuation K of which is a function of temperature considered to be constant, thus:

Medium absorption: $dI_\alpha = -KIds$

Self-radiation: $dI_s = -KI_b ds$

where I_b is the radiation intensity of a blackbody at the medium's temperature. The total variation of the radiation intensity is expressed as follows:

$dI = dI_s + dI_\alpha = (I_b - I)Kds$

Define optical thickness δ as $\delta = Ks$, then:

$$\frac{dI}{d\delta} + I = I_b \qquad (2.26)$$

which is the differential form of the radiative transfer equation. By integrating it from $\delta = 0$ to δ_1, we obtain, of course, the integral form of the radiative transfer equation:

$$I = I_1 e^{-\delta_1} + e^{-\delta_1} \int_0^{\delta_1} I_b e^\delta \, d\delta \qquad (2.27)$$

where I_1 is irradiation. For nongray media, the monochromatic radiative transfer equations can be written as:

$$\frac{dI_\lambda}{d\delta_\lambda} + I_\lambda = I_{b\lambda} \tag{2.28}$$

$$I_\lambda = I_{1\lambda} e^{-\delta_{\lambda 1}} + e^{-\delta_{\lambda 1}} \int_0^{\delta_{\lambda 1}} I_{b\lambda} e^{\delta_\lambda} d\delta_\lambda \tag{2.29}$$

The effect of scattering is not considered in the above equations. The following results after introducing the scattering term to the equations for absorption-scattering media:

$$\begin{aligned}\frac{dI_\lambda}{ds} &= -K_\lambda I_\lambda + K_{\alpha\lambda} I_{b\lambda} + Q_\lambda = -(K_{\alpha\lambda} + K_{s\lambda}) I_\lambda + K_{\alpha\lambda} I_{b\lambda} + Q_\lambda \\ &= -K_{\alpha\lambda} I_\lambda - K_{s\lambda} I_\lambda + K_{\alpha\lambda} I_{b\lambda} + Q_\lambda \end{aligned} \tag{2.30}$$

The first term on the right-hand side represents the decrease in radiant energy due to medium absorption, the second term represents the decrease in radiant energy due to medium scattering, the third term represents the increase in radiation due to the medium's own radiation, and the fourth term represents the increase in radiation due to the scattering of the irradiation from space in the s direction. Divide both sides of the equation by K_λ and substitute $K_\lambda = K_{\alpha\lambda} + K_{s\lambda}, \delta_\lambda = K_\lambda s$ to obtain:

$$\frac{dI_\lambda}{d\delta_\lambda} = -I_\lambda + \left(1 - \frac{K_{s\lambda}}{K_\lambda}\right) I_{b\lambda} + \frac{Q_\lambda}{K_\lambda} \tag{2.31}$$

2.4.2.3 Energy Balance Equation

Why establish an equation to calculate energy balance? As shown in Eq. (2.26), I_b is a function of temperature; strictly speaking, δ is also a function of temperature, so the radiation intensity I cannot be calculated using only Eq. (2.26). An energy balance equation must be established for simultaneous solution with Eq. (2.26) to determine temperature and radiation intensity.

According to the energy balance equation $dQ_s = dQ_\alpha$ and the derivation of differential volume radiation, the following equation can be obtained:

$$K \int_0^{4\pi} I \, d\Omega + Q - 4K\sigma T^4 = 0 \tag{2.32}$$

where Q is the heat obtained per unit volume of the medium due to factors other than radiation (ie, heat conduction, convection, chemical reaction).

There is an important reason for establishing the two equations and their physical meaning. The radiative transfer equation shows that radiation intensity I is weakened due to absorption, scattering, and radiation in the medium, but is enhanced due to radiation throughout the medium. If the irradiation intensity is known, any change in I can be determined by solving the radiation heat transfer equation. The emitted energy of the medium is dependent on its temperature, however, which must be determined according to the energy balance equation—so, naturally, the energy balance equation must be established.

2.4.2.4 Approximate Solution for Radiation Transfer Equation

The main difficulty in solving the above two equations lies in solving the radiative transfer equation. The possible solutions can be classified into three categories: the analytical method (which is only applicable to simple one-dimensional cases), the numerical method, which is more accurate and feasible (but is not discussed here due to space limitations), and the approximation method, which simplifies the equation by approximating it before solving it.

See the following approximate solution for transparent media:

$$I = I_1 \tag{2.33}$$

where there is no absorption or emission in the medium.

The following is an example of an emission-approximate solution (an approximate solution for optically thin media).

When optical thickness δ is small and T is relatively large, the medium's absorption can be neglected, thus:

$$I = I_1 + \int_0^\delta I_b \, d\delta = I_1 + \int_0^\delta K I_b \, ds \tag{2.34}$$

If the temperature of the medium is low, the medium's self-radiation can be neglected. See the following approximate solution for low-temperature media:

$$I = I_1 e^{-\delta} = I_1 e^{-Ks} \tag{2.35}$$

Example 2.1
Consider two infinite gray panels with temperatures T_1, T_2 and emissivity ε_1, ε_2. Between them is an almost transparent medium. The medium temperature can be determined using the approximate solution of radiation heat transfer.

Solution
When the medium is almost transparent, the radiation heat transfer equation $\frac{dI}{d\delta} + I = I_b$ becomes $I = I_b$. Assuming that there is a differential volume in the medium, the radiant energy projected to the differential volume is:

$$\int_0^{4\pi} I \, d\Omega = \int_0^{2\pi} I_1 \, d\Omega + \int_0^{2\pi} I_2 \, d\Omega$$

Because the medium is almost transparent, $I_1 = I_{b1}$ and $I_2 = I_{b2}$; I_{b1} and I_{b2} are determined as follows:

$$I_{b1} = \frac{E_{R1}}{\pi}, I_{b2} = \frac{E_{R2}}{\pi}$$

Thus, the radiosity of the two surfaces E_{R1}, E_{R2} can be solved by the network method:

$$\left.\begin{array}{l} \dfrac{E_{b1}-E_{R1}}{\dfrac{1-\varepsilon_1}{\varepsilon_1 A}} = \dfrac{E_{b1}-E_{b2}}{\dfrac{1-\varepsilon_1}{\varepsilon_1 A}+\dfrac{1}{A}+\dfrac{1-\varepsilon_2}{\varepsilon_2 A}} \\[4ex] \dfrac{E_{R2}-E_{b2}}{\dfrac{1-\varepsilon_2}{\varepsilon_2 A}} = \dfrac{E_{b1}-E_{b2}}{\dfrac{1-\varepsilon_1}{\varepsilon_1 A}+\dfrac{1}{A}+\dfrac{1-\varepsilon_2}{\varepsilon_2 A}} \end{array}\right\} \Rightarrow \begin{cases} E_{R1} = \dfrac{\varepsilon_1 E_{b1}+\varepsilon_2(1-\varepsilon_1)E_{b2}}{\varepsilon_1+\varepsilon_2-\varepsilon_1\varepsilon_2} \\[3ex] E_{R2} = \dfrac{\varepsilon_2 E_{b2}+\varepsilon_1(1-\varepsilon_2)E_{b1}}{\varepsilon_1+\varepsilon_2-\varepsilon_1\varepsilon_2} \end{cases}$$

Because the medium has weak absorbability, its total absorbing energy equals its total radiant energy, thus:

$$E_{bm} = \frac{1}{4}\int_0^{4\pi} I \, d\Omega$$

Then, substituting $E_{bm} = \sigma T_m^4$ into the above equations yields the medium temperature:

$$T_m = \left[\frac{\varepsilon_1 T_1^4 + \varepsilon_2 T_2^4 - \varepsilon_1\varepsilon_2 \dfrac{T_1^4+T_2^4}{2}}{\varepsilon_1+\varepsilon_2-\varepsilon_1\varepsilon_2}\right]^{\frac{1}{4}}$$

2.4.3 Mean Beam Length, Absorptivity, and Emissivity of Media

2.4.3.1 Average Transmissivity and Absorptivity

As shown in Fig. 2.8, A_j, A_k are isothermal diffusion surfaces between which is an isothermal absorbing medium. The blackbody radiation intensity of the medium I_{bm} and the extinction coefficient K are constant. According to the radiative transfer equation, the radiation intensity projecting from surface A_j to surface A_k can be expressed as follows:

$$I_{jk} = I_j e^{-Ks} + I_{bm}(1-e^{-Ks}) \tag{2.36}$$

The radiation heat transfer between surface A_j and surface A_k can then be derived:

$$Q_{jk} = (A_j\varphi_{jk}\tau_{jk})E_j + (A_j\varphi_{jk}\alpha_{jk})E_{bm} \tag{2.37}$$

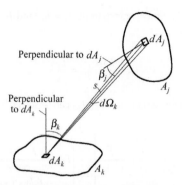

FIGURE 2.8 Radiation heat transfer between two surfaces separated by an isothermal medium.

where E_{bm} is the black emissive power of the medium. Eq. (2.37) demonstrates that if there is an absorbing medium, τ_{jk}, α_{jk} must be calculated first to solve the radiation heat transfer equation. τ_{jk} is the average transmissivity from A_j to A_k, defined as follows:

$$\tau_{jk} = \frac{1}{A_j \phi_{jk}} \int_{A_k} \int_{A_j} \frac{\tau \cos \beta_k \cos \beta_j}{\pi R^2} dA_j dA_k \qquad (2.38)$$

α_{jk} is the average absorptivity from A_j to A_k:

$$\alpha_{jk} = \frac{1}{A_j \phi_{jk}} \int_{A_k} \int_{A_j} \frac{\alpha \cos \beta_k \cos \beta_j}{\pi R^2} dA_j dA_k \qquad (2.39)$$

In the two equations above, φ_{jk} is the configuration factor without an absorbing medium, and τ, α are the medium's own transmissivity and absorptivity, respectively. See the following definition of τ, α according to Eqs. (2.38) and (2.39):

$$\alpha_{jk} + \tau_{jk} = 1 \qquad (2.40)$$

Thus, if τ_{jk} is known, α_{jk} can be obtained.

According to the definition of τ_{jk}, it is rather difficult to calculate τ_{jk} using the analytical method, because it requires complex double integral calculation. The following describes how to solve τ_{jk} in two simple cases.

1. Hemisphere to differential area on bottom center

Assume there is a hemisphere filled with an absorbing medium whose extinction coefficient is K, as shown in Fig. 2.9. Average transmissivity f can be obtained as follows:

$$\tau_{jk} = e^{-KR} \qquad (2.41)$$

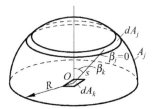

FIGURE 2.9 **Hemisphere filled with an absorbing medium.**

This is an example of the simplest case of solving for average transmissivity. Eq. (2.41) will be used again when discussing mean beam length.

2. Sphere to any part of the spherical surface

Consider a sphere filled with an absorbing medium. The average transmissivity from the entire sphere to any part of the spherical surface can be calculated as follows:

$$\tau_{jk} = \frac{2}{(2KR)^2}\left[1 - (2KR+1)e^{-KR}\right] \tag{2.42}$$

where $2KR$ is the light diameter of the sphere.

2.4.3.2 Mean Beam Length

Furnace heat transfer is mainly a radiation process between the water wall and the flame. First, let's discuss the radiation heat transfer process from a medium with uniform temperature to its boundary.

When describing average transmissivity, the second term in Eq. (2.37) represents the radiation heat from the medium to the wall. See the following:

$$Q_{mk} = A_j\varphi_{jk}\alpha_{jk}E_{bm} \tag{2.43}$$

where subscript m denotes the medium. Substituting A_k with differential area dA_k gives us:

$$dA_k E_{m,dk} = A_j\varphi_{j,dk}\alpha_{j,dk}E_{bm} \tag{2.44}$$

For a case where the radiation process is from the upper hemisphere to the differential area dA_k on the bottom center, Eq. (2.41) can be transformed into the following:

$$\alpha_{j,dk} = 1 - \tau_{j,dk} = 1 - e^{-KR}d \tag{2.45}$$

Because $\varphi_{j,dk} = \dfrac{dA_k}{A_j} = 1$, the following can be derived from Eq. (2.44):

$$E_{m,dk} = \alpha_{j,dk}E_{bm} = (1 - e^{-KR})E_{bm} \tag{2.46}$$

The definition of emissivity is:

$$\varepsilon = 1 - e^{-KR} \tag{2.47}$$

Thus:

$$E_{m,dk} = \varepsilon E_{bm} \tag{2.48}$$

Because the radiation ray between dA_k and A_j goes through the entirety of the medium's hemisphere, $E_{m,dk}$ represents the radiation power from all the medium in the hemisphere to dA_k, and the emissivity of the medium depends only on the optical thickness of the hemisphere KR.

You may find it convenient to calculate the radiation from the medium to the wall using Eq. (2.46); however the equation is limited only to cases involving radiation from the hemisphere to the bottom center of the medium.

Now, let's define an equivalence relationship. Consider the radiation from a medium with arbitrary volumetric shape to its boundary as the radiation from the medium's hemisphere to its bottom center – the radius of imaginary hemisphere R can then be called the mean beam length (represented by s, also called "effective radiation layer thickness"). Thus, we can use Eq. (2.46) to calculate emissive power for actual conditions as follows:

$$E_{mk} = (1 - e^{-Ks})E_{bm} \tag{2.49}$$

Due to the definition of an equivalent radius R (ie, s,) the calculation of emissive power from the medium (with any volumetric shape) to its boundary can be converted to the calculation of the mean bean length s of the medium.

2.4.3.3   Calculating the Mean Beam Length

Calculating the emissivity, absorptivity, and transmissivity of media are simplified considerably after learning to calculate the mean beam length, which is introduced in this section. Let's first consider a case where the optical thickness of a medium is thin, and determine the radiation from the entire volume to the entire boundary.

When the optical thickness Ks is very small, the transmissivity becomes:

$$\tau = \lim_{ks \to 0} e^{-Ks} = 1 \tag{2.50}$$

It is known from the derivation of differential volume radiation that the radiant energy emitting from the isothermal medium dV is $4KE_{bm}dV$. Because $\tau = 1$, that is, the medium has almost no absorption, all the emitting energy reaches almost to the boundary. The emitting energy from the entire volume V nearly reaching the boundary is $4KE_{bm}V$, so the emissive power can be averaged according to the entire boundary as follows:

$$E_m = 4KE_{bm}\frac{V}{A} \tag{2.51}$$

According to the definition of mean beam length:

$$E_m = (1 - e^{-Ks})E_{bm} \tag{2.52}$$

When Ks is very small, s can be replaced by s_b. Compare Eq. (2.51) with Eq. (2.52) and expand the first-order term e^{-Ks} to obtain:

$$s_b = 4\frac{V}{A} \tag{2.53}$$

Some examples of calculating s_b are as follows.

1. For a sphere with diameter D:

$$s_b = \frac{4\left(\frac{4}{3}\pi\left(\frac{D}{2}\right)^3\right)}{\left[4\pi\left(\frac{D}{2}\right)^2\right]} = \frac{2}{3}D \tag{2.54}$$

2. For a cylinder with diameter D:

$$s_b = \frac{4\left(\frac{1}{4}\pi D^2\right)}{(\pi D)} = D \tag{2.55}$$

3. For a medium between parallel infinite slabs separated by distance D:

$$s_b = 4D/2 = 2D \tag{2.56}$$

Example 2.2
When high-temperature gas covers a cross tube bundle as shown in Fig. 2.10, what is the mean beam length of the cross tube bundle?

Solution
The flue gas volume between the per-unit length of the bundle is:

$$V = (2D)^2 - 2 \cdot \frac{\pi D^2}{4} = \left(4 - \frac{\pi}{2}\right)D^2$$

The heating area corresponding to the flue gas volume between the per-unit length of the bundle is:

$$A = 4 \cdot \frac{1}{4}\pi D + \pi D = 2\pi D$$

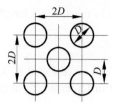

FIGURE 2.10 Mean beam length of a cross tube bundle.

So the mean bean length of the cross tube bundle can be calculated as follows:

$$s_b = 4\frac{V}{A} = 4\frac{\left(4-\frac{\pi}{2}\right)D^2}{2\pi D} = 1.546D \tag{2.57}$$

When $\delta = Ks$ is not as thin, coefficient C can be corrected as $s = Cs_b$. As listed in Table 2.1, for C of approximately 0.9:

$$s = 3.6\frac{V}{A} \tag{2.58}$$

Eq. (2.58) is applicable for accurate calculation of any geometry even when an exact geometric shape is not available.

In most cases, the effective radiation layer thickness is required to calculate the radiation heat transfer between two surfaces filled with an absorbing medium. The effective radiation layer thickness between surface A_k and surface A_j is:

$$s_{kj} = \frac{1}{A_k\phi_{kj}}\int_{A_j}\int_{A_k}\frac{\cos\beta_j\cos\beta_k}{\pi r}dA_kdA_j \tag{2.59}$$

A few details are particularly relevant here. First, shape factor $\varphi_{k,j}$ can be calculated when there is no absorbing medium, then $s_{k,j}$ can be obtained by solving Eq. (2.59). Because $\alpha_{k,j}$ is given by $\alpha = 1-e^{-Ks}$, $A_k\varphi_{k,j}\alpha_{k,j}$ can be obtained. Finally, heat transfer quantity can be determined according to Eq. (2.37).

Here are two equations for special cases:

1. Two identical rectangular slabs arranged oppositely parallel to each other, where the length of one side is $b \to \infty$:

$$\frac{\varphi_{jk}s_{jk}}{C} = \frac{4}{\pi}\left[\arctan l + \frac{1}{l}\ln\frac{1}{\sqrt{1+l^2}}\right] \tag{2.60}$$

where $l = a/c$, and c is the distance between the two rectangular slabs.

2. Two rectangular slabs vertically arranged with a common side, where length $b \to \infty$:

$$\frac{\varphi_{jk}s_{jk}}{\alpha} = \frac{1}{\pi}\left[\frac{a}{c}\ln\sqrt{\left(\frac{a}{c}\right)^2+1} + \frac{a}{c}\ln\sqrt{\left(\frac{c}{a}\right)^2+1}\right] \tag{2.61}$$

TABLE 2.1 Calculation of Mean Beam Length of Entire Volume of Medium

Volume forms	Characteristic length	s_b	s	$C = \dfrac{s}{s_b}$
Hemisphere to bottom center	Radius, R	R	R	1
Sphere to surface	Diameter, D	$\dfrac{2}{3}D$	$0.65D$	0.97
Cylinder with equal height and diameter to bottom center	Diameter, D	$0.77D$	$0.71D$	0.92
Infinite cylinder to side surface	Diameter, D	D	$0.95D$	0.95
Cylinder with equal height and diameter to surface	Diameter, D	$\dfrac{2}{3}D$	$0.60D$	0.90
Infinite medium layer to boundary	Thickness, D	$2D$	$1.8D$	0.90
Parallelepiped with side length of 1:1:4	Side length, L	$\dfrac{2}{3}L$	$1.8D$	0.90
Cube to surface	Length of short side, L	$0.89L$	$0.6L$	0.91
Medium between bundle to the tube wall				
Arranged in an equilateral triangle $s = 2D$	Tube diameter, D	$3.4(s\text{-}D)$	$0.81L$	0.88
Arranged in an equilateral triangle $s = 3D$	Tube spacing, s	$4.45(s\text{-}D)$	$3.0(s\text{-}D)$	0.85
Arranged in a square $s = 2D$		$4.1(s\text{-}D)$	$3.5(s\text{-}D)$	0.85

2.4.4 Gas Absorptivity and Emissivity [11,12,14,26,29–31]

The components of flue gas that have radiation capacity are mainly CO_2 and H_2O (gas), which both show obvious selectivity during radiation and absorption processes—in other words, they only emit and absorb radiant energy within certain light bands. The main absorbing bands of CO_2 and H_2O (gas) are listed in Table 2.2.

1. Calculation of α_{CO_2} and α_{H_2O} with a nomograph

A nomograph is a powerful calculation tool that can be applied in the case of computational backwardness (or lack of computers/calculators), which lends valuable convenience to engineering and technical personnel as they conduct scientific, reliable design based on a large number of experimental and engineering research

TABLE 2.2 Main Absorbing Bands of CO_2 and H_2O (Gas)

Gas	CO_2		H_2O	
Band range/μm	$\lambda' \sim \lambda''$	$\Delta\lambda$	$\lambda' \sim \lambda''$	$\Delta\lambda$
Band no.				
1	2.65~2.8	0.15	2.3~3.4	1.1
2	4.15~4.46	0.30	4.4~8.5	4.1
3	13~17	4.0	12~30	18

results. The typical nomographs of CO_2 and H_2O are shown in Figs. 2.11 and 2.12, respectively. The curves in Figs. 2.11a and 2.12a, which were drawn according to the reference pressure of $p_0 = 9.81 \times 10^4$ Pa, show the correction coefficients K corresponding to the total pressure p and partial pressure, respectively.

Modern engineers rarely use nomographs, because their function has gradually been replaced by empirical equations and engineering computational packages. Especially because computers are available for calculation of heat transfer in a furnace (including thermodynamic calculation), nomographs are generally not necessary. It is important to note, however, that the nomograph is still a quite accurate method of calculating absorptivity. The gas emissivity equations described below only apply to blackbodies, that is, only when a gas is assumed to be a graybody can the nomograph be used to calculate its absorptivity.

2. The effect of temperature on absorptivity

When the gas temperature equals the temperature of the blackbody (or graybody) radiation source, the gas absorptivity is also equal to its emissivity. If the temperature of the radiation source changes, and fails to match the gas temperature again, then the gas absorptivity also changes and fails to equal the emissivity. Fig. 2.13 shows the relation among CO_2 absorptivity α_{CO_2}, air temperature T_g, and the wall temperature of blackbody radiation source T_w. Fig. 2.14 shows the relation among steam absorptivity α_{H_2O}, air temperature T_g, and the wall temperature of blackbody radiation source T_w.

The following can be concluded based on Figs. 2.13 and 2.14: when ps is small, gas absorptivity decreases as gas temperature T_g increases. (This phenomenon grows more pronounced as the wall temperature of radiation source T_w increases.) When ps is large, conversely, gas absorptivity increases alongside T_g, and becomes more significant as T_w decreases.

For further analysis, please refer to the related monograph.

3. Absorptivity when there are two absorbing gases

Generally, the absorptivity of a medium containing two absorbing gases whose light bands overlap with each other is smaller than the sum of the

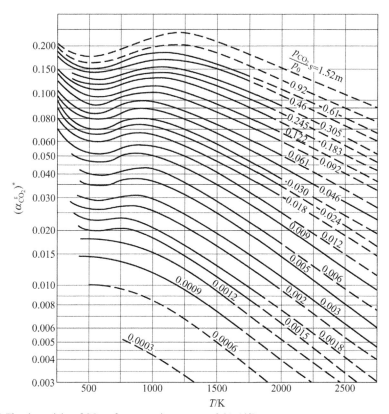

(a) The absorptivity of CO_2, reference total pressure $p_0=9.81\times10^4$Pa

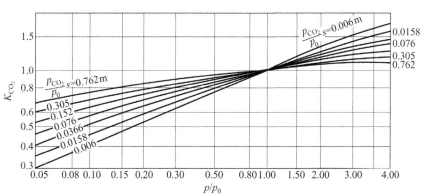

(b) The amending coefficient of the total pressure p of mixture with $\overset{\smile}{C}O_2$ and transparent gas

FIGURE 2.11   Typical nomograph of CO_2.

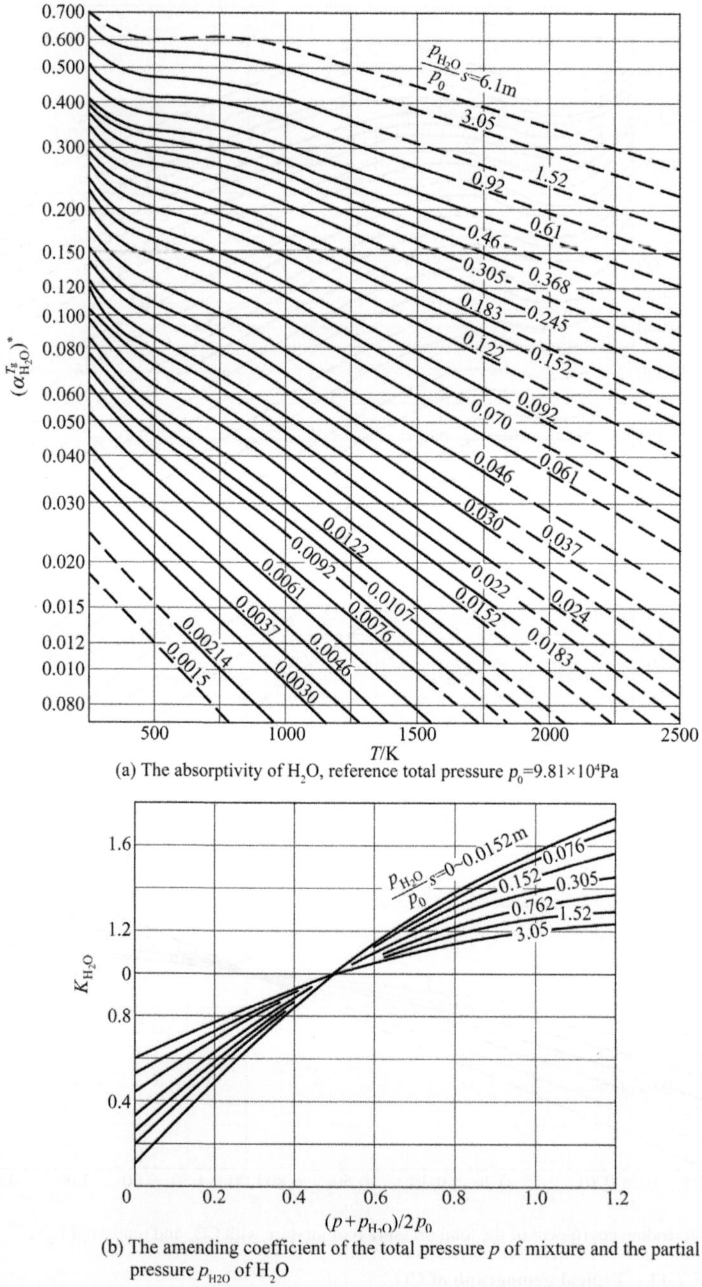

(a) The absorptivity of H_2O, reference total pressure $p_0 = 9.81 \times 10^4 Pa$

(b) The amending coefficient of the total pressure p of mixture and the partial pressure p_{H2O} of H_2O

FIGURE 2.12   **Typical nomograph of H_2O.**

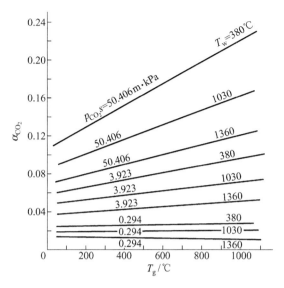

FIGURE 2.13 Relation of CO_2 absorptivity α_{CO_2}, air temperature T_g, and wall temperature of blackbody radiation source T_w.

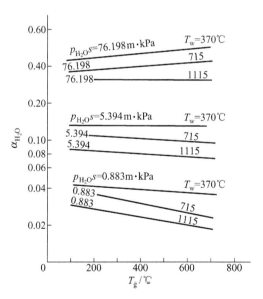

FIGURE 2.14 Relation of steam absorptivity α_{H_2O}, air temperature T_g, and wall temperature of blackbody radiation source T_w.

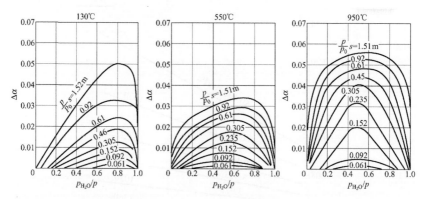

FIGURE 2.15 Correction coefficients for absorptivity of mixtures of CO_2 and H_2O.

respective absorptivity of the two gases. This phenomenon can be explained by applying quantum mechanics and statistical physics theory.

For a mixture of CO_2 and H_2O, the absorptivity is:

$$\alpha = \alpha_{CO_2} + \alpha_{H_2O} - \Delta\alpha \tag{2.62}$$

where $\Delta\alpha$ is a correction coefficient (Fig. 2.15). Because $\Delta\alpha$ is generally fairly small, it can be neglected.

Gas emissivity was once calculated primarily in the form of charts, but is now typically calculated by coding related equations into a computer, which can conveniently run several repeated calculations. The following section introduces the equations for calculating the emissivity of a triatomic gas.

According to the former Soviet Union's proposed 1973 standard for boiler thermal calculation, the emissivity of a triatomic gas can be calculated according to Bouguer's law. The gas emissivity equation, which was defined according to analysis of a large amount of industrial data, can be expressed as follows:

$$\varepsilon = 1 - e^{-K_g s} = 1 - e^{-k_g r_n ps} \tag{2.63}$$

in which the gas extinction coefficient is given by $K_g = k_g r_n ps$, $r_n = \dfrac{p_n}{p}$, where p_n is the partial pressure of triatomic gas (Pa), r_n is the volume fraction of triatomic gas, p is the total gas pressure (Pa), and s is the effective radiation layer thickness. Coefficient k_g is given by:

$$k_g = \left(\frac{0.78 + 1.6 r_{H_2O}}{\sqrt{\dfrac{p_n}{p_0} s}} - 0.1 \right) \left(1 - 0.37 \frac{T}{1000} \right) \times 10^{-5} \tag{2.64}$$

where $r_{H_2O} = \dfrac{p_{H_2O}}{p}$ is the volume fraction of H_2O, p_{H_2O} is the partial pressure of H_2O (Pa), $p_0 = 9.81 \times 10^4$ Pa is reference pressure, and T is gas temperature (K). The scope of Eq. (2.64) is limited to: $\dfrac{p_{CO_2}}{p_{H_2O}} = 0.2 \sim 2$; $p_n s = (1.2 \sim 200) \times 10^3$ m Pa; $T = 700 \sim 1800$ K.

It is important to note that it is only possible to calculate emissivity rather than absorptivity from the above two equations, so a nomograph is still needed to calculate absorptivity. Hottel proposed another calculation method which assumed that gas emissivity consists of several parts with differing extinction coefficients; his method has been used successfully for numerical calculation, as well.

2.4.5 Flue Gas and Flame Emissivity

2.4.5.1 Emissivity of Flue Gas Containing Ash

When a heat ray projects onto a particle, some of the energy is absorbed and converted into heat energy and the rest is scattered (due to the effects of refraction, reflection, diffraction, and so on). A flue gas containing fly ash particles is not a graybody, and its monochromatic extinction coefficient is related to its particle size.

Ash gas emissivity (considering triatomic gases) is expressed as follows:

$$\varepsilon = 1 - e^{-Ks} = 1 - e^{-(K_g + K_{fa})s} \tag{2.65}$$

where is the extinction coefficient, K_g is the extinction coefficient of triatomic gas, and K_{fa} is the extinction coefficient of fly ash [which can be calculated as $K_{fa} = k_{fa}\mu_{fa}p$, where p is the gas total pressure (Pa)].

Fly ash concentration μ_{fa} and flue gas density G_g are:

$$\mu_{fa} = \frac{a_{fa}[A]_{ar}}{100G_g}$$

$G_g = 1 - \dfrac{[A]_{ar}}{100} + 1.306\alpha V_a^o$ (note that G_g is the flue gas mass excluding the fly ash)

where $[A]_{ar}$ is the fuel ash content on an as-received basis, α_{fa} is the fly ash fraction dependent on combustion type (such as grate, PC, or CFB), α is the excess air coefficient, and V_a^0 is the theoretical air volume.

k_{fa} can be calculated as follows:

$$k_{fa} = \frac{4300\rho_g}{\sqrt[3]{T^2 d_m^2}} \times 10^{-5} \tag{2.66}$$

where $\rho_g = G_g/V_g$ (generally $\rho_g = 1.3$ kg/m^3 at normal state), d_m is the mean diameter of a fly ash particle (μm) (See Table D.13), and T is the flue gas temperature. The related calculation process and explanation can be found in a previous study [17].

2.4.5.2 Luminous Flame Emissivity

When combusting liquid or gas fuel, the flame radiation comes mainly from carbon black, which can continuously emit radiant energy in the range of the visible spectrum and infrared spectrum. When combusting solid fuel, the flame is all luminous.

According to the same 1973 standard referenced above, extinction coefficient K_c can be calculated as follows:

$$K_c = k_c p \qquad (2.67)$$

and k_c is given by:

$$k_c = 0.03(2 - \alpha_F)\left(1.6 \times 10^{-3} T_F'' - 0.5\right)\frac{[C]_{ar}}{[H]_{ar}}\frac{1}{P_0} \qquad (2.68)$$

where $\dfrac{[C]_{ar}}{[H]_{ar}}$ is the carbon-hydrogen ratio of the fuel as-received, and α_F is the excess air coefficient of the furnace. If $\alpha_F > 2$, then $K_c = 0$. T_F'' is the flue gas temperature at the outlet of the furnace (K), p is the flue gas temperature in the furnace, and $p_0 = 9.81 \times 10^4$ Pa is the reference pressure.

The extinction coefficient of a luminous flame is the sum of those of tri-atomic gas and carbon black:

$$K_{lf} = K_g + K_c = (k_g r_n + k_c)p \qquad (2.69)$$

In the same 1973 standard, the flame emissivity of a furnace combusting liquid or gas fuel can be calculated as follows (assuming that the furnace is filled with luminous flame and nonluminous triatomic gas):

$$\varepsilon = m\varepsilon_{lf} + (1-m)\varepsilon_g \qquad (2.70)$$

in which luminous flame emissivity is given by $\varepsilon_{lf} = 1 - e^{-K_{lf}s} = 1 - e^{-(k_g r_n + k_c)ps}$ and nonluminous flame emissivity is given by $\varepsilon_g = 1 - e^{-K_g s} = 1 - e^{-(k_g r_n)ps}$, where m is a parameter related to furnace volumetric heat load q_v, the value of which can be determined in a few different ways.

1. When $q_v \leq 400 \times 10^3$ W/m^3, m has nothing to do with the load – $m = 0.1$ and $m = 0.55$ for a gas furnace and oil furnace, respectively.
2. When $q_v \geq 1.16 \times 10^6$ W/m^3, $m = 0.6$ and $m = 1.0$ for a gas furnace and oil furnace, respectively.
3. When 400×10^3 W/m$^3 < q_v < 1.16 \times 10^6$ W/m^3, m can be calculated by interpolation according to the load.

2.4.5.3   Coal Flame Emissivity

In a pulverized coal flame, triatomic gases, ash particles, coke, and carbon black particles all have radiation capacity. Research has shown that triatomic gases, ash particles, and coke play major roles in flame radiation.

According to the 1973 Soviet Union standard, the extinction coefficient of coke K_{CO} can be calculated as follows:

$$K_{co} = k_{co} x_1 x_2 p \tag{2.71}$$

where k_{CO} is generally 10^{-5}. $x_1 = 1$ for anthracite and lean coal fuels, and $x_1 = 0.5$ for bitumite and lignite fuels. $x_2 = 0.1$ for pulverized coal furnace combustion and $x_2 = 0.03$ for grate furnace combustion. The extinction coefficient of the coal flame is then:

$$K = K_g + K_{fa} + K_{co} = \left(k_g r_n + k_{fa} \mu_{fa} + x_1 x_2 \times 10^{-5} \right) p \tag{2.72}$$

Coal flame emissivity is:

$$\varepsilon = 1 - e^{-Ks} \tag{2.73}$$

Example 2.3

Find the pulverized coal flame emissivity in a bituminous coal fired system under the following given conditions:

Theoretical air volume V^0 (normal state) = 4.81 m³/kg, excess air coefficient $\alpha = 1.25$.

Theoretical (when $\alpha = 1$) flue gas volume V_g^0 (normal state) = $V_{RO_2} + V_{N_2}^0 + V_{H_2O}^0$ = 5.218 m³/kg.

Triatomic gas volume V_{RO_2} (normal state) = 0.882 m³/kg.

Theoretical H_2O volume $V_{H_2O}^0$ (normal state) = 0.529 m³/kg.

Theoretical N_2 volume $V_{N_2}^0$ (normal state) = 3.807 m³/kg.

Coal ash as-received $[A]_{ar} = 32.48[\%]$.

Fly ratio $\alpha_{fa} = 0.9$; Flame temperature $T_{fl} = 1473$ K.

Effective radiation layer thickness $s = 5$ m.

Flue gas pressure $p = 10^5$ Pa.

Ring-roll or ball-race mills with medium speed are used for coal grinding.

Solution

V_{H_2O}(normal state) = $V_{H_2O}^0 + 0.0161(\alpha - 1)V^0$ = $0.529 + 0.0161 \times 0.25 \times 4.81$
$$= 0.548 (m^3/kg)$$

Flue gas volume V_g (normal state):

$$= V_g = V_g^0 + (\alpha - 1)V^0 + 0.0161(\alpha - 1)V^0$$

$$= 5.218 + 0.25 \times 4.81 + 0.0161 \times 0.25 \times 4.81 = 6.44 (m^3/kg)$$

$$\left.\begin{array}{l} r_{RO_2} = \dfrac{V_{RO_2}}{V_g} = \dfrac{0.882}{6.44} = 0.137 \\[3mm] r_{H_2O} = \dfrac{V_{H_2O}}{V_g} = \dfrac{0.548}{6.44} = 0.085 \end{array}\right\} = 0.222$$

$$G_g = 1 - \frac{A_{ar}}{100} + 1.306\alpha V^0$$

$$= 1 - \frac{32.48}{100} + 1.306 \times 1.25 \times 4.81 - 8.53 \,(kg/(kg\,fuel))$$

$$\mu_{fa} = \frac{[A]_{ar}\,\alpha_{fh}}{100G_g} = \frac{32.48 \times 0.9}{100 \times 8.53} = 0.0343 \,(kg/(kg\,gas))$$

$$k_g = \left(\frac{0.78 + 1.6 r_{H_2O}}{\sqrt{\dfrac{p_n}{p_0}}s} - 0.1\right)\left(1 - 0.37\frac{T_{fl}}{1000}\right)\times 10^{-5}$$

where $\dfrac{p_n}{p_0} \approx \dfrac{p_n}{p} = r_n = 0.222$, thus:

$$k_g = \left(\frac{0.78 + 1.6 \times 0.085}{\sqrt{0.222 \times 5}} - 0.1\right)\left(1 - 0.37\frac{1473}{1000}\right)\times 10^{-5}$$
$$= 0.357 \times 10^{-5}\,(m/Pa)$$

$$K_g = k_g r_n p = 0.357 \times 10^{-5} \times 0.222 \times 10^5 = 0.079\,(m^{-1})$$

$$K_{fa} = k_{fa}\mu p = \frac{4300 \times \rho_g}{\sqrt[3]{T^2 \cdot d_m^2}} \times 10^{-5}\mu_{fa}$$

As listed in Table D.13, for a mill with medium speed, $d_m = 16\,(\mu m)$, thus: $K_{fa} = \dfrac{4300 \times 1.3}{(1473 \times 16)^{2/3}} \times 10^{-5} \times 0.0343 \times 10^5 = 0.233\,(m^{-1})$

Because this is a pulverized bituminous coal boiler, $x_1 = 0.5$, $x_2 = 0.1$. Therefore, the extinction coefficient of the coal flame is: $K = K_g + K_{fa} + x_1 x_2 p \times 10^{-5} = 0.079 + 0.233 + 0.5 \times 0.1 \times 10^5 \times 10^{-5} = 0.362\,(m^{-1})$ and coal flame emissivity is:

$$\varepsilon_g = 1 - e^{-Ks} = 1 - e^{-0.362 \times 5} = 0.836$$

Chapter 3

Radiation Heat Exchange Between Isothermal Surfaces

Chapter Outline

This chapter will first describe the radiation heat transfer between solid surfaces without considering the effect of the medium among them, that is, the medium is transparent. Transparent objects or media are those of which the transmissivity $\tau = 1$. Then consideration will be given to the radiation heat transfer between the medium and the surfaces.

When the medium is transparent, the radiative heat transfer among solid surfaces is influenced by the following three factors: (1) the temperature of each surface; (2) the emissivity and absorptivity of each surface; (3) the areas, shapes, and relative position of the surfaces, which is called the configuration factor under simplified conditions used for engineering calculations. Commonly, these three factors are interrelated and interact with each other, so it becomes overly complicated for engineering calculations and needs to be simplified appropriately. Chapter: Theoretical Foundation and Basic Properties of Thermal Radiation has introduced the simplified conditions for engineering calculations, and

Theory and Calculation of Heat Transfer in Furnaces. http://dx.doi.org/10.1016/B978-0-12-800966-6.00001-6

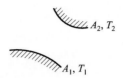

FIGURE 3.1 Closed system composed of two surfaces.

assumes that all discussions in this book, except where specifically explained, satisfy the simplified conditions.

3.1 RADIATIVE HEAT EXCHANGE BETWEEN SURFACES IN TRANSPARENT MEDIA [12,14]

According to heat transfer theory, the energy conservation principle is the basis for the calculation of all energy exchange and transfer. When there is only a single thermal energy form, it becomes the heat balance principle. There are different methods for establishing radiation heat transfer balance equations, such as the network method, the net heat flux method, the Monte Carlo simulation method, and so on.

3.1.1 Radiative Heat Transfer of a Closed System Composed of Two Surfaces

A closed system composed of two surfaces represents simplified conditions for engineering calculations—the surface is isothermal, diffuse gray, and the radiation properties and radiosity of the surface are uniform. In this section, we'll find the radiation heat transfer between the two surfaces shown in Fig. 3.1.

The problem can be solved with either the network method or the net heat flux method. Let's take a look at the network method first.

1. Network method

Fundamental heat transfer theory asserts that there are two radiation heat transfer resistances for a graybody surface: surface heat resistance and space heat resistance, as shown in Fig. 3.2.

The surface heat resistance of the graybody can be calculated as $\dfrac{1-\varepsilon}{\varepsilon A}$, thus, the surface heat resistances for surface 1 and surface 2 are $\dfrac{1-\varepsilon_1}{\varepsilon_1 A_1}$ and $\dfrac{1-\varepsilon_2}{\varepsilon_2 A_2}$, respectively. The space heat resistance is $\dfrac{1}{A_1\varphi_{12}}$ or $\dfrac{1}{A_2\varphi_{21}}$.

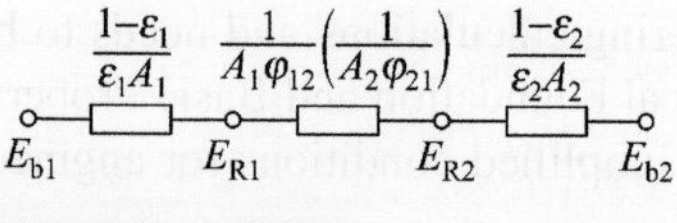

$$E_{b1} \quad \frac{1-\varepsilon_1}{\varepsilon_1 A_1} \quad E_{R1} \quad \frac{1}{A_1\varphi_{12}}\left(\frac{1}{A_2\varphi_{21}}\right) \quad E_{R2} \quad \frac{1-\varepsilon_2}{\varepsilon_2 A_2} \quad E_{b2}$$

FIGURE 3.2 Radiation heat transfer resistance.

Similar to the series circuit principle, the radiation heat transfer between surface 1 and surface 2 can be directly written as follows:

$$Q_{12} = \frac{E_{b1} - E_{b2}}{\dfrac{1-\varepsilon_1}{\varepsilon_1 A_1} + \dfrac{1}{A_1 \varphi_{12}} + \dfrac{1-\varepsilon_2}{\varepsilon_2 A_2}} \tag{3.1}$$

where $E_{b1} = \sigma T_1^4, E_{b2} = \sigma T_2^4$.

2. Net heat flux method

The following is obtained by establishing the heat exchange relation between surface 1 and surface 2 using radiosity concepts:

$$Q_{12} = Q_{R1,2} - Q_{R2,1} \tag{3.2}$$

where $Q_{R1,2}$ is the radiosity heat from surface 1 to surface 2, and $Q_{R2,1}$ is the radiosity heat from surface 2 to surface 1.

According to the relation between radiosity heat Q_R and emissive power E_R, the following two equations are obtained:

$$Q_{R1,2} = \int_{A_1} E_{R1} \varphi_{d_1,2} \, dA_1 \tag{3.3}$$

$$Q_{R2,1} = \int_{A_2} E_{R2} \varphi_{d_2,1} \, dA_2 \tag{3.4}$$

Because the emissive power E_R can be expressed as $E_R = E_b + \left(\dfrac{1}{\varepsilon} - 1\right) q$ for a graybody:

$$E_{R1} = E_{b1} + \left(\frac{1}{\varepsilon_1} - 1\right) q_1 \tag{3.5}$$

$$E_{R2} = E_{b2} + \left(\frac{1}{\varepsilon_2} - 1\right) q_2 \tag{3.6}$$

Consider the following under simplified condition #4 (see chapter: Theoretical Foundation and Basic Properties of Thermal Radiation):

$$\varphi_{d_1,2} = \varphi_{12}$$

$$\varphi_{d_2,1} = \varphi_{21}$$

and substitute Eqs. (3.3)–(3.6) into Eq. (3.2) to obtain:

$$Q_{1,2} = \left[E_{b1} + \left(\frac{1}{\varepsilon_1} - 1\right) q_1\right] \phi_{12} A_1 - \left[E_{b2} + \left(\frac{1}{\varepsilon_2} - 1\right) q_2\right] \phi_{21} A_2 \tag{3.7}$$

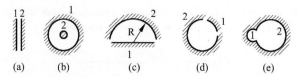

FIGURE 3.3   Surfaces with uniform radiation. (a) Infinite-length parallel plane, (b) concentric sphere or cylinder, (c) hemisphere to base plane or infinite-length semicircular cylinder to strip, (d) two separate surfaces locating at the same sphere or infinite-length cylinder, (e) two inner surfaces of spherical crowns.

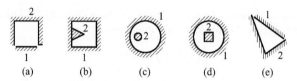

FIGURE 3.4   Surfaces with nonuniform radiation. (a) Three surfaces of a rectangle and fourth surface, (b) a cone in a rectangle cavity, (c) eccentric sphere or cylinder, (d) rectangle in a cylinder, (e) one side of triangle and other sides.

Because the closed system is composed of two surfaces, $-Q_{1,2} = Q_{2,1}$, that is, $-q_1 A_1 = q_2 A_2$. Eq. (3.7) then becomes:

$$Q_{1,2} = \frac{E_{b1}\varphi_{12}A_1 - E_{b2}\varphi_{21}A_2}{\left(\dfrac{1}{\varepsilon_1}-1\right)\varphi_{12}+1+\left(\dfrac{1}{\varepsilon_2}-1\right)\varphi_{21}} \tag{3.8}$$

Clearly, the calculation results from the above two methods are consistent. Now the heat exchange amount $Q_{1,2}(Q_{2,1})$ and the temperature of surface 1 are known, thus the temperature of surface 2 can be calculated.

Chapter: Theoretical Foundation and Basic Properties of Thermal Radiation contains a discussion on simplified condition #4 for engineering calculation, which assumes that the effective surface radiation (incident radiation) is uniform, that is, $\varphi_{d_1 2} = \varphi_{12}$, $\varphi_{d_2 1} = \varphi_{21}$. In practice, only a few cases satisfy these conditions (Fig. 3.3); the majority do not (Fig. 3.4).

So can the above equations be used for radiation heat transfer calculation in situations that fail to satisfy simplified condition #4?

Example 3.1
Find the radiation heat transfer from the inner surface of a whole sphere to the surface of the hemisphere (including the circle plane), as shown in Fig. 3.5. Consider two different calculation methods: one that strictly satisfies the simplified condition (that is, considers the system as two parts), and another that considers the system as a whole, and compare the calculation differences between them. Assume that $\varepsilon_1 = \varepsilon_2 = \varepsilon$.

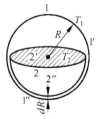

FIGURE 3.5 **Radiation heat transfer from inner surface of a sphere to the surface of the hemisphere.**

Solution

$$\varphi_{d_1 2} \neq \varphi_{12}$$

The original system 1-2 is replaced by two parts labeled 1'-2' and 1"-2" (Fig. 3.5). According to the energy conservation principle:

$$Q_{12} = Q'_{12} + Q''_{12}$$

in which Q'_{12} and Q''_{12} can be calculated as follows:

$$Q'_{12} = \frac{\sigma\left(T_1^4 - T_2^4\right)}{\dfrac{1-\varepsilon_1}{\varepsilon_1}\dfrac{1}{A'_1} + \dfrac{1}{\varphi'_{12}A'_1} + \dfrac{1-\varepsilon_2}{\varepsilon_2}\dfrac{1}{A'_2}}$$

$$Q''_{12} = \frac{\sigma\left(T_1^4 - T_2^4\right)}{\dfrac{1-\varepsilon_1}{\varepsilon_1}\dfrac{1}{A''_1} + \dfrac{1}{\varphi''_{12}A_1} + \dfrac{1-\varepsilon_2}{\varepsilon_2}\dfrac{1}{A'_2}}$$

where $A'_1 = A''_1 = A'_2 = \dfrac{1}{2}(4\pi R^2)$ (neglecting the dR term), and $A'_2 = \pi R^2$. Because $\varphi'_{12}A'_1 = \varphi'_{21}A'_2$, $\varphi'_{21} = 1$, so $\varphi'_{12}A'_1 = A'_2$. Similarly, $\varphi''_{12} = \varphi''_{21} = 1$ means $\varphi''_{12}A''_1 = A'_2$, so Q'_{12} and Q''_{12} become:

$$Q'_{12} = \frac{2}{\dfrac{3}{\varepsilon} - 1}\sigma\pi R^2\left(T_1^4 - T_2^4\right)$$

$$Q''_{12} = \frac{2}{\dfrac{2}{\varepsilon} - 1}\sigma\pi R^2\left(T_1^4 - T_2^4\right)$$

After defining $F = \sigma\pi R^2\left(T_1^4 - T_2^4\right)$:

$$Q_{12} = \left(\frac{2\varepsilon}{3-\varepsilon} + \frac{2\varepsilon}{2-\varepsilon}\right)F = \frac{2\varepsilon(5-2\varepsilon)}{\varepsilon^2 - 5\varepsilon + 6}F$$

Now consider the system as a whole:

$$Q''_{12} = \frac{\sigma\left(T_1^4 - T_2^4\right)}{\dfrac{1-\varepsilon_1}{\varepsilon_1 A_1} + \dfrac{1}{\varphi_{12}A_1} + \dfrac{1-\varepsilon_2}{\varepsilon_2 A_2}}$$

Because $A_1 = A_1' + A_1'' = 4\pi R^2$, $A_2 = A_2' + A_2'' = 3\pi R^2$, $\varphi_{12}A_1 = A_2$:

$$Q_{12}^0 = \frac{12\varepsilon}{7 - 3\varepsilon} F$$

The calculation error is:

$$\delta = \frac{Q_{12} - Q_{12}^0}{Q_{12}} = 1 - \frac{Q_{12}^0}{Q_{12}}$$

$$= \frac{2\varepsilon(\varepsilon - 1)}{2\varepsilon(6\varepsilon^2 - 29\varepsilon + 35)}$$

$$= \frac{\varepsilon - 1}{6\varepsilon^2 - 29\varepsilon + 35}$$

By considering the maximum value of δ when $\varepsilon \in (0,1)$, the stationary point can be calculated as follows:

$$\frac{d\delta}{d\varepsilon} = \frac{-6(\varepsilon^2 - 2\varepsilon - 1)}{(6\varepsilon^2 - 29\varepsilon + 35)^2}$$

Let $\dfrac{d\delta}{d\varepsilon} = 0$, then $\varepsilon = 1 \pm \sqrt{2}$, then there is no stationary point when $\varepsilon \in (0,1)$.

Obviously, when $\varepsilon = 1$, then $\delta = 0$, $\lim_{\varepsilon \to 0} \delta = \dfrac{1}{35} = 2.857\%$.

All calculation errors δ for different ε can then be listed as follows:

$$\varepsilon = 0.01, \delta = -2.852\%$$
$$\varepsilon = 0.1, \delta = -2.80\%$$
$$\varepsilon = 0.5, \delta = -2.27\%$$
$$\varepsilon = 0.9, \delta = -0.73\%$$
$$\varepsilon = 0.99, \delta = -0.08\%$$

Thus, $\delta_{max} < 3\%$.

The calculation error is small (below 3%) when using the simplified method to solve the above problem; therefore, simplified condition #4 is often neglected during engineering calculations.

3.1.2 Radiation Transfer of a Closed System Composed of Multiple Surfaces

In this section, we'll consider a system with several surfaces labeled 1, 2, ..., n, as shown in Fig. 3.6 and determine the amount of radiation heat transfer between the surfaces using the network method.

For any intermediate node i, the balance equation is:

$$\frac{\varepsilon_i A_i}{1 - \varepsilon_i}(E_{bi} - E_{Ri}) + \sum_{j \neq i} A_i \varphi_{ij}(E_{Rj} - E_{Ri}) = 0 \tag{3.9}$$

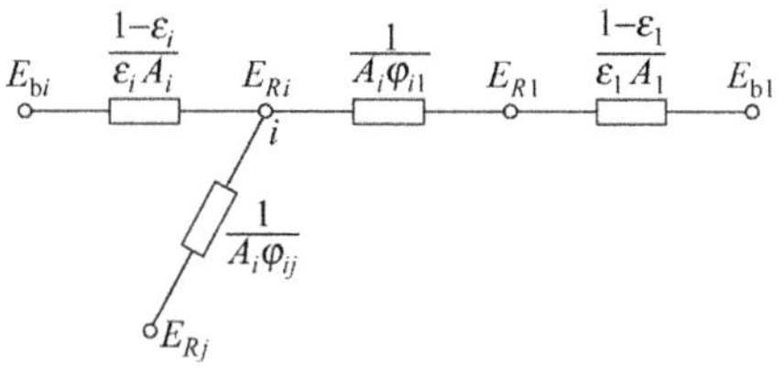

FIGURE 3.6 The network method for calculating radiation heat transfer between several surfaces.

Because $\sum_{j\neq i}\varphi_{ij}+\varphi_{ii}=1$, $\sum_{j\neq i}\varphi_{ij}=1-\varphi_{ii}$. By substituting this into Eq. (3.9):

$$E_{Ri}=\frac{1}{1-\varphi_{ii}(1-\varepsilon_i)}\left[(1-\varepsilon_i)\sum_{j\neq i}\varphi_{ij}E_{Rj}+\varepsilon_i E_{bi}\right] \tag{3.10}$$

Surface i is isothermal, so $q_i = 0$. Assume that $q_i > 0$ when losing heat, then:

$$\begin{aligned} q_i &= E_{Ri}-E_{Ii}=E_{Ri}-\sum_j \frac{1}{A_i}E_{Rj}A_j\varphi_{ji} \\ &= E_{Ri}-\sum_j \frac{1}{A_i}E_{Rj}A_i\varphi_{ij} \\ &= E_{Ri}-\sum_j E_{Rj}\varphi_{ij} \\ &= 0 \end{aligned}$$

Thus E_{Ri} becomes:

$$E_{Ri}-\left(\sum_{j\neq i}E_{Rj}\varphi_{ij}+\varphi_{ii}E_{Ri}\right)=0$$

and:

$$E_{Ri}=\frac{1}{1-\varphi_{ii}}\sum_{j\neq i}E_{Rj}\varphi_{ij} \tag{3.11}$$

In fact, Eq. (3.11) can be directly derived from Eq. (3.10) when $\rho_i = 1, \alpha_i = 0$, $\varepsilon_i = 0$.

When q_i is known and E_{bi} is unknown, assume that $q_i > 0$ when losing heat. Due to the following:

$$E_{Ri}=E_{bi}-\frac{1-\varepsilon_i}{\varepsilon_i}q_i$$

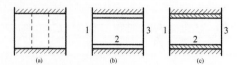

FIGURE 3.7   Common hole walls. (a) Brick setting hole wall. (b) Water cooling hole wall. (c) Metal hole wall.

then:

$$E_{bi} = E_{Ri} + \frac{1-\varepsilon_i}{\varepsilon_i} q_i$$

Substituting the above equation into Eq. (3.10) yields:

$$E_{Ri} = \frac{1}{1-\varphi_{ii}} \left(\sum_{j \neq i} \varphi_{ij} E_{Rj} + q_i \right) \tag{3.12}$$

3.1.3   Hole Radiative Heat Transfer

Next, let's analyze an example of calculating the heat transfer from a door, a hole, or a gap to the outer environment according to the results presented above—being able to do so successfully is very important in real-world engineering.

There are three types of hole walls commonly used, as shown in Fig. 3.7.

The hole conditions can be classified into two categories according to their thermal conductivity: $\lambda_w = 0$ and $\lambda_w = \infty$. First, we'll assume that the surface is isothermal, (thus, $\lambda_w = 0$).

Example 3.2

The temperature of a circular hole is 1000°C, and the emissivity is 0.6. The environment is a large room with a temperature of 20°C, which can be considered a blackbody. Let's calculate the radiation heat transfer from the circular hole. The simplified solution to this question is to regard the radiosity of the entire inner heating surface as uniform, and work through the problem integrally.

The radiosity on the surface is variable. The inner surface can be divided into four parts (hole bottom 1, and side surfaces 2, 3, and 4) for analysis (Fig. 3.8). Let's assume an imaginary black surface 5 covers the hole, and solve the radiosity of every part before calculating the final amount of heat transfer—as such, this is a precise, stepwise solution.

The following works through the problem first stepwise, then in an integral manner for the sake of illustration.

1. Stepwise solution

For precalculation preparation, first consider the emissive power of each part of the surface:

$$E_{b1} = E_{b2} = E_{b3} = E_{b4} = \sigma T_1^4$$

FIGURE 3.8 **Radiation heat transfer from a circular hole to the outer environment.**

$$= (5.67 \times 10^{-8}) \times (1000 + 273)^4 = 1.4887 \times 10^5 [W/m^2]$$

$$E_{b5} = \sigma T_5^4 = (5.67 \times 10^{-8}) \times (20 + 273)^4 = 417.8 [W/m^2]$$

and its emissivity:

$$\varepsilon_1 = \varepsilon_2 = \varepsilon_3 = \varepsilon_4 = 0.6$$
$$\varepsilon_5 = 1.0$$

its area:

$$A_1 = A_5 = A_6 = A_7 = \pi [cm^2]$$
$$A_2 = A_3 = A_4 = 2\pi [cm^2]$$

and the configuration factors, which can be calculated based on the example circular plate surface shown in Fig. 1.25. For the configuration factors between two parallel circular planes with the same radius, utilize Eq. (1.93):

$$\Phi = \frac{1 + 2R^2 - \sqrt{1 + 4R^2}}{2R^2}, \quad R = r/h$$

where r is the radius of the circular plane, and h is the distance between the two planes.

All the configuration factors can be calculated as follows:

$$\Phi_{11} = \Phi_{55} = 0$$
$$\Phi_{16} = 0.37, \Phi_{17} = 0.175, \Phi = 0.1$$
$$\Phi_{12} = \Phi_{16} = 0.63 = \Phi_{54}$$
$$\Phi_{13} = \Phi_{16} - \Phi_{17} = 0.195 = \Phi_{53}$$
$$\Phi_{14} = \Phi_{17} - \Phi_{15} = 0.075 = \Phi_{52}$$
$$\Phi_{21} = \Phi_{26} = \Phi_{12} A_1/A_2 = \Phi_{12}/2 = 0.315 = \Phi_{45} = \Phi_{36} = \Phi_{37}$$
$$\Phi_{22} = 1 - \Phi_{21} - \Phi_{26} = 1 - 0.315 - 0.315 = 0.37 = \Phi_{33} = \Phi_{44}$$
$$\Phi_{31} = \Phi_{13} A_1/A_3 = \Phi_{13}/2 = 0.0975 = \Phi_{27} = \Phi_{46}$$
$$\Phi_{32} = \Phi_{36}\Phi - 31 = 0.2175 = \Phi_{34} = \Phi_{43} = \Phi_{23}$$
$$\Phi_{41} = \Phi_{14} A_1/A_4 = \Phi_{14}/2 = 0.0375 = \Phi_{25}$$
$$\Phi_{42} = \Phi_{46} - \Phi_{41} = 0.06 = \Phi_{24}$$

Next, calculate the emissive power of each part of the surface using Eq. (3.12):

$$E_{R1} = (1 - \varepsilon_1)(\Phi_{12} E_{R2} + \Phi_{14} E_{R4} + \Phi_{15} E_{b5}) + \varepsilon_1 E_{b1}$$

$$E_{R2} = \frac{1}{1 - \Phi_{22}(1 - \varepsilon_2)} [(1 - \varepsilon_2)(\Phi_{21} E_{r1} + \Phi_{23} E_{R3} + \Phi_{24} E_{R4} + \Phi_{25} E_{b5} + \varepsilon_2 E_{b2})]$$

$$E_{R3} = \frac{1}{1-\Phi_{33}(1-\varepsilon_3)}[(1-\varepsilon_3)(\Phi_{31}E_{R1}+\Phi_{32}E_{R2}+\Phi_{34}E_{R4}+\Phi_{35}E_{b5}+\varepsilon_3E_{b3})]$$

$$E_{R4} = \frac{1}{1-\Phi_{44}(1-\varepsilon_4)}[(1-\varepsilon_4)(\Phi_{41}E_{R1}+\Phi_{42}E_{R2}+\Phi_{43}E_{R3}+\Phi_{45}E_{b5}+\varepsilon_4E_{b4})]$$

Substitute all the values to obtain:

$$E_{R1} = 0.252E_{R2}+0.078E_{R3}+0.03E_{R4}+89341$$

$$E_{R2} = 0.1479E_{R1}+0.1021E_{R3}+0.2817E_{R4}+104848$$

$$E_{R3} = 0.04577E_{R1}+0.1021E_{R2}+0.1021E_{R4}+104848$$

$$E_{R4} = 0.01761E_{R1}+0.02817E_{R2}+0.1021E_{R3}+104848$$

Then the equation set can be solved by the Gauss Iteration Method:

$$E_{R1} = 1.4003\times10^5[W/m^2]$$

$$E_{R2} = 1.4326\times10^5[W/m^2]$$

$$E_{R3} = 1.3872\times10^5[W/m^2]$$

$$E_{R4} = 1.2557\times10^5[W/m^2]$$

Finally, calculate the total amount of radiation heat transfer. The outward radiation heat transfer values of each part are:

$$Q_i = \frac{E_{oi}-E_{Ri}}{(1-\varepsilon_i)/\varepsilon_i A_i} = \frac{\varepsilon_i A_i}{1-\varepsilon_i}(E_{bi}-E_{Ri})$$

$$Q_1 = \frac{\varepsilon_1 A_1}{1-\varepsilon_1}(E_{b1}-E_{R1}) = \frac{0.6\times\pi\times10^{-4}}{1-0.6}(1.4887-1.4003)\times10^5 = 4.1658[W]$$

$$Q_2 = \frac{\varepsilon_2 A_2}{1-\varepsilon_2}(E_{b2}-E_{R2}) = \frac{0.6\times2\pi\times10^{-4}}{1-0.6}(1.4887-1.4326)\times10^5 = 5.2873[W]$$

$$Q_3 = \frac{\varepsilon_3 A_3}{1-\varepsilon_3}(E_{b3}-E_{R3}) = \frac{0.6\times2\pi\times10^{-4}}{1-0.6}(1.4887-1.3872)\times10^5 = 9.5661[W]$$

$$Q_4 = \frac{\varepsilon_4 A_4}{1-\varepsilon_4}(E_{b4}-E_{R4}) = \frac{0.6\times2\pi\times10^{-4}}{1-0.6}(1.4887-1.2557)\times10^5 = 21.959[W]$$

and the total amount of outward, radiation heat transfer of the circular hole is:

$$\Sigma Q = Q_1+Q_2+Q_3+Q_4 = 40.979[W]$$

2. Integral solution

Assuming the radiosity is uniform, consider the inner surface in its entirety:

$$A_1 = \pi+3\times2\pi = 7\pi[cm^2], A_5 = \pi[cm^2]$$

$$\Phi_{51} = 1.0, \varepsilon_1 = 0.6, \varepsilon_5 = 1.0$$

$$Q = \frac{E_{b1} - E_{b5}}{\dfrac{1-\varepsilon_1}{\varepsilon_1 A_1} + \dfrac{1}{A_5} + \dfrac{1-\varepsilon_5}{\varepsilon_5 A_5}} = \frac{A_5(E_{b1} - E_{b5})}{\left(\dfrac{1}{\varepsilon_1} - 1\right)\dfrac{A_5}{A_1} + \dfrac{1}{\varepsilon_5}} \tag{3.13}$$

$$= \frac{\pi \times 10^{-4}(1.4887 \times 10^{-5} - 417.8)}{\left(\dfrac{1}{0.6} - 1\right)\dfrac{1}{7} + 1}$$

$$= 42.581[W]$$

Comparing the above two solutions, the relative error is $\Delta = \dfrac{42.581 - 40.979}{40.979} =$ 3.9%.

Therefore, the integral solution can be utilized to simplify the calculation in practice.

If the circular hole surface 5 is small enough compared to hole depth, surface 5 can be considered a blackbody surface and the outward radiation of the circular hole can be replaced by the outward radiation of surface 5 at the same temperature. In this case, because the size of surface 5 is not small compared to hole depth, this solution is not appropriate.

$$Q_1 = A_5 \sigma T_1^4 = \pi \times 10^{-4} \times 5.67 \times 10^{-8} \times 1273^4 = 46.78[W]$$

Compared to the stepwise calculation result for the same scenario, the error is:

$$\Delta = \frac{46.78 - 40.98}{40.98} = 14.15\%$$

Now, let's consider a case where $\lambda_w = \infty$. For a water cooling hole wall, $T_1 > T_2 \approx T_3$. Because the hole wall approximates a blackbody (due to ash deposition), and flame surface A_1 and outer surface A_1 both can be considered blackbodies, we have the following:

$$\begin{cases} Q_{13} = E_{b1}\varphi_{13}A_1 \\ Q_{12} = E_{b1}(1-\varphi_{13})A_1 \end{cases} \tag{3.14}$$

where $E_{b1} = \varepsilon T_1^4$. The configuration of a circular hole is:

$$\varphi_{13} = \frac{2\left(1 + \dfrac{d^2}{2h^2} - \sqrt{1 + \dfrac{d^2}{h^2}}\right)}{\dfrac{d^2}{h^2}}$$

where d and h are the diameter and length of the circular hole, respectively.

Let's consider a metal hole wall, now – hole wall 2 is an insulating surface with uniform temperature, thus we obtain the following by applying the network method:

$$Q = \frac{E_{b1} - E_{b3}}{\dfrac{1-\varepsilon_1}{\varepsilon_1} + R\dfrac{1-\varepsilon_3}{\varepsilon_1 A_3}} \tag{3.15}$$

where R is the thermal resistance between node 1 and node 3.

The detailed calculation process is not listed here. If we consider the system to be three blackbodies, the calculation is simpler.

3.1.4 Radiative Heat Transfer of Hot Surface, Water Wall, and Furnace Wall

To conclude this section, let's briefly discuss the radiation heat transfer between a flame, water wall, and furnace wall – this is a common concern in real-world engineering. Assume that the water wall is smooth and comprised of a number of discrete tubes.

First, we'll determine the configuration factor of the water wall as shown in Fig. 3.9.

First, define the configuration factor from the flame to the water wall φ_{gt}:

$$\varphi_{gt} = 1 + \frac{d}{s}\arctan\sqrt{\left(\frac{s}{d}\right)^2 - 1} - \sqrt{1 - \left(\frac{d}{s}\right)^2} \tag{3.16}$$

Then, find the configuration factor from the water wall to the flame φ_{tg}:

$$\varphi_{tg} = \frac{s}{\pi d}\varphi_{gt} = \frac{1}{\pi}\left(\frac{s}{d} + \arctan\sqrt{\left(\frac{s}{d}\right)^2 - 1} - \sqrt{\left(\frac{s}{d}\right)^2 - 1}\right) \tag{3.17}$$

followed by the configuration factor between the water wall tubes φ_{tt}—because $\varphi_{tg} = \varphi_{tw}$ and $\varphi_{tt} + 2\varphi_{tg} = 1$, φ_{tt} can be calculated as follows:

$$\varphi_{tt} = 1 - 2\varphi_{tg} = \frac{2}{\pi}\left(\sqrt{\left(\frac{s}{d}\right)^2 - 1} - \frac{s}{d} + \arcsin\frac{d}{s}\right) \tag{3.18}$$

Next, we'll determine the configuration factor from the flame to the furnace wall φ_{gw}:

$$\varphi_{gw} = 1 - \varphi_{tg} = \sqrt{1 - \left(\frac{d}{s}\right)^2} - \frac{d}{s}\arctan\sqrt{\left(\frac{s}{d}\right)^2 - 1} $$

$$\varphi_{wg} = \varphi_{gw} \tag{3.19}$$

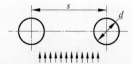

FIGURE 3.9 Configuration factor from the flame to the water wall.

Notably, the above discussion takes place under the premise that the temperatures of the flame, water wall tube, and furnace wall are uniform. Of course, this is not always true in practice, where it should be calculated according to the actual conditions. For the membrane water walls in large and medium-sized boilers, for example, the above relation becomes simpler because there is no direct radiation heat transfer between the flame and furnace wall. Water wall tubes are still commonly used in small-sized industrial boilers, for which the configuration factor from the flame to the furnace wall have to be involved.

Next, let's look at effective configuration factor x. In a furnace, it is generally assumed that the flame is the hot surface, the water wall is the heating surface, and the furnace wall is the insulating surface. Assume that the emissivity of the flame ε_g and water wall ε_t are 1, and the temperature of each surface is uniform. Based on configuration factor properties, the heat balance relation, and the fourth-power law, the heat absorbed by the water wall can be derived as follows:

$$Q_t = \sigma\left(T_g^4 - T_t^4\right)A\varphi_{gt}(1+\varphi_{gw}) \tag{3.20}$$

For calculation in practice, Eq. (3.20) can be rewritten as:

$$Q = \sigma\left(T_g^4 - T_t^4\right)H \tag{3.21}$$

in which:

$$H = Ax$$
$$x = \varphi_{gt}(1+\varphi_{gw}) = \varphi_{gt}+\varphi_{gt}\varphi_{gw} \tag{3.21a}$$
$$= \varphi_{gt}+\varphi_{wt}\varphi_{gw}$$

$$= \varphi_{gt} + \varphi_{wt}(1-\varphi_{gt}) \tag{3.21b}$$

where H is the area of the effective heating surface, A is the area of the furnace wall arranged with water walls, and x is the effective configuration factor of the water wall.

It is worth noting that x differs from the configuration factor defined by heat transfer—x is a compound quantity including two parts: the configuration factor from the flame to the tube directly, and the configuration factor from the reflected flame from the furnace wall to the tube. By definition, x only considers the effect of furnace wall reflections and not the effects of tube emissivity.

Let's look now at reducing the emissivity of the furnace surface. While calculating the heat transfer in a furnace, the water wall and the furnace wall are sometimes considered a whole, called a "furnace surface" (Fig. 3.10).

A furnace surface is generally considered to be a graybody. The ratio between the absorbed radiation q_α and irradiation q_I of the furnace surface can be defined as the effective absorptivity of the furnace surface α_{fw}, that is, the effective emissivity ε_{fw}:

$$\varepsilon_{fw} = \frac{q_\alpha}{q_I} \tag{3.22}$$

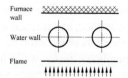

FIGURE 3.10 **Furnace surface radiation.**

where the absorbed radiation heat q_α consists of five parts:

1. The heat absorbed by the water wall:

$$q_1 = q_1 \varphi_{gt} \varepsilon_t \tag{3.23}$$

2. The heat absorbed by the furnace wall:

$$q_2 = q_1 (1 - \varphi_{gt}) \varepsilon_w \tag{3.24}$$

3. The heat reflected from the furnace wall and absorbed by the water wall:

$$q_3 = q_1 (1 - \varphi_{gt})(1 - \varepsilon_w) \varphi_{wt} \varepsilon_t \tag{3.25}$$

4. The heat reflected from the adjacent water wall tubes and absorbed by the water wall tube:

$$q_4 = q_1 \varphi_{gt} (1 - \varepsilon_t) \varphi_{tt} \varepsilon_t \tag{3.26}$$

5. The heat reflected from the water wall and then again absorbed by the water wall:

$$q_5 = q_1 \varphi_{gt} (1 - \varepsilon_t) \varphi_{tw} \varepsilon_w \tag{3.27}$$

Thus, q_α is:

$$q_\alpha = q_1 + q_2 + q_3 + q_4 + q_5 = \sum_{i=1}^{5} q_i$$

Then ε_{fw} can be calculated using Eq. (3.22).

In short, the use of appropriate engineering approaches and necessary simplified methods allows us to derive simple algorithms that are applicable for engineering calculations in practice.

3.2 RADIATIVE HEAT EXCHANGE BETWEEN AN ISOTHERMAL MEDIUM AND A SURFACE

Chapter: Emission and Absorption of Thermal Radiation contains a discussion on the radiation characteristics of absorption-scattering media; the following section further describes the characteristics of radiation heat transfer between the medium and the furnace wall.

3.2.1 Heat Transfer Between a Medium and a Heating Surface

The main purpose of studying the heat transfer in a furnace is to determine the radiation heat transfer between the high-temperature medium (flue gas) generated from combustion, and the surfaces (eg, heating surface, furnace surface). It is typically assumed that the temperature, emissivity, and absorptivity of the medium and the surfaces are uniform. In real-world conditions in which the parameters are nonuniform, partition refinement can be applied to the system until the parameters of each part can be considered uniform. The following discussion complies with these assumptions.

The heat transfer between the medium and heating surface can be calculated as follows. The emissive power of the heating surface and the medium are $\varepsilon_t \sigma T_t^4 A_r$ and $\varepsilon_g \sigma T_g^4 A_r$. The emissive power of the medium (ie, the radiation projected onto the heating surface) is the sum of the medium's own radiation and the transmissive radiation of the heating wall, $\varepsilon_g \sigma T_g^4 A_r + Q_{Rt}(1-\alpha_g)$, so the reflected radiation of the heating surface is $(1-\alpha_t)\left[\varepsilon_g \sigma T_g^4 A_r + Q_{Rt}(1-\alpha_g)\right]$. Thus, the radiosity equation for the heating surface is:

$$Q_{Rt} = \varepsilon_t \sigma T_t^4 A_r + (1-\alpha_t)\left[\varepsilon_g \sigma T_g^4 A_r + Q_{Rt}(1-\alpha_g)\right] \tag{3.28}$$

Solving the above equation gives us:

$$Q_{Rt} = \frac{\varepsilon_t T_t^4 + \varepsilon_g (1-\alpha_t) T_g^4}{1-(1-\alpha_t)(1-\alpha_g)}\sigma A_r \tag{3.29}$$

Considering the heat, the medium's loss equals the amount of radiation heat transfer, then:

$$
\begin{aligned}
Q &= Q_g - \alpha_g Q_{Rt} \\
&= \varepsilon_g \sigma T_g^4 A_r - \alpha_g Q_{Rt} \\
&= \frac{\dfrac{\varepsilon_g}{\alpha_g} T_g^4 - \dfrac{\varepsilon_t}{\alpha_t} T_t^4}{\dfrac{1}{\alpha_t} + \dfrac{1}{\alpha_g} - 1}\sigma A_r
\end{aligned}
\tag{3.30}
$$

Let's take a closer look at Eq. (3.30). If the heating surface is a graybody, then $\varepsilon_t = \alpha_t$, thus the equation can be simplified as follows:

$$Q = \frac{\alpha_t}{\alpha_g + \alpha_t - \alpha_g \alpha_t}\left(\varepsilon_g T_g^4 - \alpha_t T_t^4\right)\sigma A_r \tag{3.31}$$

and because $\alpha_g + \alpha_t(1-\alpha_g) < 1, \alpha_t + \alpha_g(1-\alpha_t) > \alpha_t$, then:

$$1 > \frac{\alpha_t}{\alpha_g + \alpha_t - \alpha_g \alpha_t} > \alpha_t$$

The absorptivity of the cooling water tube α_t is approximately 0.8. Let $\dfrac{\alpha_t}{\alpha_g + \alpha_t - \alpha_g \alpha_t} \approx \dfrac{1 + \alpha_t}{2} = \dfrac{1 + \varepsilon_t}{2}$, thus Eq. (3.31) becomes:

$$Q = \frac{1 + \varepsilon_t}{2} \varepsilon_g T_g^4 \left[1 - \frac{\alpha_g}{\varepsilon_g} \left(\frac{T_t}{T_g} \right)^4 \right] \sigma A_r \qquad (3.32)$$

Eq. (3.32) is commonly used to calculate the radiation heat transfer of boiler back-end surfaces.

If the heating surface and the medium are both graybodies, then $\varepsilon_t = \alpha_t$, $\varepsilon_g = \alpha_g$, so Eq. (3.30) can be simplified as follows:

$$Q = \alpha_F \sigma \left(T_g^4 - T_t^4 \right) A_r \qquad (3.33)$$

where $\alpha_F = \dfrac{1}{\dfrac{1}{\alpha_g} + \dfrac{1}{\alpha_t} - 1}$ is the furnace emissivity, namely, the system emissivity.

3.2.2 Heat Transfer Between a Medium and a Furnace

A furnace consists of a heating surface and a furnace wall. Two different cases are investigated here: one in which the heating surface and furnace wall are separate, and another in which the heating surface and furnace wall can be considered one unit (such as a furnace surface consisting of a bare tube wall and a furnace wall).

To determine the heat transfer between the medium, furnace wall, and heating surface, define cooling factor X as $X = \dfrac{A_r}{A_r + A_w}$ and let $\omega = \dfrac{A_r}{A_w}$, thus $\omega = \dfrac{X}{1 - X}$. $A_r = XA$. Cooling factor X and effective configuration factor $x = \varphi_{gt} + (1 - \varphi_{gt}) \varphi_{wt}$ are the same, though they are expressed from different perspectives and forms.

See the following configuration factors: from heating surface to heating surface $\varphi_{tt} = 1 - \varphi_{tw}$, from heating surface to furnace wall $\varphi_{tw} = 1 - \varphi_{tt}$, from furnace wall to heating surface $\varphi_{wt} = (1 - \varphi_{tt}) \omega$, and from furnace wall to furnace wall $\varphi_{ww} = 1 - \varphi_{wt}$.

Medium radiation Q_g is determined as follows:

$$Q_g = \varepsilon_g T_g^4 A$$

where T_g is medium temperature, A is the area, and $A = A_r + A_w$.

The emissive power of a heating surface consists of the following parts: the heating surface's own radiation $\varepsilon_t \sigma A_r T_t^4$, the reflected radiation of the flame radiation $Q_g X (1 - \alpha_t)$, the reflected radiation of the radiosity of the furnace wall $Q_{Rw} [(1 - \varphi_{tt}) \omega] (1 - \alpha_g)(1 - \alpha_t)$, and the reflected radiation of the radiosity of the heating surface $Q_{Rw} \varphi_{tt} (1 - \alpha_g)(1 - \alpha_t)$.

The emissive power of the furnace wall is equal to the radiation projected onto it, because the furnace wall is insulated. Furnace radiosity consists of the following parts: the irradiation caused by flame radiation $Q_g(1-x)$, irradiation caused by heating surface radiation $Q_{Rt}(1-\varphi_{tt})(1-\alpha_g)$, and irradiation caused by furnace wall radiation $Q_{Rw}[1-(1-\varphi_{tt})\omega](1-\alpha_g)$.

According to the heat balance principle, the heat that the heating surface receives is equal to that the flame loses, thus:

$$Q = Q_g - Q_{Rt}\alpha_g - Q_{Rw}\alpha_g \qquad (3.34)$$

Combining all the above equations yields the detailed expression of Q.

Next, let's determine the heat transfer between the medium and a surface considered to be one unit that includes the heating surface and furnace wall. There are two cases that demonstrate this particularly well: the suspension-firing furnace (Fig. 3.11) and the grate furnace (Fig. 3.12).

The overall peripheral area of the suspension-firing furnace is $A_{fw} = A_r + A_w$, $x = \dfrac{A_r}{A_w}$, where A_r is the area of the heating surface and A_w is the area of the furnace wall. (The furnace wall is originally denoted by subscript fw, but for simplicity, subscript τ is used to denote it in the following derivation.)

The emissive power of the flame is $Q_{Rg} = Q_g + (1-\alpha_g)Q_{Rt}$, where Q_g is the flame radiation itself. The emissive power of the furnace wall is $Q_{Rt} = Q_g + (1-x\alpha_t)Q_{Rg}$, where Q_t is the furnace wall radiation itself. Thus:

$$Q_{Rg} = \frac{Q_g + (1-\alpha_g)Q_t}{1-(1-\alpha_g)(1-x\alpha_t)} \qquad (3.35)$$

FIGURE 3.11   Suspension-firing furnace.

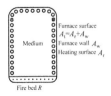

FIGURE 3.12   Grate furnace.

$$Q_{Rg} = \frac{Q_t + (1 - x\alpha_t)Q_g}{1 - (1 - \alpha_g)(1 - x\alpha_t)} \qquad (3.36)$$

By considering the heat the flame loses as the radiation transfer heat, we obtain:

$$Q = Q_{Rg} - Q_{Rt} = \frac{\dfrac{Q_g}{\alpha_g} - \dfrac{Q_t}{x\alpha_t}}{\dfrac{1}{x\alpha_t} + \dfrac{1}{\alpha_g} - 1} \qquad (3.37)$$

Substituting $Q_g = \alpha_g A_t \sigma T_g^4$, $Q_t = \alpha_t A_r \sigma T_t^4$ into Eq. (3.37) yields:

$$Q = \sigma \alpha_F A_r \left(T_g^4 - T_t^4 \right) \qquad (3.38)$$

where α_F is furnace emissivity, namely system emissivity, which can be calculated as follows:

$$\alpha_F = \frac{1}{\dfrac{1}{\alpha_t} + x\left(\dfrac{1}{\alpha_g} - 1\right)} = \frac{1}{\dfrac{1}{\varepsilon_t} + x\left(\dfrac{1}{\varepsilon_g} - 1\right)} \qquad (3.39)$$

Define $r = \dfrac{R}{A_t}$ for the grate furnace, where R is the fire bed area, A_t is the furnace area and $x = \dfrac{A_r}{A_t} = \dfrac{A_r}{A_w + A_r}$. Similar to the suspension-firing furnace, the following two formulas can be derived:

$$Q_{Rg} = Q_g + Q_{Rt}\varphi_{tt}(1 - \alpha_g) + Q_R = Q_g + Q_{Rt}(1 - \gamma)(1 - \alpha_g) + Q_R \qquad (3.40)$$

$$Q_{Rt} = Q_t + (1 - x\alpha_t)Q_{Rg} \qquad (3.41)$$

where Q_R is the emissive power of the fire bed, $Q_g = \alpha_g A_t \sigma T_g^4$ is the flame radiation itself, and $Q_t = \alpha_t A_r \sigma T_t^4$ is the water wall radiation itself. Thus, the emissive power of the fire bed is:

$$Q_R = (1 - \alpha_g)R\sigma T_g^4 = (1 - \alpha_g)xA\sigma T_g^4 \qquad (3.42)$$

Substituting Eq. (3.42) into Eqs. (3.40) and (3.41) yields:

$$Q_{Rt} = \frac{x\alpha_t A_t \sigma T_t^4 + (1 - x\alpha_t)MA_t \sigma T_g^4}{x\alpha_t + (1 - x\alpha_t)M} \qquad (3.43)$$

$$Q_{Rg} = \frac{MA_t \sigma T_g^4 + (1 - M)x\alpha_t A_t \sigma T_t^4}{x\alpha_t + (1 - x\alpha_t)M} \qquad (3.44)$$

where $M = \alpha_g + r(1 - \alpha_g)$ is an intermediate variable. Therefore, the heat exchange quantity is:

$$Q = Q_{Rg} - Q_{Rt} = \frac{M\alpha_t x A_t \sigma \left(T_g^4 - T_t^4\right)}{x\alpha_t + (1 - x\alpha_t)M} = \alpha_F A_r \sigma \left(T_g^4 - T_t^4\right) \quad (3.45)$$

where system emissivity is:

$$\alpha_F = \frac{M\alpha_t}{x\alpha_t + (1 - x\alpha_t)M} = \frac{1}{\dfrac{1}{\alpha_t} + x\left(\dfrac{1}{M} - 1\right)} \quad (3.46)$$

The above derivation is based on the fourth-power law, which attributes the effect of medium properties and geometric characteristics to a system emissivity.

3.2.3 Calculating Radiative Heat Transfer According to Projected Heat

The former Soviet Union's standard uses the Gurvitch method to perform the boiler thermal calculation, with slight differences between the 1957 version and the 1973 version. In the Gurvitch method, radiation heat transfer is calculated based on irradiation.

1. 1957 standard
The 1957 standard contains the following fundamental assumptions: that the heating surface (ie, the water wall) is a blackbody with temperature of 0 K (thus $\alpha_t = 1, T = 0\,K$), that the water walls are arranged uniformly, and that the flame is a graybody.

Based on these assumptions, derivation from Eqs. (3.38) and (3.39) yields the following:

$$Q = \alpha_F \sigma A_r T_g^4 \quad (3.47)$$

$$\alpha_F = \frac{1}{1 + x\left(\dfrac{1}{\alpha_g} - 1\right)} \quad (3.48)$$

In actuality, the water wall is not a blackbody, so in practice a coefficient of 0.82 is introduced into furnace emissivity α_F to amend the result. In addition, there may be fouling on the surface of the water wall, and the temperature of the wall is quite high, thus fouling factor ζ is also introduced for correction. α_F can then be expressed as:

$$\alpha_F = \frac{0.82}{1 + \left(\dfrac{1}{\alpha_g} - 1\right)\zeta x} \quad (3.49)$$

The fouling factor will be discussed below in detail. Eq. (3.49) has fallen out of common use; instead, modern engineers utilize the equations in the 1973 standard.

2. 1973 standard

The assumption necessary for calculation under the 1957 standard is obviously not reasonable. Although introducing correction factors improves the calculation to a certain extent, the results are still not sufficient. The thermal efficiency coefficient obtained from experiments under the 1973 standard can be used to amend the heat exchange quantity.

The thermal efficiency coefficient is the ratio between the absorbed radiation energy Q by the water wall and incident radiant energy Q_I onto the water wall. See the following:

$$\psi = \frac{Q}{Q_I} = \frac{Q_I - Q_{Rt}}{Q_I} \tag{3.50}$$

where Q_{Rt} is the radiosity of the water wall. Q_I and Q_{Rt} can be measured with a radiometer, thus ψ is an empirical coefficient. Under the 1973 standard, the assumption that the temperature of the water wall is 0 K is removed and instead it is assumed that ψ is constant, so the heat exchange quantity becomes:

$$Q = \psi Q_I \tag{3.51}$$

Replacing x with ψ and using the 1957 standard form yields:

$$Q = \tilde{\alpha}_F \psi A_t \sigma T_g^4 \tag{3.52}$$

where $\tilde{\alpha}_F$ is the furnace emissivity for calculating incident radiant energy Q_I, which is different from the system emissivity discussed in heat transfer theory [see Eq. (3.46)].

For a suspension-firing furnace:

$$\tilde{\alpha}_F = \frac{1}{1 + \psi \left(\dfrac{1}{\alpha_g} - 1 \right)} \tag{3.53}$$

For a grate furnace:

$$\tilde{\alpha}_F = \frac{1}{1 + \psi \left(\dfrac{1}{M} - 1 \right)}, M = \alpha_g + \left(1 - \alpha_g\right)r \tag{3.54}$$

Next, let's explore the relation between ψ, x, and ζ. By definition, thermal efficiency coefficient ψ is the ratio between the radiant energy absorbed by heating surface Q and the radiant energy projected onto furnace surface Q_I; the effective configuration factor x is the ratio between the radiant energy projected onto heating surface Q_{Ir} and that projected onto furnace surface Q_I. Fouling factor ζ is the ratio between the radiant energy absorbed by heating surface Q and the radiant energy projected onto heating surface Q_{Ir}, that is:

$$\begin{cases} \psi = \dfrac{Q}{Q_1} \\[2mm] x = \dfrac{Q_{1r}}{Q_1} \\[2mm] \zeta = \dfrac{Q}{Q_{1r}} \end{cases}$$

Thus:

$$\psi = x\zeta \tag{3.55}$$

The effects of fouling on both surface emissivity and temperature are included in ζ. Table D14 in the appendix lists the values of ζ under different conditions.

3.3 RADIATIVE HEAT EXCHANGE BETWEEN A FLUE GAS AND A HEATING SURFACE WITH CONVECTION [12,17]

Heating surfaces in the back-end surface, superheater, reheater, economizer, and air preheater are convection heat surfaces, thus there is not only radiation heat transfer between the flue gas and the heating surface, but also convective heat transfer. In addition, convective heat transfer is significant in low-temperature areas (such as the air preheater). The following provides a brief introduction to calculating the radiation heat transfer combined with a few forms of heat transfer.

Eq. (3.32) gives us the radiation heat transfer formula for the medium surrounded by the heating surface. By rewriting it in heat flux form, we obtain:

$$q_r = \frac{Q_r}{A_r} = \sigma \frac{1 + \varepsilon_t}{2} \varepsilon_g T_g^4 \left[1 - \frac{\alpha_g}{\varepsilon_g} \left(\frac{T_t}{T_g} \right)^4 \right] \tag{3.56}$$

Assuming that the medium is a graybody ($\alpha_g = \varepsilon_g$), the sum of convective heat transfer q_c and radiation heat transfer q_r is:

$$\begin{aligned}
q &= q_c + q_r \\
&= h_c(T_g - T_t) + \sigma \frac{1 + \varepsilon_t}{2} \varepsilon_g T_g^4 \left[1 - \left(\frac{T_t}{T_g} \right)^4 \right] \\
&= h_c(T_g - T_t) + \sigma \frac{1 + \varepsilon_t}{2} \varepsilon_g T_g^3 \frac{1 - \left(\dfrac{T_t}{T_g} \right)^4}{1 - \dfrac{T_t}{T_g}} (T_g - T_t) \\
&= h_c(T_g - T) + h_r(T_g - T_t) \\
&= h_1(T_g - T_t)
\end{aligned} \tag{3.57}$$

where $h_1 = h_c + h_r$ is the overall heat transfer coefficient at the flue gas side, in which the radiation heat transfer coefficient is:

$$h_r = \sigma \frac{1+\varepsilon_t}{2} \varepsilon_g T_g^3 \frac{1-\left(\dfrac{T_t}{T_g}\right)^4}{1-\dfrac{T_t}{T_g}} \tag{3.58}$$

Considering that the flue gas flow field within the heating surface is nonuniform, thus utilization coefficient ξ is necessary to amend the heat transfer coefficient:

$$h_1' = \xi h_1 \tag{3.59}$$

ξ can be obtained from the related charts or tables in application, which are usually proposed based on experimental or operational data of the boiler.

The overall heat transfer quantity is:

$$Q = K \Delta t A_r \tag{3.60}$$

in which the heat transfer coefficient is given by:

$$K = \frac{1}{\dfrac{1}{h_1} + \rho + \dfrac{1}{h_2}} \tag{3.61}$$

where $\rho = \dfrac{\delta_a}{\lambda_a}$ is the ash deposit coefficient (ie, fouling heat resistance), δ_a is the thickness of the fouling layer, h_2 is the heat transfer coefficient of the working medium (water or steam) side, Δt is the temperature difference between the flue gas and the working medium, and A_r is the heat transfer area, which can be determined in two different ways—when one side is steam or water and the other side is flue gas, then A_r is the area of the flue gas side; when the two sides are both gas (eg, flue gas and air), then A_r is the average area of both sides.

A few supplementary explanations are necessary for h_r. Let's first determine the meaning of each term in Eq. (3.58): ε_t is the emissivity of the heating surface, ($\varepsilon_t = 0.68$ for the slag screen and $\varepsilon_t = 0.80$ in any other case), ε_g is the emissivity of the flue gas (see chapter: Emission and Absorption of Thermal Radiation), T_g is the temperature of the flue gas, which is the average of the temperature at the inlet and outlet of the heating surface (when the temperature difference is greater than 300°C, use the logarithmic mean value; otherwise, use the arithmetic average value). T_t is the temperature of the furnace surface (in K), which is calculated as follows:

$$T_t = t_t + 273$$

$$t_t = t + \left(\rho + \frac{1}{h_2}\right) \frac{B_{cal}}{A_r}(Q + Q_r) \tag{3.62}$$

where t is the average temperature of the working medium, and ρ is the ash deposition coefficient. Eq. (3.62) is applicable for platen superheaters and convection superheaters. For economizers and air preheaters, the equation can be simplified as follows:

$$t_t = t + \Delta t$$

where $\Delta t = 80°C$ at high-temperature sections, $\Delta t = 60°C$ at mid-temperature sections, and $\Delta t = 25°C$ at low-temperature sections.

Next, correct the front space of the tube bundle as follows:

$$h'_r = h_r \left[1 + A \left(\frac{T_g^0}{1000} \right)^{0.25} \left(\frac{l_t^0}{l_t} \right)^{0.07} \right] \tag{3.63}$$

where l_t^0 is the depth of the radiation space (in front of the tube bundle), l_t is the depth of the tube bundle, T_g^0 is the temperature of the flue gas in the front space of the tube bundle, and A is a coefficient that equals 0.2 when burning heavy oil or coal gas, 0.4 when burning anthracite, lean coal, or bituminous coal, and 0.5 when burning lignite or shale.

Example 3.3
Find the radiation heat transfer coefficient h_r, heat transfer coefficient K, and heat transfer quantity Q in a system where the final superheater of the boiler has the following properties.

The area of the heating surface $H = 122.5 \text{ m}^2$.
The flue gas effective radiation layer thickness $s = 0.252 \text{ m}$.
The average temperature of the flue gas $\theta_g = 706.8 \text{ °C}$, $T_g = 980 \text{ K}$.
The average temperature of the working medium $t = 361.8°C$.
The average temperature difference $\Delta t = 344.6°C$.
The convective heat transfer coefficient of the flue gas $h_c = 72.8 \text{ W/(m}^2 \text{ °C)}$.
The convective heat transfer coefficient of the working medium $h_2 = 1357 \text{ W/}$ $(\text{m}^2 \text{ °C})$.
The flue gas component (volume fraction) $r_{H_2O} = 0.058, r_n = 0.1983$.
Fly ash concentration (mass fraction) $\mu_{fa} = 0.0109 \text{ kg/kg}$.
The diameter of fly ash particles $d_{fa} = 13 \text{ μm}$.
The ash depositon coefficient of the heating surface $\rho = 0.00537 \text{ m}^2 \cdot °\text{C/W}$.
Flue gas density $\rho_g = 1.33 \text{kg/m}^3$.
The emissivity of fouling on the tube wall $\varepsilon_w = 0.82$.

Solution
By analyzing the known conditions, it becomes clear that if h_r is obtained, then K and Q are easily calculated. In order to obtain h_r, two unknown variables (ε_g and T_w) should be determined first; ε_g can be determined according to calculation equations for flue gas containing ash.

The overall extinction coefficient of flue gas K is the sum of K_g and K_{fa}. K_g and K_{fa} are expressed as follows:

$$K_g = k_g r_n p = \left(\frac{0.78 + 1.6 r_{H_2O}}{\sqrt{\frac{p_n}{p_0}} s} - 0.1 \right) \left(1 - 0.37 \frac{T_g}{1000} \right) r_n p$$

$$= \left(\frac{0.78 + 1.6 \times 0.058}{\sqrt{0.1983 \times 0.252}} - 0.1 \right) \left(1 - 0.37 \frac{980}{1000} \right) \times 0.1983 \times 1$$

$$= 0.481 (\text{m}^{-1})$$

$$K_{fa} = K_{fa} \mu_{fa} p = \frac{4300 \rho_g \mu_{fa} p}{\left(T_g^2 d_{fa}^2 \right)^{1/3}} = \frac{4300 \times 1.33 \times 0.0109 \times 1}{(980^2 \times 13^2)^{1/3}} = 0.114 (\text{m}^{-1})$$

Thus, the overall extinction coefficient of flue gas K can be calculated as:

$$K = K_g + K_{fa} = 0.481 + 0.114 = 0.595 (\text{m}^{-1})$$

$$\varepsilon_g = 1 - e^{-Ks} = 1 - e^{-0.595 \times 0.252} = 1 - e^{-0.150} = 0.139$$

So $t_w = t + \dfrac{Q}{H} \left(\rho + \dfrac{1}{h_2} \right)$, because $\dfrac{Q}{H} = \dfrac{t_w - t}{\varepsilon + 1/h_2}$ (obtained from comparing it with equation $t_w = t + \dfrac{B_{cal}}{H} \left(\rho + \dfrac{1}{h_2} \right) (Q + Q_r)$, which is an expression for heat transfer and heat balance in the boiler thermal calculation and is not given here in detail. Since there is no Q_r in the final superheater, thus $Q_r = 0$; also, because the unit of Q is kJ/kg, B_{cal} is required).

Obviously, t_w is related to heat transfer quantity Q, but both are unknown. Therefore, we can only assign a value to t_w or Q and perform trial and error calculations until the error between the assumed value and the calculated value is acceptably small.

First, assume that $Q = 2.46 \times 10^6$ W and calculate the following:

$$t_w = t + Q/H(\rho + 1/h_2) = 361.8 + 2.46 \times 10^6/122.5(0.00537 + 1/1357) = 484.4(^\circ\text{C})$$
$$T_w = t_w + 273 = 757.4(\text{K})$$

Then the radiation heat transfer coefficient is:

$$h_r = \sigma(1 + \varepsilon_w)/2 \varepsilon_g T_g^3 \frac{1 - (T_w/T_g)^4}{1 - (T_w/T_g)}$$

$$= 5.67 \times 10^{-8} \times (1 + 0.82)/2 \times 0.139 \times 980^3 \times \frac{1 - (757.4/980)^4}{1 - (757.4/980)}$$

$$= 19.12 \left(\text{W/m}^2 \cdot {}^\circ\text{C} \right)$$

So the overall heat transfer coefficient is:

$$K = \frac{1}{1/(h_c + h_r) + \rho + 1/h_2} = \frac{1}{1/(72.8 + 19.12) + 0.00537 + 1/1357}$$
$$= 58.872 \left(W/m^2 \cdot °C \right)$$
$$Q = H\Delta t K = 122.5 \times 344.6 \times 58.872 = 2.49 \times 10^6 (W)$$

The relative error between the calculation value and the assumed value is 1.2%, which is within ±15%, so there is no need to recalculate.

Let's assign a value to t_w for trial and error calculation next.

For example, assume that $t_w = 500 °C$, $T_w = 773$ K, then:

$$h_r = \sigma[(1 + \varepsilon_w)/2]\varepsilon_g T_g^3 \frac{1 - (T_w/T_g)^4}{1 - (T_w/T_g)}$$
$$= 5.67 \times 10^{-8}(1 + 0.82)/2 \times 0.139 \times 980^3 \times \frac{1 - (773/980)^4}{1 - (773/980)}$$
$$= 19.59 (W/m^2 \cdot °C)$$

$$K = \frac{1}{1/(h_c + h_r) + \rho + 1/h_2} = \frac{1}{1/(72.8 + 19.59) + 0.00537 + 1/1357}$$
$$= 59.06 \left(W/m^2 \cdot °C \right)$$

$$Q = H\Delta t K = 122.5 \times 344.6 \times 59.06 = 2.493 \times 10^6 (W)$$
$$t_w = t + (\rho + 1/\alpha_2)Q/H = 361.8 + (0.00537 + 1/1357)2.493 \times 10^6/122.5$$
$$= 486.1 °C$$

The relative error between the calculation value and the assumed value is 2.8%. Because it is also within ±15%, there is no need to recalculate.

Chapter 4

Heat Transfer in Fluidized Beds

Chapter Outline

4.1 FUNDAMENTAL CONCEPTS OF FLUIDIZED BEDS

4.1.1 Definition and Characteristics of Fluidized Beds

The first-generation fluidized bed boiler is the bubbling fluidized bed boiler (BFB boiler). The circulating fluidized bed boiler (CFB boiler) is an updated version of the BFB boiler, thus it is the second-generation fluidized bed boiler. The CFB boiler is a device that burns solid fuel to produce steam or hot water, and the boiler furnace works with special fluid dynamic characteristics. Fine particles (Class A particles or Class B particles in Geldart classification, ie, particles 30~500 μm in size) are circulated through the entire furnace. The gas velocity is higher than the average terminal velocity of an average-size particle; at the same time, enough particles return to help maintain uniform temperature distribution in the furnace. In brief, the flow pattern at the lower part of the CFB boiler furnace is fluidized, and the upper part is characterized by material transportation. "Solid fuels" are usually fossil fuels such as coal, and sometimes biomass or solid waste.

Theory and Calculation of Heat Transfer in Furnaces. http://dx.doi.org/10.1016/B978-0-12-800966-6.00004-1
101

Most particles that leave the CFB boiler furnace are separated by a gas-solid separation device (the "separator") and sent back to the furnace at a high enough flow rate from the return port near the furnace bottom to sustain the particle backmixing circulation in the furnace at whatever level is necessary. A schematic diagram of the structure of a CFB boiler is shown in Fig. 4.1. Part of the air needed by the fuel is suppied into the furnace by the air distributor at the bottom of the furnace—this is the "primary air." The amount of primary air is usually lower in practice than the theoretical air demand, usually 40~60% of the total air required. "Secondary air," then, is supplied into the furnace at a certain height above the air distributor, either from all around or from a certain direction. Fuel burns in the furnace to release heat, part of the heat is absorbed by the water wall or steam-cooling heating surface (including the superheater and reheater), and the remaining heat is absorbed by the convective heating surface at the back-end ductwork of the boiler or carried away by exhaust flue gas.

Facilitating necessary material circulation through special hydrodynamic properties by fast fluidization or dilute phase backmixing is a very crucial part of running a CFB boiler successfully—this process is also how the CFB boiler was named. A careful combination of bed sectional gas velocity, material circulating flow rate, particle characteristics, material size distribution, and boiler

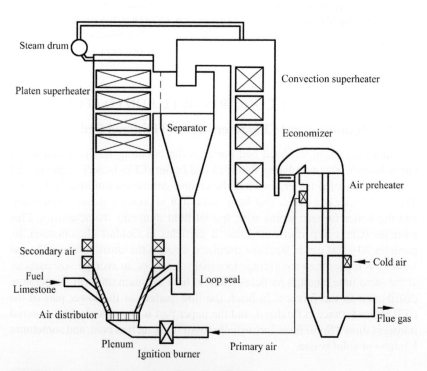

FIGURE 4.1   CFB boiler structure schematic diagram.

geometric shape can produce special hydrodynamic characteristics. Under these hydrodynamic characteristics, solid material is fluidized by the gas flow at a velocity faster than the terminal velocity of single-particle material. At the same time, solid material is not directly entrained by the gas flow as occurs in a gas transportation system. Conversely, material moves up and down in uneven-sized particle clusters, forming high-level internal material backmixing. These small clusters move upward, downward, and around as they are formed, broken, and regenerated constantly. These special hydrodynamic characteristics can carry a certain amount of large-particle material with terminal velocity far higher than the sectional average gas velocity. This gas-solid motion mode forms a large gas-solid slip velocity. From the viewpoint of hydrodynamics the particles in the zone other than the dense-phase zone of the CFB boilers are in a fast fluidized zone. These remarkable hydrodynamic properties distinguish the CFB boiler furnace from other types of boilers (such as the furnace of the grate-firing boiler or the furnace of the pulverized coal suspension-firing boiler). Fig. 4.2 and Table 4.1 provide a detailed comparison of different types of boilers.

4.1.2 Basic CFB Boiler Structure

A typical CFB boiler can be divided into two parts. The first part contains the following components: the furnace or fast fluidized bed (combustion chamber), the gas–solid separation device (cyclone separator or inertial separator), and the solid material recycling device (standpipe, loop seal, or external heat exchanger for some CFB boilers). These components facilitate the circulation of solid material with the fuel burned inside. Like a pulverized coal boiler, the furnace of a CFB boiler usually has a water wall, where a portion of the combustion heat is absorbed by water wall tubes.

The second part is the convective flue gas passage, which includes a superheater, reheater, economizer, and air preheater, where the remaining heat from the gas is absorbed. Other auxiliary parts of the CFB boiler include the

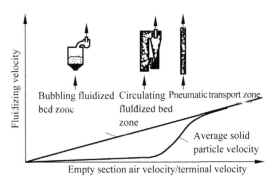

FIGURE 4.2   Gas dynamic characteristics of different combustion types.

TABLE 4.1 Comparison of CFB Boiler and Other Boilers

Characteristics	Fire grate combustion	Bubbling fluidized bed	Circulating fluidized bed boiler (CFB boiler)	Pulverized coal furnace
Bed height or fuel combustion zone height/m	0.2	1.2	15–40	27–45
Sectional air velocity/(m/s)	1.2	1.5–2.5	4–8	4–6
Excess air coefficient at furnace exit	1.2–1.3	1.2–1.25	1.15–1.3	1.15–1.3
Sectional thermal load/(MW/m^2)	0.5–1.5	0.5–1.5	3–5	4–6
Coal particle size (mm)	6–32	< 6	< 6	< 0.1
Load turn-down ratio	4:1	3:1	(3–4):1	2:1
Combustion efficiency (%)	85–90	90–96	95–99	99
NO$_x$ emission volume fraction (10^{-6})	400–600	300–400	50–150	400–700
Desulfurization efficiency in furnace (%)	Low	80–90	80–95	Low

bottom ash cooler, bottom ash conveyor, fuel supply device, and limestone supply device.

The cross section area of the lower part of a CFB boiler furnace is small and expands upward. The area of the air distributor at the bottom of the furnace is usually the smallest of all the components in order to maintain a good fluidized state even for segregated particles. The furnace wall below (even sometimes above) the secondary air inlet is covered by a refractory layer, and the covered area is determined by fuel characteristics.

The cross section area of the upper part of the furnace is uniform and larger than that of the lower part, and the evaporation heating surface is usually arranged at this part of the furnace wall. The gas-solid separation device and non-mechanical material loop seal are arranged outside the furnace, and these parts are also covered by abrasion-resistant and refractory layers. In some boiler designs, a portion of the hot material circulating between the separator and furnace is bypassed to an external heat exchanger, where some heat is absorbed. The external heat exchanger is a bubbling fluidized bed comprised of a heating surface with immersed pipes. There is only a small amount of air in the external heat exchanger, thus the combustion fraction is very small. At present, except for the Lurgi CFB boiler, CFB boilers do not have external heat exchangers.

To increase the fuel's residence time in the furnace, fuel is usually fed from the furnace bottom. Fuel can be fed through an independent fuel feeding port, or from the loop seal below the standpipe with hot bed material. Primary air enters the furnace through the air distributor, and secondary air is supplied into the furnace from a certain height above the air distributor to ensure the fuel burns completely. Though heat is absorbed along the furnace height continuously, due to good mixing of material throughout the furnace height, the temperature along the furnace height is basically uniform.

4.1.3   Different Types of CFB Boilers

At present, the CFB boiler has a large capacity (more than 1000 t/h) and is developing toward even larger capacity and the supercritical parameters. The most common CFB boilers in use worldwide include the Lurgi (Lurgi Ltd., Germany), Pyroflow (Ahlstrom Ltd., Finland), FW (Foster Wheeler Ltd., USA), Circofluid (Babcock Ltd., Germany), and Inner Recycle (B&W Ltd., USA). Schematic diagrams of these different types of CFB boilers are shown in Fig. 4.3.

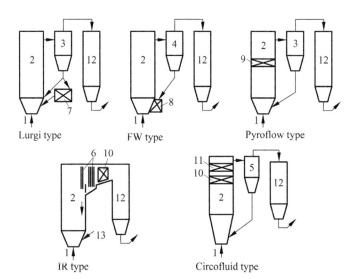

FIGURE 4.3   CFB boiler furnace types. Note: (1) air distributor, (2) furnace, (3) adiabatic high-temperature cyclone separator, (4) high-temperature steam-cooling cyclone separator, (5) mid-temperature cyclone separator, (6) internal separation unit, either U-shape or shutter, (7) external bed heat exchanger, (8) INTREX heat exchanger, (9) Ω tube panel superheater, (10) superheater, (11) high-temperature economizer, (12) back-end ductwork, (13) fly ash recirculating port.

The fuel particle size entering the furnace of a Lurgi CFB boiler is 50~500 μm, the fluidized velocity is 3~10 m/s, the primary air ratio is 40%, and the bulk density on the air distributor in the normal state, 273.15 K (0°C), 101.325 KPa, is 300 kg/m³. The bulk density (normal state) above the secondary air inlet port is approximately 30~50 kg/m³. High-temperature adiabatic separation is adopted. The most significant characteristics of the Lurgi CFB boiler are its external bubbling fluidized bed heat exchanger (EHE), which has a special function suited to the large scale of CFB boilers, as well as the ability to control load and combustion temperature, desulfurization in the furnace, and fuel adaptability. The disadvantages of the EHE are its complex structure and large inertia of load change. A typical Lurgi CFB boiler is illustrated in Fig. 4.4.

The Pyroflow CFB boiler, of Finnish organization Ahlstrom Ltd. (Fig. 4.3), is similar to the Lurgi but does not have an EHE. Both of them adopt high-temperature adiabatic cyclone separation, and are high-rate CFB boilers (with a circulation rate of approximately 70~80). The fluidized velocity of the Pyroflow is 5~5.5 m/s, and the primary air ratio is 40~70%. The bulk density on the air distributor (normal state) is over 100 kg/m³, and the bulk density above the secondary air inlet port is approximately 30~50 kg/m³.

The FW CFB boiler, of Foster Wheeler Ltd. (USA), is not intrinsically different from the Pyroflow CFB boiler. Its most notable characteristic is that it employs a high-temperature, steam-cooling cyclone separator. The Inner Recycle (IR) CFB boiler (B&W, USA) has a U-shaped separator at its furnace outlet,

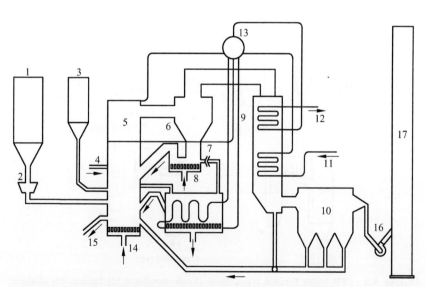

FIGURE 4.4   Systematic diagram of a Lurgi CFB boiler. Note: (1) coal bunker, (2) coal feeder, (3) limestone bunker, (4) secondary air, (5) furnace, (6) cyclone separator, (7) hot ash control valve, (8) external bubbling bed heat exchanger, (9) back-end ductwork, (10) baghouse filter, (11) water feeding inlet, (12) steam outlet, (13) steam drum, (14) primary air, (15) slag, (16) induced draft fan, (17) stack.

and the separated fine ash flows downward along the furnace's rear wall to form the internal recycle. The separation efficiency of the IR circulation system is low, and generally the secondary-separated fly ash in the back-end ductwork must be recovered back to the furnace to burn again.

The Circofluid CFB boiler (Babcock Ltd., Germany), as shown in Fig. 4.5a, adopts low-rate (10~20) and mid-temperature (400~500°C) separation. Its fluidized velocity is 3.5~5 m/s, and the bulk density above the suspension stage (normal state) is 1.5~2 kg/m³. The primary air ratio is approximately 60%. Fig. 4.5b shows a sketch of a 130 t/h Circofluid CFB boiler manufactured by Beijing Boiler Works with imported patent technology.

4.1.4   CFB Boiler Characteristics

There is a certain amount of solid particles (bed material) in the furnace of any CFB boiler, with particle sizes usually 0.1~0.5 mm. The solid particles include one or several of the following:

1. Sand or gravel (when burning low-ash fuel such as biomass).
2. Fresh or spent desulfurizer (when burning high-sulfur coal or when desulfurization is needed).

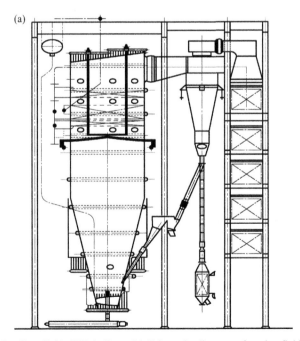

FIGURE 4.5   Circofluid CFB boilers. (a) Schematic diagram of a circofluid CFB boiler, (b) general drawing of 130t/h Circofluid CFB boiler manufactured by China (front view), (c) general drawing of 130t/h Circofluid CFB boiler manufactured by China (top view).

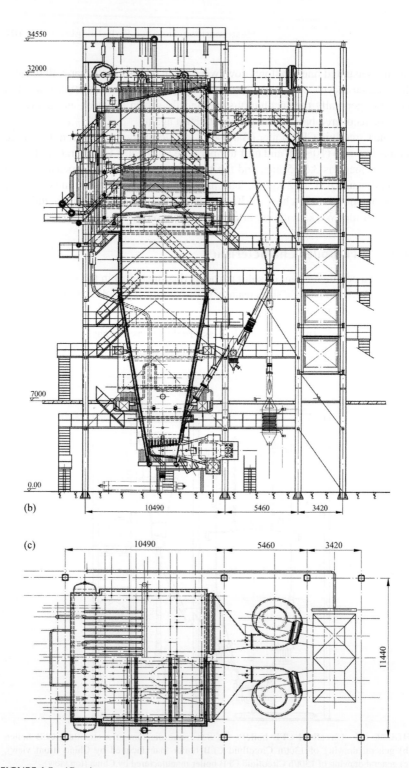

34550

32000

7000

0.00

(b)

10490 | 5460 | 3420

(c)

10490 | 5460 | 3420

11440

FIGURE 4.5 (*Cont.*)

3. Coal ash (when burning high-ash or medium-ash coal, where desulfurization is not needed).

Sometimes, the bed material is a combination of the above. The size distribution of the fuel entering the furnace does not always determine the particle size of the bed material (especially for low-ash fuel), because the fuel only occupies a very small percentage ($1\sim3\%$) of the total amount of bed material in the CFB boiler. In addition, fracture and cracks in the bed material inevitably occur during the fuel combustion process.

The CFB boiler has many unique merits that lend it better environmental and energy-saving properties than other solid-fuel boilers.

1. Wide flexibility for fuel

   In a CFB boiler, fuel only accounts for $1\sim3$ wt.% of the bed material, and the remainder is incombustible solid particles (such as desulfurizer, ash, or sand). The bed hydrodynamic characteristics of the CFB boiler allow gas/solid and solid/solid materials to mix very well, thus the fuel can mix with large amounts of high-temperature bed material rapidly after entering the furnace. Fuel is heated rapidly above ignition temperature, while the bed temperature does not drastically decrease. As long as the fuel heating value is higher than the heat needed to heat the fuel and air to ignition temperature, the CFB boiler can use any fuel without needing auxiliary fuel. The CFB boiler can run on various kinds of solid or semi-solid fuels successfully, from anthracite to lignite and from biomass to solid waste. Of course, this does not mean the boiler can burn a wide range of fuels without needing some amount of modification. Many commercial CFB boilers burn low-grade, high-ash content ($40\sim60\%$) coal fuel such as gangue.

2. High combustion efficiency

   In addition, the combustion efficiency of a CFB boiler is higher than that of a BFB boiler. The combustion efficiency of CFB boilers ranges from $95\sim99\%$, sometimes even above 99.5%. The high combustion efficiency of CFB boilers is mainly due to their highly effective gas-solid mixing ability and longer residence time. CFB boiler efficiency is similar to that of a pulverized coal boiler, which is higher than a grate-firing furnace.

3. High desulfurization efficiency

   CFB boilers can also remove sulfur in the furnace, which is an extraordinary characteristic that other furnaces (such as PC boilers or grate-firing boilers) do not have. Desulfurization in-furnace requires low initial investment, and provides high efficiency, convenient operation, and cost-effectiveness compared to postcombustion desulfurization. In-furnace desulfurization in the CFB boiler is more efficient than that in the BFB boiler—to reach 90% desulfurization efficiency, a typical CFB boiler needs a molar ratio of limestone to sulfur Ca/S of $1.5\sim2.5$, while a BFB boiler needs a ratio to Ca/S of $2.5\sim3$ or even higher to reach the same efficiency.

4. Low NO_x emission

Low NO_x emission is another prominent characteristic of CFB boilers. Industrial operation data have shown that the NO_x emission volume fraction of a typical CFB boiler reaches $150 \times 10^{-6} \sim 50 \times 10^{-6}$ due to the special combustion condition of the dense-phase zone, low-temperature combustion, and staged air supply, especially for highly volatile fuel. Primary air less than the amount of stoichiometric combustion is supplied from the furnace bottom, thus nitrogen in the fuel cannot fully react with oxygen to form oxynitrides, and a small amount is reduced to N_2. Secondary air is supplied from a certain height above the furnace bottom air distributor, leaving excess air of about 20%. Because the nitrogen in the fuel has transformed into molecular nitrogen at this point, there is only a small change in the NO_2 formed above the reduction area. In the low-temperature combustion zones of a CFB boiler (800~950°C), generally, nitrogen in the air does not react with oxygen to form thermal NO_x.

5. Small furnace cross section area

Fig. 4.6 compares the sectional thermal load of several combustion types. The sectional thermal load of an atmospheric CFB boiler is $3 \sim 5$ MW/m², nearly equal to or higher than that of a PC boiler. As shown in the figure, for the same thermal load, a BFB boiler needs a furnace cross-section area $2 \sim 3$ times larger than a CFB boiler.

It is important to mention that because pressurized fluidized bed boiler technology is still relatively unmatured, it has not yet been widely applied in the field at present.

CFB boilers have high sectional thermal load, mainly due to the high sectional gas velocity ($4 \sim 8$ m/s) in their furnaces. Intense gas–solid mixing promotes rapid heat release and transfer, which enhances sectional thermal load. Its small furnace cross section area makes the CFB boiler suitable for renovation of existing coal- or oil-fired boilers.

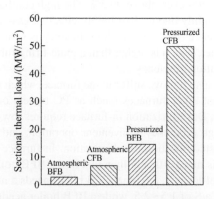

FIGURE 4.6 Sectional thermal loads of different combustion types.

6. Simple fuel system

Compared to PC boilers, CFB boilers have simple fuel systems due to the fact that they do not necessitate coal pulverization, and instead only preliminary crushing of the fuel (the crushing is often completed by the fuel supplier), saving the considerable effort necessary for a coal pulverization system. Because its fuel-feeding ports are relatively few, the fuel supply system of the CFB boiler is quite simple. At the given heat duty, the furnace cross section area of the CFB boiler is small, and meanwhile, due to effective mixing and extension of its combustion area, one feeding port can meet the fuel demand of a much larger area than that of the BFB boiler. Every fuel-feeding port of the CFB boiler can supply a much larger bed area and its sectional thermal load is much more than that of the BFB boiler.

7. Wide turn-down ratio and fast load response

Since its sectional gas velocity is high and its absorption of heat is easy to control, load response in a CFB boiler is very quick. Generally, the load changing rate of a CFB boiler can be 2~4%, sometimes even 10%. The load changing range of the CFB boiler is much wider than that of the PC boiler, as well—up to (3~4):1, and it is able to burn stably at 20~30% of the rated output.

8. Low corrosion of boiler back-end heating surface

Due to in-furnace desulfurization and complete combustion, the back-end heating surface (especially the low-temperature air preheater) of the CFB boiler shows less corrosion than that of the PC boiler or grate-firing boiler with the same coal and the same exhaust gas temperature. In other words, the exhaust gas temperature can be reduced appropriately through reliable operation to enhance boiler efficiency.

There are also a few important disadvantages of CFB boiler technology.

1. The structure is quite complicated. For example, because a high-temperature cyclone separator (which is very large and expensive) is used, the steel consumption of the whole CFB boiler jumps by about 20%.

2. The flow resistance of the bed material circulation system of the entire boiler is quite large, which increases the power consumption of the system—its power consumption is approximately 7% of its own power generation. That said, due to doing without the pulverization system, the CFB boiler system consumes about as much electricity as, or only slightly more than, the PC boiler.

3. Additionally, the heating surface and refractory layer in the CFB boiler furnace are washed by solid particles, so inappropriate design may cause some extent of abrasion.

4. Finally, the N_2O emission is higher than that of other combustion types. The N_2O emission range of the CFB boiler is $50 \times 10^{-6} \sim 200 \times 10^{-6}$, considerably higher than that of the PC boiler ($<20 \times 10^{-6}$).

The primary development direction of the CFB boiler, on the one hand, is to simplify boiler structure, save cost, and modify or adopt a new low-resistance, small-scale separator; on the other hand, developers are interested in taking advantage of the CFB boiler's high-efficiency and energy-saving properties on a larger scale. At present, 1500 t/h CFB boilers have been applied in industries worldwide, and a 2000 t/h CFB boiler has been put into operation in China. CFB boilers with larger capacity and higher parameters (such as supercritical) are currently being developed.

4.2 CONVECTIVE HEAT TRANSFER IN GAS–SOLID FLOW

During the design and operation of grate-firing boilers and suspension firing boilers, the heat transfer on the heating surface depends on the combustion situation, and is usually self-adaptive. In a CFB boiler, however, the situation is different—the heat transfer of the gas–solid flow is more complicated than that of pure gas, but the heat transfer is closely related to combustion and can even be considered to fully determine the combustion situation and boiler load under certain conditions. Therefore, it is necessary to research and analyze the heat transfer of the gas-solid flow in the CFB boiler to design and operate the system as efficiently as possible. Fig. 4.7 shows the distribution and arrangement of the heating surface in a typical CFB boiler.

In the CFB boiler, heat transfer modes of the gas-solid flow can be divided into the following five situations:

1. Heat transfer between gas and solid particles, within two-phase flow in the furnace.

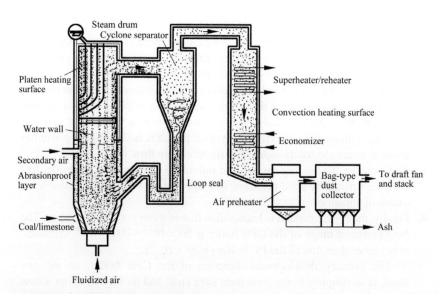

FIGURE 4.7   A CFB boiler heating surface.

2. Heat transfer between the bed material and water wall, within two-phase flow in the furnace's heating surface.
3. Heat transfer between the bed material and immersed pipes in the furnace, within the dense-phase zone in the bubbling bed.
4. Heat transfer between the bubbling bed material and immersed pipes in the external heat exchanger (EHE).
5. Heat transfer in the cyclone separator (or another type of separator).
6. The above five situations can be classified into two classes: heat transfer between gas and particles, or heat transfer between the gas-solid flow and the wall or tubes. Table 4.2 provides the orders of magnitude of heat transfer coefficients during the above heat transfer situations.

Compared to the heat transfer coefficient between a bed material and heating surface, heat transfer coefficient calculation between gas and solid particles is rarely necessary, but in some specific cases this calculation is also needed. When the gas-solid is near the air distributor, solid inlet port, or secondary air inlet port, the gas-solid temperature is different than the average bed material temperature and thus thermal inertia is very critical to the combustion process. In coal-burning conditions, the heating rate of coal particles has a significant impact on the release of volatile matter; combustion, abrasion, and cracks are also influenced by the heat transfer situation of coal particles. In addition, heat transfer between gas and solid particles controls the response characteristics of

TABLE 4.2 Heat Transfer Coefficients in CFB Boilers

Location	Heating surface type	Typical heat transfer coefficient/(W/(m²·K))
Water wall tube above refractory coating in furnace (890–950°C)	Evaporation heating surface	110–220
Platen superheater in furnace (890–950°C)	Evaporating heating surface, reheater, or superheater	50–150
Cyclone separator (890–900°C)	Superheater	20–50
Horizontal tube of external heat exchanger (600–850°C)	Evaporating heating surface, reheater, or superheater	280–450
Tube bundles in cross flue gas flow within back-end heating surface (200–800°C)	Economizer, superheater, or reheater	40–85
Between flue gas and bed material (177–420 μm, 50°C)	Furnace	30–200

the transitional process in a CFB boiler and influences the characteristics of the automatic control system. Heat transfer between gases and solids, such as heat transfer of the dense-phase zone near the air distributor or the dilute-phase zone, is also important, but due to space limitations are not discussed here. Interested readers can refer to the related monograph.

At present, heat transfer processes in CFB boilers are not well-understood. There yet lacks sufficient data for industrial CFB boilers, in particular. Experimental observations could be applied to the information provided in this section to understand the effects of related parameters on heat transfer, and to discuss the heat transfer mechanisms related to design and operation variables.

4.2.1 Two-Phase Flow Heat Transfer Mechanism

In a CFB boiler with fine particles, solid particles usually form clusters in the rising gas flow. The flow with dispersed solid particles is called the dispersed phase, and the remainder is called the cluster phase. Most of the bed material particles rise along the bed central area, while clusters near the furnace wall fall down along the wall surface, as shown in Fig. 4.8. Clusters that have coalesced from solid particles form, disappear, and reform in cycles. Therefore, heat transfer to the heating surface includes the radial heat conduction of solid particle clusters, the convective heat transfer of the dispersed phase, and the radiative heat transfer of the two phases.

Clusters falling along the wall surface experience an unsteady-state heat conduction process to the wall surface. Through conduction and radiation, clusters transfer heat to the wall and are cooled down. In boilers, the heating surface is very high above the air distributor—as such, heat conduction to clusters forms a thermal boundary layer over the course of a very long travel range.

4.2.2 Factors Impacting Two-Phase Heat Transfer

Heat transfer from the bed material to the wall surface in a CFB boiler is affected by certain design and operation parameters including material concentration,

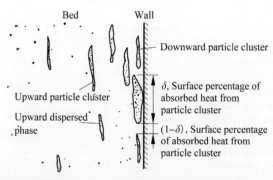

FIGURE 4.8 Schematic diagram of surface heat transfer in a CFB boiler.

bed velocity (bed sectional gas velocity), bed temperature, and geometric structure. We'll discuss these separately.

1. Influence of material concentration and particle size

In a fast bed, the average material concentration on the surface has the most
important impact on the heat transfer coefficient between the bed material
and the surface. The material concentration on the surface is in direct proportion to the average material concentration of the bed. Fig. 4.9 shows the
influence curve of average bed material concentration on the heat transfer
coefficient at ambient temperature. The fourth-power law states that the
radiative factor is very important at high temperatures. Fig. 4.10 shows different results based on Fig. 4.9; the differences between them are due to
radiation. As shown in Figs. 4.9 and 4.10, the heat transfer coefficient increases as material concentration increases. The slopes of the curves in both
figures show that the heat transfer coefficient is directly proportional to the
square root of the material concentration.

   The heat from the bed to the heating surface is transferred through sliding solid particle clusters and rising gas flow with dispersed-phase particles.
Experiments have shown that heat transfer by conduction from solid particle
clusters to a heating surface is much larger than heat transfer by convection.
The coverage rate of solid particle clusters on a dense-phase bed heating surface is larger than that on a dilute-phase zone heating surface, therefore, the
heat transfer coefficient between clusters and the heating surface in a dense-
phase zone is larger than that in a dispersed-phase zone. In a CFB boiler, the
heat transfer coefficient from the furnace bottom to the outlet varies and is

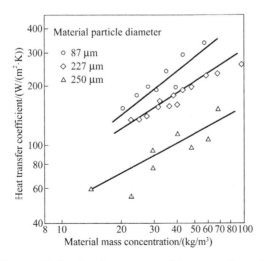

FIGURE 4.9 **Influence of bed sectional average material concentration on the heat transfer
coefficient at ambient temperature.**

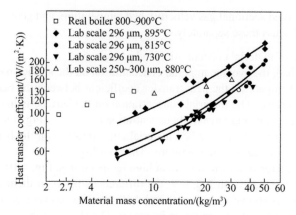

FIGURE 4.10 **Influence of material concentration on the heat transfer coefficient at elevated temperature.**

influenced by many factors including fuel characteristics, total air supply, primary and secondary air ratios, bed material circulation rate, bed material inventory, material particle size distribution, and temperature distribution.

Fig. 4.10 shows the heat transfer coefficients measured in both a real CFB boiler and a lab-scale experiment. The heat transfer coefficient of the boiler was measured based on the real heating surface of a power-generating, 110 MWe (steam output 420 t/h) CFB boiler. Though the error in industrial tests is large, we can still assert that the measured heat transfer coefficient of a real boiler is larger than that measured in the lab, as the solid particle flow rate falling along the furnace wall is much larger in a real boiler. Because the average material concentration measured by the static pressure difference method cannot reflect the solid particle flow rate completely, it should be noted that the heat transfer comparison shown in Fig. 4.10 was made with data from two devices of quite disparate sizes.

As shown in Fig. 4.9, the smaller the particle diameter is, the larger the heat transfer coefficient is.

2. Influence of fluidized velocity

As opposed to BFB boilers, the fluidized velocity of CFB boilers influences the material concentration but does not significantly affect the heat transfer. As shown in Fig. 4.11, when material concentration is constant, the heat transfer coefficient varies slightly at different fluidized velocities. In many situations, if the material circulation rate is constant, when the fluidized velocity increases, the decrease in the heat transfer coefficient caused by the decreasing material concentration is more pronounced than the increase in the heat transfer coefficient caused by the increasing fluidized velocity. The influence of fluidized velocity on heat transfer requires further analysis for beds with large particle size or low density.

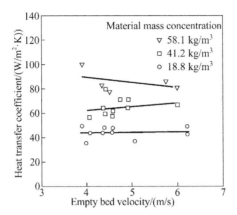

FIGURE 4.11   Influence of fluidized velocity on heat transfer when material concentration keeps constant.

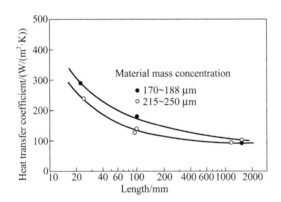

FIGURE 4.12   Influence of heating surface length on the heat transfer coefficient.

3. Influence of heating surface vertical length

Fig. 4.12 shows the influence curve of the vertical length of the heating surface on the heat transfer coefficient. As shown, the heat transfer coefficient measured along the heating surface decreases as vertical length increases. Though the heat transfer coefficient decreases continuously as length increases, the rate at which it decreases grows slower. As clusters fall along the heating surface, the temperature gradually comes near the surface temperature—as the temperature difference between the surface and clusters decreases, the calculated heat transfer coefficient decreases as heating surface length increases. In real boilers, any one specific cluster cannot move along the surface endlessly—when clusters fall to a certain height, they return to the bed center or break and are replaced by new clusters. Therefore, after going through a certain vertical length of the heating surface, the heat transfer coefficient tends toward a stable value.

4. Influence of bed temperature

Research has shown that at constant material concentration, the total heat transfer coefficient increases as bed temperature increases (Fig. 4.13). Due to the increase in temperature, the thermal resistance between clusters near the wall layer and the surface decreases, causing an increased heat transfer coefficient. In addition, the increased radiation of particles in the bed and gas leads to an increased heat transfer coefficient. In a real CFB boiler, the impact of bed temperature on the heat transfer coefficient is more complicated, and the thermal resistance of the whole heat transfer process requires further discussion; we do not analyze it here due to space limitations, but urge interested readers to consult our references.

4.2.3 Two-Phase Flow Convective Heat Transfer

The main difficulty inherent to establishing the heat transfer model from the bed to the surface in a fast bed is the lack of knowledge of gas-solid flow characteristics in the fast bed. In general, it is believed that heat is transferred to the unsteady thin layer of solid particles sliding along the surface to form the thermal boundary layer. The larger the boiler capacity is, the thicker the boundary layer is. Analysis of the mass balance, momentum balance, and energy balance of the gas-solid flow near the wall allows us to obtain details regarding the heat transfer from the bed to the wall surface—said analysis is rather complex, as you'll discover later.

There are two flow patterns in a fast bed: dispersing solid particles (solid-particle dispersed phase) and clusters. The clusters and solid-particle dispersed phase simultaneously contact different parts of the bed wall. Supposing δ_c is the average percentage of the surface area covered by clusters, h_c is the convective heat transfer coefficient, and h_r is the radiative heat transfer coefficient, the

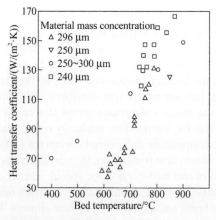

FIGURE 4.13 Change of the total heat transfer coefficient with bed temperature at constant material concentration (20 kg/m³).

oveall heat transfer coefficient of the heating surface can be denoted as the sum of h_c and h_r, that is:

$$h = h_c + h_r \tag{4.1}$$

$$h_c = \delta_c h_{cc} + (1 - \delta_c) h_{dc} \tag{4.1a}$$

$$h_r = \delta_c h_{cr} + (1 - \delta_c) d_{dr} \tag{4.1b}$$

where h_{cc} is the convective heat transfer coefficient from the clusters to the heating surface, h_{dc} is the convective heat transfer coefficient from the solid-particle dispersed phase to the heating surface, h_{cr} is the radiative heat transfer coefficient from the clusters to the heating surface, and h_{dr} is the radiative heat transfer coefficient from the solid-particle dispersed phase to the heating surface.

At any time, the wall surface of the CFB boiler is partially covered by clusters while the rest of the wall is exposed to the solid-particle dispersed phase. As shown in Fig. 4.8, the average time coverage rate of clusters on the surface is:

$$\delta_c = \frac{1}{2} \left[\frac{1 - v_w - Y}{(1 - v_c)} \right]^K \tag{4.2}$$

where coefficient K = 0.5, v_w is voidage at the wall surface, v_c is voidage in the clusters, and Y is the percentage of solid particles in the particle dispersed phase, which increases from the bed center to the heating surface and reaches its maximum on the heating surface. Research shows that the distribution of vertical voidage is related only to the average of vertical dimensionless distance (r/R) and sectional average voidage, thus, the empirical equation for surface voidage is:

$$v(R) = v_w = v^n \tag{4.3}$$

where n has been determined through experimentation to be 3.811.

Convective heat transfer includes convective heat transfer of clusters and the particle dispersed phase, that is:

$$h_c = \delta_c h_{cc} + (1 - \delta_c) h_{dc} \tag{4.4}$$

1. Convective heat transfer of clusters

In a CFB boiler furnace, clusters move along the heating surface. After contacting the heating surface, the clusters either break and disappear or bounce to another place in the furnace. When clusters contact the heating surface, there is unsteady heat conduction between the clusters and heating surface. At the initial stage of heat transfer, only the first adjacent layer of clusters transfers heat to the surface and its temperature falls close to the surface temperature. When clusters remain on the surface long enough, internal particles of the clusters begin to participate in unsteady heat conduction to the surface. Analyzing the unsteady heat transfer between the surface and clusters allows us to obtain the instant local heat transfer coefficient h_t:

$$h_t = \sqrt{\frac{\lambda_c c_c \rho_c}{\tau \pi}} \qquad (4.5)$$

where λ_c, c_c, and ρ_c are the heat conductivity coefficient, specific heat capacity, and density, respectively, and τ is time.

Because the heat conduction of clusters is analyzed based on the heat conduction of small clusters in the BFB boiler, it can be similarly understood that the characteristics of clusters are the same as those of the emulsion phase in the BFB boiler. So the specific heat capacity of clusters is $C_c = [(1-\upsilon_c)C_p + \upsilon_c C_g]$, density is $\rho_c = [(1-\upsilon_c)\rho_p + \upsilon_c \rho_g]$, and υ_c is voidage of clusters. The heat conductivity coefficient of clusters λ_c can be found in the literature [16].

Supposing the adherence time between the clusters and surface is τ_c, the average heat transfer coefficient is:

$$h_{cc} = \frac{1}{\tau_c}\int_0^{\tau_c} h_t\,d\tau = \sqrt{\frac{4\lambda_c c_c \rho_c}{\pi \tau_c}} \qquad (4.6)$$

In a fast bed, the thermal resistance between the clusters and surface includes two components: the contact resistance between the clusters and surface, and the average thermal resistance of the clusters. Contact resistance can be calculated according to the thermal resistance of the thin gas layer thickness ($d_p/10$). After analyzing the unsteady heat conduction process of the clusters and surface, the heat transfer component h_{cc} can be calculated using the following equation:

$$h_{cc} = \frac{1}{\dfrac{d_p}{10\lambda_g} + \left[\dfrac{\tau_c \pi}{4\lambda_c c_c \rho_c}\right]^{0.5}} \qquad (4.7)$$

where τ_c is the average adherence time of the clusters, and λ_g is the heat conductivity coefficient of the gas, which can be determined according to the average temperature of the thin gas layer.

If thermal time constant J is less than the cluster adherence time, then other conditions besides heat conduction of the cluster layer must be considered. The thermal time constant J is:

$$J = \frac{c_p d_p^2 \rho_p}{36\lambda_g} < \tau_c \qquad (4.8)$$

For a long, continuous heating surface, such as the actual situation in a real CFB boiler, the adherence time of clusters is quite long. At this time, compared with contact thermal resistance, the unsteady heat conduction of clusters is very important, and the influence of particle size on the heat transfer coefficient less so.

So when cluster adherence time is short, the first item of the denominator in Eq. (4.7) is important. Under these circumstances, heat transfer is restricted by the cluster adherence layer. From this, for the case that particle size is large and adherence time is short, the convective heat transfer component of clusters h_{cc} can be calculated from the following equation:

$$h_{cc} = \frac{10\lambda_g}{d_p} \tag{4.9}$$

However, when adherence time is long, the heat conduction of clusters affects the convective heat transfer component. At this point, Eq. (4.7) should not be used to calculate the convective heat transfer component.

The heat conduction of clusters and surfaces depends on the duration of time that the clusters adhere on the wall surface, in other words, the time needed for clusters to move along the heating surface before breaking apart. Adhered clusters accelerate their slide assisted by gravity and impeded by surface resistance and upward gas flow drag force—the final result of these forces causes clusters to reach a maximum speed U_m. When clusters pass through the entire heat surface without breaking, adherence time can be calculated by the following equation of motion:

$$L = \frac{U_m^2}{g}[\exp(-g\tau_c/U_m)-1]+U_m\tau_c \tag{4.10}$$

where L is the vertical length of the heating surface, and τ_c is the adherence time of the clusters to the surface. The maximum velocity U_m can be measured through experimentation or calculated according to the empirical relationship of surface shear stress and cluster thickness. Approximately, the value of U_m can be considered 1.2~2.0 m/s.

As shown in Fig. 4.12, the influence of cluster adherence time (or vertical heating length) on the heat transfer coefficient decreases gradually. Therefore, for large boilers, the sensitivity of the heating transfer coefficient to adherence time is small. The influence of adherence time on the calculation of the total heat transfer coefficient in a real CFB boiler is negligible.

2. Convective heat transfer of particle dispersed phase
   In a fast bed, the wall surface contacts not only the clusters, but also the rising gas flow. The heat transfer between the dispersed phase and wall surface also requires analysis. Previous studies have shown that the following heat

transfer calculation equation derived from the dilute-phase gas-solid mixture can be applied to calculate the heat transfer coefficient of the dispersed phase:

$$h_{dc} = \frac{\lambda_g}{d_p} \frac{c_p}{c_g} \left[\frac{\rho_{dis}}{\rho_p} \right]^{0.3} \left[\frac{U_t^2}{g d_p} \right]^{0.21} Pr \tag{4.11}$$

where ρ_{dis} is the density of the particle dispersed phase, $\rho_{dis} = [\rho_p Y + \rho_g (1 - Y)]$, λ_g and c_g are the thermal conductivity and specific heat capacity of gas, respectively, U_t is the terminal velocity of an average-size particle, and Pr is the Prandtl number.

At ambient temperature, the radiative component can be ignored. Eq. (4.11) provides the lower limit of the heat transfer coefficient, while Eq. (4.9) provides its upper limit; these limits are very important for control and change of the load in the CFB boiler, and have reference value for the design and operation of real CFB boiler systems.

A notable uncertainty factor in Eq. (4.11) is the particle volume concentration Y in the dispersed phase—when Y is 0.001%, the calculated values are consistent with experimental data. However, the overall heat transfer coefficient is not sensitive to Y in many cases.

4.3 RADIATIVE HEAT TRANSFER IN GAS–SOLID FLOW

Convection and radiation are modes of heat transfer in the CFB boiler furnace that must be accounted for. Radiative heat transfer is an important route for the heat transfer in the fast bed, especially in a situation with a high temperature ($>700°C$) and low bulk density (<30 kg/m^3). This section adopts the simplified engineering calculation conditions discussed in chapter: Theoretical Foundation and Basic Properties of Thermal Radiation, assuming that the two-phase flow and surface are both isothermal graybodies.

The radiative heat transfer in the fast bed includes two parts: one is mainly from the radiation of clusters contacting the surface, and the other is the radiation from the particle dispersed phase to the surface. The total radiative heat transfer coefficient from the bed to the surface can be expressed as follows:

$$h_r = \delta_c h_{cr} + (1 - \delta_c) h_{dr} \tag{4.12}$$

where δ_c is the average time coverage rate of clusters on the surface, h_{cr} is the radiative component of the clusters, and h_{dr} is the radiative component of the dispersed phase.

1. Radiation of particle dispersed phase
   For a working medium with low particle concentration, the following equation can be used to evaluate reduced emissivity of the particle suspended

phase to a surface with a cover of clusters, that is, suspended particle emissivity in the bed:

$$\varepsilon_p = 1 - e^{-1.5\varepsilon_{p,s}Ys/d_p} \tag{4.13}$$

where $\varepsilon_{p,s}$ is average emissivity of the particle surface, and Y is the volume percentage of solid particles in the furnace. When there is an abundance of experimental and industrial data for the radiative heat transfer component, the value of Y in Eq. (4.13) can be determined approximately. Effective radiation layer thickness s can be calculated using the following equation:

$$s = \frac{3.5V}{A} \tag{4.14}$$

where V is the radiative volume of the bed, and A is the coverage area of bed radiative volume V. In a medium with low particle concentration, such as in the suspension (dilute bed) of a BFB boiler, the emissivity of the dilute bed without accounting for the impact of diffuse reflection is considered as follows:

$$\varepsilon_d = \varepsilon_g + \varepsilon_p - \varepsilon_g\varepsilon_p \tag{4.15}$$

where ε_p and ε_g are the emissivity of particles and gas, respectively. Research has shown that if the values of s and Y make ε_p more than 0.5~0.8, the influence of diffuse reflection must be considered. For large CFB boilers, the emissivity of the dilute bed can be calculated according to the equation proposed by Brewster [16]:

$$\varepsilon_d = \sqrt{\frac{\varepsilon_{p,s}}{(1-\varepsilon_{p,s})B}\left(\frac{\varepsilon_{p,s}}{(1-\varepsilon_{p,s})B} + 2\right)} - \frac{\varepsilon_{p,s}}{(1-\varepsilon_{p,s})B} \tag{4.16}$$

where $\varepsilon_{p,s}$ is the average emissivity of a particle surface, $\varepsilon_{p,s} = 1 - e^{-c_\varepsilon \cdot c_p^n}$, coefficient c_ε is 0.1~0.2, c_p is material concentration, index n is 0.2~0.4, and B is a coefficient; for isotropous diffuse reflection, $B = \frac{1}{2}$, for diffuse reflection particles, $B = \frac{2}{3}$.

The emissivity of the particle dispersed phase ε_d can be calculated using Eq. (4.15) or Eq. (4.16), which introduces the definition of diffuse reflection proposed by Brewster. The radiative heat transfer coefficient of the particle dispersed phase can be calculated by the following equation:

$$h_{dr} = \varepsilon_{d,sys}\sigma\frac{T_b^4 - T_s^4}{T_b - T_s} \tag{4.17}$$

$$\varepsilon_{d,sys} = \frac{1}{\frac{1}{\varepsilon_d} + \frac{1}{\varepsilon_s} - 1} \tag{4.17a}$$

where ε_s is the emissivity of the heating surface, $\varepsilon_{d,sys}$ is the particle dispersed phase system emissivity, and T_s is the temperature of the heating surface.

2. Radiation of particle clusters

The emissivity of particle clusters ε_c can be determined based on the multiple reflection of particle clusters from the following equation:

$$\varepsilon_c = \frac{1+\varepsilon_{p,s}}{2} \tag{4.18}$$

where $\varepsilon_{p,s}$ is the emissivity of the bed material.

The heat transfer coefficient of clusters h_{cr} can be calculated by changing the ε_d in Eq. (4.17) to ε_c:

$$h_{cr} = \varepsilon_{c,sys}\sigma\frac{T_b^4 - T_s^4}{T_b - T_s} \tag{4.19}$$

$$\varepsilon_{c,sys} = \frac{1}{\dfrac{1}{\varepsilon_c}+\dfrac{1}{\varepsilon_s}-1} \tag{4.19a}$$

In a typical CFB boiler, the heating surface is formed by tube panels. The particle concentration in the bed is small, especially in the upper heating surface zone of the furnace; therefore, dilute-phase radiative heat transfer is predominant. It is appropriate, to this effect, to use a projected heating surface area to evaluate the radiative heat transfer component and to employ the total peripheral area to evaluate the convective heat transfer component. Eqs. (4.17) and (4.18) are suitable for large CFB boilers, but are not for small CFB boilers or the radiative heat transfer of lab-scale fire-resistant surfaces.

4.4   HEAT TRANSFER CALCULATION IN A CIRCULATING FLUIDIZED BED

A major difference between CFB boilers and PC boilers is that the material (including coal ash, desulfurization agent, and so on) concentration in CFB boilers is much higher than that in PC boilers, and that the material concentration along the heating surface is uneven. Material concentration is critical for heat transfer in a furnace, so the material concentration at the furnace outlet should be calculated first—this can be evaluated according to the external circulation rate. The material concentration at different furnace heights is determined by the internal circulation rate, which varies along the furnace height. The material concentration within the lower part is high, while that within the upper part is low. In a large-scale CFB boiler, different calculation equations must be adopted to calculate the water wall, dividing water wall, platen superheater, and platen reheater separately. Material concentration and radiative heat transfer both have significant impact on convective heat transfer.

The structure/size of the furnace's heating surface, including width and thickness of fin, also significantly affects the average heat transfer coefficient. The width of fins affects the aggregation of material clusters, and is related to the utilization coefficient of the extended heating surface. Similar to the PC boiler, the temperature level in the CFB boiler furnace significantly impacts radiative heat transfer. The calculation methods presented here were established according to previously published work and methods proposed by Tsinghua University, with careful consideration of engineering convenience and feasibility.

The absorbed heat of the heating surface in a CFB boiler can be calculated by the following equation:

$$Q = K \cdot A_g \cdot \Delta T \tag{4.20}$$

where Q is the heat transfer rate, K is the heat transfer coefficient based on the total area of the gas side, ΔT is the temperature difference, and A_g is the total peripheral area of the gas side.

4.4.1 Influence of Heating Surface Size on Heat Transfer

The heat transfer coefficient of a furnace's heating surface can be calculated using Eq. (4.21). The thermal resistance includes four parts: gas side thermal resistance $\frac{1}{h_b^0}$, working medium side resistance $\frac{1}{h_m} \cdot \frac{A_g}{A_m}$, the heating surface's own thermal resistance $\frac{\delta_1}{\lambda}$, and additional thermal resistance ρ_{as}. See the following:

$$K = \frac{1}{\frac{1}{h_b^0} + \frac{1}{h_m}\frac{A_g}{A_m} + \rho_{as} + \frac{\delta_1}{\lambda}} \tag{4.21}$$

where h_b^0 is the heat transfer coefficient based on the total wall surface of the gas side, h_m is the heat transfer coefficient of the working medium side (which can be calculated according to thermal calculation of the former Soviet Union's 1973 standard), A_g is the total area of the gas side, A_m is the total area of the working medium side, δ_1 is tube thickness, and λ is the metal thermal conductivity of the heating surface.

The heat transfer coefficient based on the total wall surface of the gas side is expressed as follows:

$$h_b^0 = [P(\eta - 1) + 1]\frac{h_b}{1 + \rho_s \cdot h_b} \tag{4.22}$$

where P is the fin area factor, $P = \dfrac{A_{\text{fin}}}{A_{\text{g}}}$, A_{fin} is fin area, η is fin utilization, h_{b} is the gas side heat transfer coefficient shown in Eq. (4.34), and ρ_{s} is the ash deposition coefficient of the heating surface (0.0005).

The fin area coefficient is:

$$P = \frac{A_{\text{fin}}}{A_{\text{g}}} = \frac{s-d}{s-\delta+\left(\dfrac{\pi}{2}-1\right)d} \tag{4.23}$$

where s, δ, and d are tube pitch, fin thickness, and tube diameter, respectively, which are depicted in Fig. 4.14.

The fin utilization coefficient is:

$$\eta = \frac{\tan h(\beta \cdot w'')}{\beta \cdot w''} \tag{4.24}$$

where β is related to the heating situation of the heating surface, fin structure/size, and membrane wall fin material, which can be denoted as follows:

$$\beta = \sqrt{\frac{Nh_{\text{b}}(w+\delta)}{\delta\lambda(1+\rho_{\text{s}}h_{\text{b}})}} \tag{4.25}$$

where λ is the metal thermal conductivity, δ is fin thickness, and N represents the heating situation—for single-face heating, $N = 1$, and for double-face heating, $N = 2$.

The real fin width is:

$$w = \frac{s-d}{2} \tag{4.26}$$

Effective width in Eq. (4.24) is:

$$w'' = \frac{w'}{\sqrt{N}} \tag{4.27}$$

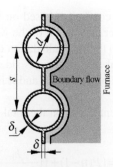

FIGURE 4.14 Schematic diagram of membrane wall.

The reduced width of this equation is:

$$w' = \frac{w}{\mu} \tag{4.28}$$

According to experimental and operational data, the relationship of fin width coefficient μ and structure/size can be obtained as follows:

$$\mu = f_\mu\left(\frac{s}{d}\right) \tag{4.29}$$

When $\dfrac{s}{d} = 1.3$, $\mu = 0.97$; when $\dfrac{s}{d} = 1.7$, $\mu = 0.9$.

In Eq. (4.21), ρ_{as} is additional thermal resistance, which can be expressed as follows:

$$\rho_{as} = \frac{\delta_a}{\lambda_a} \tag{4.30}$$

where δ_a is the refractory layer thickness of the heating surface and λ_a is the thermal conductivity of the fire-resistance layer, which can be calculated using the following equation:

$$\lambda_a = a_0 + a_1\overline{T}_a \tag{4.31}$$

where a_0 and a_1 are coefficients and T_a is the average temperature of the refractory layer, which can be calculated using the following equation:

$$\overline{T}_a = (T_b + T_w)/2 \tag{4.32}$$

where T_b is the gas side temperature and T_w is the heating surface temperature, as shown in Eq. (4.36).

In the heating surface structure (Fig. 4.14), the ratio of external surface area and internal surface area of the heating surface is:

$$\frac{A_g}{A_m} = 1 + \frac{2}{\pi}\left(\frac{s - \delta - (2 - \pi)\delta_1}{d - 2\delta_1} - 1\right) \tag{4.33}$$

where δ_1 is tube wall thickness, s is tube pitch, and δ is fin thickness.

4.4.2 CFB Boiler Gas Side Heat Transfer Coefficient

The heat transfer from gas/material two-phase mixtures to the heating surface includes convection and radiation. According to linear superposition of these two types of heat transfer, the gas side heat transfer coefficient in the CFB boiler is:

$$h_b = h_r + h_c \tag{4.34}$$

where h_r is the radiative heat transfer coefficient and h_c is the convective heat transfer coefficient.

In principle, h_r and h_c can be obtained through the methods introduced in Sections 4.2 and 4.3. Here, we've introduced a simplified calculation method for engineering practice.

1. Radiative heat transfer coefficient

$$h_r = \varepsilon_{sys}\sigma \frac{T_h^4 - T_w^4}{T_b - T_w} \tag{4.35}$$

In this equation, σ is the Boltzmann constant, ε_{sys} is gas side system emissivity, and T_w is the water wall tube surface temperature, calculated as follows:

$$T_w = T_m + \Delta T_w \tag{4.36}$$

where T_m is the working medium temperature in the heating tubes. The difference between the internal and external temperatures of the carbon steel water wall tube is:

$$\Delta T_w = 0.7(T_b - T_m) \cdot N \cdot \left(\frac{A_{fin}}{A_g}\right)^{0.4} \cdot \frac{1000}{h_m} \tag{4.37}$$

where T_b is the gas side temperature, T_m is the working medium temperature in the heating tubes, N is a coefficient for the heating situation (either 1 or 2, as mentioned above), $\dfrac{A_{fin}}{A_g} = P$ is the fin area ratio, and h_m is the heat transfer coefficient of the working medium side.

In a superheater and reheater made from steel alloy:

$$\Delta T_w = 0.7(T_b - T_m) \cdot N \cdot \frac{A_{fin}}{A_g} \cdot \frac{1000}{h_m} \tag{4.38}$$

The system emissivity between the surface and gas side ε_{sys} can be written as follows:

$$\varepsilon_{sys} = \frac{1}{\dfrac{1}{\varepsilon_b} + \dfrac{1}{\varepsilon_w} - 1} \tag{4.39}$$

where ε_b is gas-side emissivity and ε_w is heating surface emissivity (usually 0.5~0.8). It should be noted here that the definition of this equation is different from that of Eqs. (4.17a) and (4.19a).

In gas-solid two-phases, gas side emissivity ε_b includes both particle emissivity and gas emissivity, which are calculated similarly to Eq. (4.15):

$$\varepsilon_b = \varepsilon_p + \varepsilon_g - \varepsilon_p\varepsilon_g \tag{4.40}$$

where ε_p is solid material emissivity and ε_g is gas emissivity. ε_p can be calculated using Eq. (4.16) (assume $\delta_c \approx 0$, $\varepsilon_p \approx \varepsilon_d$), and ε_g can be calculated using Eqs. (2.63) and (2.64) for gas emissivity (see chapter: Emission and Absorption of Thermal Radiation). A further simplified method can be adopted for calculation in practice.

2. Convective heat transfer coefficient

The convective heat transfer coefficient, which includes gas convection and particle convection, can be expressed as follows:

$$h_c = h_{gc} + h_{pc} \tag{4.41}$$

where h_{gc} is the gas convective heat transfer coefficient and h_{pc} is the particle convective heat transfer coefficient. The gas convective heat transfer coefficient is:

$$h_{gc} = C_{gc} \cdot \omega_g \tag{4.42}$$

where C_{gc} is the gas convection coefficient (around 4~5), and ω_g is gas velocity. The particle convective heat transfer coefficient is:

$$h_{pc} = C_{pc} w_g^{\frac{1}{2}} \cdot h_{pc}^0 \tag{4.43}$$

where ω_g is the gas velocity and h_{pc}^0 is the theoretical particle convective heat transfer coefficient in incipient fluidization, the value of which is related to particle size, particle temperature, and heating surface configuration. C_{pc} is the particle convection coefficient, which is related to the local material concentration in the vicinity of the furnace water wall.

Material concentration near the furnace water wall can be calculated using empirical equations, or according to the characteristic material concentration of the furnace.

Chapter 5

Heat Transfer Calculation in Furnaces

Chapter Outline

Calculation of heat transfer in boilers consists of two parts: calculations of heat transfer in the combustion chamber and in the convection heating surfaces. The first four sections of this chapter are devoted to the former, in which different kinds of boilers are considered, including grate-firing boilers, suspension-firing boilers and circulating fluidized bed (CFB) boilers. Section 5.5 gives a brief introduction to the latter and Section 5.6 describes the process of heat transfer calculation in brief, as a conclusion of the chapter.

Theory and Calculation of Heat Transfer in Furnaces. http://dx.doi.org/10.1016/B978-0-12-800966-6.00005-3

5.1   HEAT TRANSFER IN FURNACES [27,28]

5.1.1   Processes in the Furnace

A functioning furnace undergoes four interactive processes: flow, combustion, heat, and mass transfer. Heat transfer is only one process in the furnace, for which an exact solution cannot be obtained unless four groups of equations, corresponding to the four processes, are solved simultaneously. Thus, due to inherent complexity, strictly theoretical analysis is impossible.

1. Basic equations of heat transfer in flame

   Fuels burn in the combustion chamber, releasing energy and forming the flame. The basic steady-state energy equation of heat transfer from the flame to the furnace walls is as follows:

$$\nabla \cdot \left(\rho V c_p T \right) = \nabla \cdot \left(\Gamma_T \nabla T \right) + S_Q \tag{5.1}$$

   where the left-hand side is the convection form of enthalpy ($c_p T$), V is velocity, the first term on the right-hand side is the diffusion term, Γ_T is the diffusion coefficient, and S_Q is the generalized source term. See the following:

$$S_Q = Q_{ch} + Q_R \tag{5.2}$$

   (5.2) where Q_{ch} is the heat release rate of the chemical reaction and Q_R is the heat transfer rate of radiation.

   The above is only an energy conservation equation—convection and diffusion terms are usually calculated by the difference method. The equation also differs from general transport equations due to the complex spatial integral term Q_R, which is the primary difficulty of the solution.

   As discussed in previous chapters, the calculation of Q_R is a complex spatial integral because Q_R is determined by the geometry, thermodynamic state, species distribution, and other factors throughout the entire space in the chamber due to the noncontact mode of radiative heat transfer. An analytical solution is, by definition, impossible, so numerical methods are commonly utilized, among which the heat flux method, domain method, and probability simulation method are most popular.

2. Mathematical model of heat transfer in the furnace

   The heat flux method, domain method, and probability simulation method are all numerical methods of calculating the heat transfer rate of radiation in a furnace. The models these methods are based on are incomplete, however, as they only describe the principles the temperature field must obey and require information to be given such as flow field, heat source distribution, and physical parameters.

   As a matter of fact, Eq. (5.1) alone is insufficient for full mathematical formulation of the interaction of flow, combustion, heat, and mass transfer in the furnace. These interactive processes, collectively referred to as the "combustion process," consist of turbulent flow and combustion, heat transfer in the flame, and multiphase flow and combustion, all of which obey basic physical and chemical laws represented below as conservation equations.

Continuity equation: law of conservation of mass.
Momentum equation: Newton's second law.
Energy equation: law of conservation of energy.
Chemical balance equation: law of conservation and transformation of chemical species.

All these equations can be represented in a unified form as follows:

$$\frac{\partial}{\partial \tau}(\rho \Phi) + \nabla \cdot (\rho u \Phi + J_\Phi) = S_\Phi \tag{5.3}$$

where Φ represents physical flux, such as the constant, velocity, enthalpy, or concentration corresponding separately to continuity, momentum, energy, and reaction equations. u is velocity, J_Φ is diffusion flux, S_Φ is a source term, and, τ is time.

Once detailed expressions of J_Φ and S_Φ are obtained, the governing equation of combustion can be represented as follows:

$$\frac{\partial}{\partial \tau}(\rho \Phi) + \nabla \cdot (\rho u \Phi) = \nabla \cdot (\Gamma_\Phi \nabla \Phi) + S_\Phi \tag{5.4}$$

where the first term on the left-hand side is time-varying, the second convection term and the first on the right-hand side are diffusion terms, and S_Φ is the source term. Eq. (5.4) reduces to (5.1) taking Φ as $c_p T$.

Turbulent flow is popular in real-world engineering, but mathematical formulation is not closed for time-averaging turbulent equations. Additional equations are needed to this effect, resulting in different turbulent models such as the k-ε model. Each model may provide solutions for distributions of temperature, velocity, pressure, and chemical species concentration in the furnace, fully describing the characteristics of heat and mass transfer and chemical reactions. However, as mentioned above, the process of solving these equations is so complex that numerical methods are used instead to obtain an acceptable approximation. Simplified models based on data gathered through engineering experience (empirical coefficients, in particular) are usually adopted for technical applications.

5.1.2 Classification of Heat Transfer Calculation Methods

Essentially speaking, calculations for heat transfer are all semitheoretical approaches based on experience; certain parameters such as thermal conductivity, thermal diffusivity, diffusion coefficient, viscosity coefficient, and emissivity are all determined by measurement, during which an accurate relationship between these coefficients and temperature or pressure is mostly unavailable. Empirical methods also attribute uncertainty to one or several factors, including the heat transfer coefficient, thermal effective coefficient, and others. Calculation is always based on some basic theories, of course (postulates, assumptions, laws, theorems, etc.), making it a semiempirical or semitheoretical method of analysis. Furnace heat transfer calculations are more empirical. The following section outlines a handful of useful calculations for classification based on spatial dimensions.

There are zero-dimensional, one-dimensional, two-dimensional, and three-dimensional models available for application to furnace heating calculation. In a zero-dimensional model, all physical quantities within the furnace are uniform and the results are averaged. This method is the one most often used for engineering design, and is the standard method for thermal calculation in China. One-dimensional models are used to study changes in the physical quantities along the axis (height) of the furnace, where the physical quantity in the perpendicular plane is uniform. This model has practical value for engineering projects such as large-capacity boilers.

The two-dimensional model is mainly used for axisymmetric cylindrical furnaces, such as vertical cyclone furnaces. The three-dimensional model describes the furnace process (flow, temperature, chemical species fields, and so on), using three-dimensional coordinates (x, y, z). In principle, only a three-dimensional model can correctly describe the furnace process – in reality, all the equations used so far for describing the furnace process fail to obtain analytical solutions, and only the numerical methods can reach approximate solutions. Even for an approximate solution the amount of calculation is very large—slow or small-capacity computers are not up to the task.

The experience method was previously most commonly applied to zero-dimensional models due to a lack of adequate understanding of the furnace process and related mechanisms. Currently, the semiempirical method is growing in popularity—this method is based on fundamental equations such as the thermal balance equation and radiative heat transfer equation, as well as certain coefficients or factors obtained through experimentation.

5.1.3 Furnace Heat Transfer Calculation Equation

Here, the furnace heat transfer calculation method is introduced using a zero-dimensional model. This method is primarily based on the energy conservation equation and the radiative heat transfer equation.

The thermal balance equation is as follows:

$$Q = \varphi B_{cal}(I_a - I_F'') = \varphi B_{cal} V\bar{C}(T_a - T_F'') \tag{5.5}$$

where φ is the heat preservation coefficient (which accounts for the heat loss due to the cooling effect of the furnace wall), B_{cal} is the design fuel (burnt fuel) supply rate, $B_{cal} = \dfrac{100 - q_{uc}}{100} B$, B is the fuel supply rate per boiler unit (obtained by the boiler heat balance), $I_a(T_a)$ is the enthalpy of flue gas at theoretical combustion temperature T_a, $I_F''(T_F'')$ is the enthalpy of flue gas at the temperature of the exit of the furnace T_F'', and $\overline{VC}$ is the mean overall heat capacity of the combustion products between T_a and T_F'', expressed as follows:

$$\overline{VC} = \frac{I_a - I_F''}{T_a - T_F''} \tag{5.6}$$

Note that the values of the above-mentioned quantities, I_a, T_F'', $\overline{VC}$, are calculated in terms of per kg or per m^3 of fuel.

Next, let's discuss the radiative heat transfer equation. As discussed in chapter: Theoretical Foundation and Basic Properties of Thermal Radiation, Planck's law and its corollary (Stefan–Boltzmann law) form the basis for calculation of radiation heat transfer. The basic formula for calculating radiation heat transfer is the Stefan–Boltzmann law, which can be conducted in two different ways.

1. Direct calculation of radiation heat (the Hottel method) as follows:

$$Q = \sigma a_F A \left(T_g^4 - T_w^4 \right) \tag{5.7}$$

$$a_F = \frac{1}{\dfrac{1}{\varepsilon_w} + x\left(\dfrac{1}{\varepsilon_g} - 1 \right)} \tag{5.8}$$

2. According to the projected radiation heat transfer (the Gurvich method) as follows:

$$Q = \sigma \tilde{a}_F \psi A T_g^4 \tag{5.9}$$

$$\tilde{a}_F = \frac{1}{1 + \psi\left(\dfrac{1}{\varepsilon_g} - 1 \right)} \tag{5.10}$$

The above formulas are applicable to suspension-firing furnaces, as introduced in chapter: Heat Transfer in Fluidized Beds. As long as a_F (or $\tilde{a}_F$), T_g, and T_w (or ψ) are obtained, Q can be determined effectively. During calculation in practice, Eq. (5.5) is typically used to calculate Q, but T_g is usually determined by solving Eqs. (5.5–5.10) simultaneously.

5.1.4   Flame Temperature

First, let's consider furnace temperature distribution—strictly speaking, it is a function of time and space $T = (x, y, z, \tau)$, where τ is time. Of course, this is only a phenomenological description. Speaking to the mechanism, the temperature

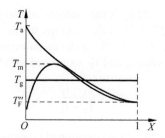

FIGURE 5.1 Furnace temperature distributions.

distribution in the furnace is affected by the furnace geometry, fuel characteristics, burner structure, boiler load level, combustion condition, and other factors.

In Fig. 5.1, T_a marks the adiabatic combustion temperature. Assuming that the combustion is complete in a very short period of time, the heat transfer process has not yet started and the combustion products are at their highest possible temperature. Curve T (along the direction of furnace height X) is the actual temperature, the maximum of T is T_m, the T of the furnace exit ($X = 1$) is T_F'', and the mean value of T along $X \in [0,1]$ is T_g. Naturally, $T_a > T_m > T_g > T_F''$, where adiabatic combustion temperature T_a, average flame temperature T_g, and furnace outlet temperature T_F'' are all important factors during heat transfer analysis and calculation.

Next, let's discuss an empirical equation that describes furnace temperature distribution. Furnace temperature distribution is a function of time and spatial coordinates, but that does not mean there isn't any other description similar to the three-dimensional description. In general, furnace temperature distribution is considered to be in a stable steady-state condition under steady working conditions, that is, $T = (x, y, z)$. See the following simple empirical formula:

$$\theta^4 = e^{-\alpha X} - e^{-\beta X} \tag{5.11}$$

where $\theta = \dfrac{T}{T_a}$ is the dimensionless temperature, α and β are the empirical factors, $X = \dfrac{x}{L}$ is the dimensionless flame position, L is the total length of the flame (ie, the furnace height), and x is the distance from the nozzle of the burner.

Let $X = 1$ in Eq. (5.11) so that the outlet temperature of the furnace is:

$$\theta_F'' = \left(e^{-\alpha} - e^{-\beta} \right)^{\frac{1}{4}} \tag{5.12}$$

where $\theta_F'' = \dfrac{T_F''}{T_a}$.

Calculate the highest flame temperature location X_m, because $\dfrac{d\theta}{dX}\big|_{X=X_m} = 0$:

$$X_m = \frac{\ln \alpha - \ln \beta}{\alpha - \beta} (\alpha > 0, \beta > 0, \alpha \neq \beta) \tag{5.13}$$

Calculate average flame temperature $\left(\theta_g = \dfrac{T_g}{T_a} \right)$:

$$\theta_g = \int_0^1 \theta \, dX \tag{5.14}$$

This equation is not particularly convenient to calculate. According to the above analysis, θ_F'', θ_g, and X_m are all functions of α and β. The relationship between θ_F'' and θ_g can be determined using Eqs. (5.12–5.14) to simplify the calculation. The typical form of Eq. (5.11) is:

$$T_g^4 = m T_a^{4(1-n)} T_F''^{4n} \tag{5.11a}$$

where m and n are empirical coefficients. In Eq. (5.11), X is a coordinating variable, suggesting that heating surfaces are arranged around the flame, the temperature field of the furnace cross section is of uniform distribution, and only radial radiative heat transfer is considered. This hypothesis is actually in approximate form—the temperature field of the furnace has different degrees of heterogeneity, and there are heating surfaces at both ends of the flame, therefore, the above empirical equation and its inference can only be applied to qualitative analysis. In other words, it is not appropriate to apply it in quantitative engineering calculations (except when validated by industrial tests).

Next, let's take a look at an analytical solution for average flame temperature. Due to the complexity of the furnace process, the model for average flame temperature must be simplified. The results can only be applied to approximate qualitative analysis, but there is practical significance for this method itself—roughly simplifying a complex process into its most valuable components is very useful in engineering practice.

Assume that fuel combusts completely instantaneously at the burner exit and reaches adiabatic combustion temperature T_a, and that heat transfer only occurs in the radial direction of the furnace axis; ignore heat transfer in the axial direction (one-dimensional model). Then, from thermal balance:

$$-B_{cal} V \overline{C} dT = \sigma \widetilde{a_F} \psi T^4 dA \tag{5.15}$$

where A is the furnace area and $V\overline{C}$ is mean overall heat capacity. Integrate Eq. (5.15) to obtain:

$$-\int_{T_a}^{T_F''} \frac{dT}{T^4} = \frac{\sigma \widetilde{a_F} \psi}{B_{cal} V \overline{C}} \int_0^A dA$$

so that:

$$\frac{1}{3}\left(\frac{1}{T_F''^3} - \frac{1}{T_a^3} \right) = \frac{\sigma \widetilde{a_F} \psi}{B_{cal} V \overline{C}} A \tag{5.16}$$

According to the basic equation of heat transfer in furnaces in Eqs. (5.5) and (5.9):

$$B_{cal}V\overline{C}(T_a - T_F'') = \sigma \widetilde{a_F} \psi T_g^4 A \tag{5.17}$$

which means:

$$\frac{T_a - T_F''}{T_g^4} = \frac{\sigma \widetilde{a_F} \psi A}{B_{cal}V\overline{C}} \tag{5.18}$$

Substitute this equation into Eq. (5.16) and transform:

$$T_g^4 = rT_F''^4 \tag{5.19}$$

where:

$$r = \frac{3}{\left(\frac{T_F''}{T_a}\right)^3 + \left(\frac{T_F''}{T_a}\right)^2 + \left(\frac{T_F''}{T_a}\right)} \tag{5.20}$$

Another method is also possible. Let maximum flame temperature be T_m and at the point near burner elevation let $A_m/A = X_m$, where A_m is the area of furnace surface below the maximum flame elevation, then:

$$r = \frac{3(1 - X_m)}{\left(\frac{T_F''}{T_a}\right)^3 + \left(\frac{T_F''}{T_a}\right)^2 + \left(\frac{T_F''}{T_a}\right)} \tag{5.21}$$

When deducing the above equation, remember that $T_m = T_a$ and Eq. (5.19) is satisfied.

Calculating average flame temperature T_g is a key issue when analyzing furnace heat transfer. Different calculation formulas for T_g are adopted for different methods of calculating furnace heat transfer. All calculation formulas for T_g are empirical equations, including some correction factors. Consider the two following examples.

1. Suppose that the gas temperature in the furnace is homogeneous, that is $T_g = T_F''$, but can be corrected in practice:

$$T_g = T_F'' \left(1 + \sum \Delta i\right) \tag{5.22}$$

where Δi is a correction factor considering fuel type, combustion type, burner tilt, water-cooling degree, and other necessary aspects.

2. Consider Tg a function of T_F'' and T_a, where $T_g = f(T_F'', T_a)$. Add a correction factor for combustion conditions to yield the following Gurvich method equation:

$$T_g^4 = M^{\frac{3}{5}} \frac{\left(\frac{T_a}{T_F''}\right)^3}{\sqrt[3]{\left(\frac{T_a}{T_F''} - 1\right)}} T_F''^4 \tag{5.23}$$

where $M = A\text{-}BX_B$ is the correction for combustion conditions, X_B is the relative height of the burner ($X_B = \dfrac{H_B}{H_F}$, H_B is burner height from furnace bottom, H_F is furnace height), and A and B are empirical coefficients determined by fuel type and combustion type.

5.2 HEAT TRANSFER CALCULATION IN SUSPENSION-FIRING FURNACES [11]

Next, let's calculate the heat transfer in a suspension-firing furnace. Keep in mind several of the methods introduced above, including the theoretical foundation of radiative heat transfer, heat radiation of absorbed scattering media, radiative heat transfer of surfaces with transparent media, and radiative heat transfer from isothermal media to surfaces. The same basic assumptions (uniform surface temperature and isothermal medium) apply, though in practice, they are not accurate; in a real-world furnace, surface temperature (including the water wall) is not uniform—even the medium temperature in the furnace is nonuniform. The following section examines the furnace process when the medium temperature is considered to be nonuniform, with special focus on radiative heat transfer.

Heat transfer calculation equations and furnace emissivity were introduced in chapter: Heat Transfer in Fluidized Beds. This chapter introduces average flame temperature T_g, which allows us to obtain the rate of heat transfer. Here, we integrate this content and provide information regarding the acquisition and utilization of some useful empirical coefficients. The Gurvich method, which is suitable for both suspension-firing and grate-firing furnaces, is introduced below.

5.2.1 Gurvich Method

According to the radiative heat transfer equation introduced in Section 5.1:

$$Q = \tilde{a}_F \sigma \psi A T_g^4$$

The Polyak-Shorin [44] formula is:

$$T_g^4 = m T_a^{4(1-n)} T_F''^{4n}$$

where generally, m≈1, and the thermal balance formula is:

$$Q = \varphi B_{cal} V \bar{C} \left(T_a - T_F'' \right)$$

where φ is the heat preservation coefficient considering heat loss due to furnace wall radiation and convection.

Then:

$$\theta_F''^{4n} - \frac{Bo}{m\tilde{a}_F}\left(1 - \theta_F''\right) = 0 \qquad (5.24)$$

where $Bo = \dfrac{\varphi B_{cal} V \bar{C}}{\sigma \psi A T_a^3}$ is the Boltzmann number. From m$\approx$1, n varies according to the combustion condition, so the above equation can be written as:

$$\theta_F'' = f\left(\frac{Bo}{\tilde{a}_F}, n\right) \tag{5.25}$$

where $\theta_F'' = \dfrac{T_F''}{T_a}$. According to Eq. (5.25), the following completely empirical equation can be obtained from experimental data:

$$\theta_F'' = \frac{\left(\dfrac{Bo}{\tilde{a}_F}\right)^{0.6}}{M + \left(\dfrac{Bo}{\tilde{a}_F}\right)^{0.6}} \tag{5.26}$$

where M is the flame center modification factor, which is dependent on flame center relative position X_m. Eq. (5.26) can be transferred into the following:

$$T_F'' = \frac{T_a}{M\left(\dfrac{\sigma \psi A \tilde{a}_F T_a^3}{\varphi B_{cal} V \bar{C}}\right)^{0.6} + 1} \tag{5.27}$$

Furnace outlet gas enthalpy I_F'' can be determined based on furnace outlet gas temperature T_F'', followed by heat transfer in the furnace $Q = \varphi B_{cal}(Q_a - I_F'')$. Eq. (5.27) can be used to calculate furnace outlet gas temperature.

5.2.2 Calculation Method Instructions [17,22,35,37,40,43,45]

First, calculate flame center modification factor M according to fuel type, combustion type, burner position, and burner tilt. The following M values are important (please also refer to Table D6.11):

$M = 0.54 - 0.2X_m$ for gas and heavy-oil-fired boilers.

$M = 0.59 - 0.5X_m$ for PC boilers buring bituminous coal and lignite   (5.28)

$M = 0.56 - 0.5X_m$ for PC boilers burning anthracite, lean coal, and high-ash bituminous coal.

$M = 0.59$ for grate-firing boilers with a thin coal layer (a wind-force coal thrower boiler).

$M = 0.52$ for grate-firing boilers with a thick coal layer (grate boiler, stoker). Flame center relative position X_m can be calculated by the following equation:

$$X_m = X_B + \Delta X \tag{5.29}$$

where $X_B = H_B/H_F$, H_B is burner height from the burner center to the furnace hopper center or furnace bottom, and H_F is furnace height from the furnace exit center to the furnace hopper center or furnace bottom. For a pulverized coal furnace, apply the following correction ΔX values:

$\Delta X = 0.1$ D≤420 t/h(116 kg/s)
$\Delta X = 0.05$ D>420 t/h(116 kg/s)
$\Delta X = 0$ Tangential fired boiler

When the combustor tilts upwards 20 degree, ΔX increases by 0.1. When the combustor tilts downwards 20 degree, ΔX decreases by 0.1. When the tilting angle is between −20 degree~20 degree, ΔX can be calculated based on ΔX linear interpolation.

Average flame temperature is not included in Eq. (5.26) (the Gurvich equation). Based on the radiative heat transfer equation, thermal balance equation, and Eq. (5.26), however, the average flame temperature can be determined through the Gurvich method. See the following:

$$T_g^4 = \frac{MT_F'' T_a^3}{\sqrt[3]{\frac{1}{M^2}\left(\frac{T_a}{T_F''}-1\right)^2}} \tag{5.30}$$

Do note that this equation is completely empirical.

Let's examine the scope of application of the Gurvich method a bit. Under the 1973 standard suggested by the former Soviet Union for boiler thermal calculation, Eq. (5.26) can be applied when $\theta_F'' \leq 0.9$. What was the foundation for this scope?

From the thermal balance equation:

$$q = \frac{\varphi B_{cal} V \overline{C}}{A}(T_a - T_F'') \tag{5.31}$$

and radiative heat transfer equation:

$$q = \sigma \tilde{a}_F T_g^4 \psi \tag{5.32}$$

as well as Eqs. (5.31) and (5.32):

$$\frac{Bo}{\tilde{a}_F} = \frac{1}{1-\theta_F''}\frac{T_g^4}{T_a^4} \tag{5.33}$$

substitute this equation into Eq. (5.26):

$$M^{\frac{5}{3}}\frac{\theta_F''^{\frac{5}{3}}}{(1-\theta_F'')^{\frac{2}{3}}} = \left(\frac{T_g}{T_a}\right)^4 \tag{5.34}$$

We know that $T_g \leq T_a$, so:

$$\frac{\theta_F''}{(1-\theta_F'')^{\frac{2}{5}}} \leq \frac{1}{M} \tag{5.35}$$

where when $M = 0.3$, $\theta_F'' \leq 0.95$, when $M = 0.44$, $\theta_F'' \leq 0.9$, and when $M = 0.5$, $\theta_F'' \leq 0.85$. Generally, M is not less than 0.44, therefore $\theta_F'' \leq 0.9$ is the scope of the Gurvich equation.

The above analysis determined the application scope based on the relationship $T_g \leq T_a$. In fact, $T_g < T_m < T_a$, thus the actual application scope is a bit narrower, usually $\theta_F'' < 0.7$.

The heat transfer in large-capacity boilers should be calculated a little differently. The Gurvich equation, Eq. (5.26), was obtained based on experimental data from a boiler with output D $\leq 200 \sim 300$ t/h. (The influence of temperature uniformity in the furnace was ignored during data analysis and modeling.) The output of a modern boiler, especially a large utility boiler, is far greater than this value (with maximum capacity above 3000 t/h). When boiler volume is large, Eq. (5.26) does not produce very accurate results—the calculated furnace outlet gas temperature is then lower than the measured value (often lower than $100 \sim 130°C$). The following is an equation for furnace heat transfer suitable for a large boiler, with correction for an uneven temperature field:

$$T_F'' = T_a \left[1 - M \left(\frac{\tilde{a}_F \psi T_a^2}{10800 q_H} \right)^{0.6} \right] \tag{5.36}$$

where the calculation methods of $\tilde{a}_F$, ψ, T_a, and M are as introduced above. Please note that thermal efficiency coefficient ψ is an average value. The heat release rate per unit of radiative heating surface area q_H, can be calculated by the following equation:

$$q_H = \frac{B_{cal} Q_{fuel}}{A_F}$$

where Q_{fuel} is the heat input by fuel of 1 kg:

$$Q_{fuel} = \frac{Q_{ar,net,p}(100 - q_{uc} - q_{ug} - q_{ph})}{100 - q_{uc}} + Q_{air}$$

Approximately:

$$Q_{fuel} = Q_{ar,net,p}$$

where the empirical value of q_H for large-volume boilers is recommended. For bituminous coal or lean coal, $q_H = 290 \sim 310$ kW/m^2; for low-grade lignite, when boiler capacity is $180 \sim 275$ kg/s ($660 \sim 990$ t/h), $q_H = 210 \sim 280$ kW/m^2;

for heavy oil, $q_H = 560 \sim 635$ kW/m^2. These data can be referred to easily during boiler design.

The furnace volume heat release rate is: $q_V = \dfrac{BQ_{ar,net,p}}{V_F}$, furnace cross-section heat release rate is $q_A = \dfrac{BQ_{ar,net,p}}{A_F}$, and furnace release rate is $q_H = \dfrac{B_{cal}Q_r}{H}$, where V_F is furnace volume, A_F is furnace cross section area, H is furnace heating surface area, B is fuel consumption, and Q_r is the furnace heat transfer rate.

A few additional instructions are necessary for furnace heat transfer calculation. There are two main goals when performing furnace heat transfer calculations: (1) determining the transfer rate and the heat distribution in each heating surface of the boiler, and (2) solving and choosing the proper furnace outlet gas temperature (prior to the slag screen) to prevent slagging and tube burst from occurring in the convection heating surface downstream of the furnace exit.

1. Design and verification calculation

   Furnace heat transfer calculation is a part of the overall thermal calculation of the boiler. Boiler thermal calculation also includes design calculation (where an original need is given to design a new boiler), and verification calculation (where a finished design or an existing boiler's ability to meet an original requirement when fuel or load changes is verified).

   Furnace heat transfer calculation is necessary for both design calculation and verification calculation. For the furnace outlet gas temperature, generally $T_F'' < T_{DT}$ is needed, T_{DT} is the deformation temperature of coal ash, $T_F'' \leq T_{ST} - 100(K)$, and T_{ST} is the softening temperature of coal ash.

2. Calculation of V_F and A_F

   While calculating furnace effective radiation layer thickness $s = 3.6\dfrac{V_F}{A_F}$, the V_F and A_F are needed. V_F is the enclosed volume in terms of the water wall center line or furnace wall when without a water wall. The horizontal plane passing through the half-height of the furnace hopper acts as the bottom boundary of V_F for suspension-firing furnaces. A_F is the area covering the volume of V_F; for grate-firing boilers the fire bed area is included in the total A_F.

5.2.3   Furnace Heat Transfer Calculation Examples

Case 5.1: for calculating heat transfer (a part of overall thermal calculation) of a 410 t/h PC boiler, refer to Appendix C.

5.3   HEAT TRANSFER CALCULATION IN GRATE FURNACES [12,13,19]

5.3.1   Heat Transfer Calculation in Grate Furnaces in China

The Gurvich method can be applied to suspension-firing furnaces and grate-firing furnaces. In China, engineers use the modified version of the Gurvich method to carry out furnace heat transfer calculation.

Furnace heat transfer calculation involves the radiative heat transfer from the flame or high-temperature combustion products around the furnace wall. Assume that furnace surface (as indicated in Fig. 3.10, the water wall and the furnace wall are called a "furnace surface") temperature is T_w, emissivity is ε_w, and its area equals the effective radiative heating surface A_r of the same side of the furnace wall. The radiative heat transfer of the flame is transferred to the water wall through the plane parallel to a water wall panel; this plane can be considered the flame's radiative plane, the temperature of which is equal to average flame temperature T_g. The emissivity equals the radiative emissivity from the flame to the furnace surface a_F. Then, the heat transfer between the flame and furnace surface can be simplified into radiative heat transfer between two parallel infinite planes. According to the heat transfer principle, radiative heat transfer between the flame and furnace surface can be expressed as follows:

$$Q = a_F \sigma A_r \left(T_g^4 - T_w^4 \right) \tag{5.37}$$

or:

$$q = a_F \sigma \left(T_g^4 - T_w^4 \right) \tag{5.37a}$$

where a_F is system emissivity for parallel planes, expressed as:

$$a_F = \frac{1}{\dfrac{1}{\varepsilon_g} + \dfrac{1}{\varepsilon_w} - 1} \tag{5.38}$$

The thermal balance equation can also be obtained from the gas side:

$$Q = \varphi B_{cal} \left(Q_{fuel} - I_F'' \right) \tag{5.39}$$

where φ is the heat preservation coefficient, B_{cal} is the design fuel supply rate, I_F'' is gas enthalpy at the furnace exit, and Q_{fuel} is the heat carried per kg of fuel and corresponding air, expressed as follows:

$$Q_{fuel} = Q_{ar,net,p} + I_{ph} + Q' \tag{5.40}$$

where I_{ph} is the fuel's sensible heat and the sensible heat carried by the air when heated by an external air heater.

The heat available to furnace Q_t can be calculated using the following equation:

$$Q_t = Q_{fuel} \left(1 - \frac{q_{ug} + q_{uc} + q_{ph}}{100 - q_{uc}} \right) + Q_{air} - Q' \tag{5.41}$$

where Q' is the air heat from an external air heater and Q_{air} is the heat carried by the air corresponding to each kg of fuel, which can be calculated as follows:

$$Q_{air} = (\alpha_F'' - \Delta\alpha_F)I_{ha}^0 + \Delta\alpha_F I_{ca}^0 \tag{5.42}$$

When the boiler is not equipped with an air preheater:

$$Q_{air} = \alpha_F'' I_{ca}^0 \tag{5.43}$$

where α_F'' is the excess air coefficient at the furnace exit, $\Delta\alpha_F$ is the furnace air leakage coefficient, I_{ha}^0 is the enthalpy of the theoretical amount of air at the exit of an air preheater, and I_{ca}^0 is the enthalpy of the theoretical amount of air at the entrance of an air preheater.

Note that, the same as I_a, I_F'', $V\overline{C}$, the values of $Q_t, Q_{fuel}, Q_{air}, Q', I_{ca}^0, I_{ha}^0$ are all calculated in terms of per kg or per m^3 of fuel.

The temperature that fuel burns at under adiabatic conditions is the theoretical combustion temperature, denoted by T_a, the value of which can be looked up on the gas enthalpy-temperature table according to the heat available to the furnace Q_t. Eq. (5.39) can then be rewritten as follows:

$$Q = \varphi B_{cal} V\overline{C}(T_a - T_F'') \tag{5.44}$$

where $V\overline{C}$ is the mean overall heat capacity of the combustion products per 1 kg of fuel, V is the generated gas volume per 1 kg of fuel, and $\overline{C}$ is the average specific heat capacity of combustion products at constant pressure with unit kJ/(m$^3 \cdot$°C).

From Eqs. (5.4) and (5.8) we obtain:

$$V\overline{C} = \frac{Q_{fuel} - I_F''}{T_a - T_F''} = \frac{Q_{fuel} - I_F''}{\theta_a - \theta_F''} \tag{5.45}$$

where T_F'' (K) and θ'' (°C) are the furnace outlet gas temperature. Therefore, the basic equation for furnace heat balance can be determined based on Eqs. (5.37) and (5.44):

$$a_F \sigma A_r(T_g^4 - T_w^4) = \varphi B_{cal} V\overline{C}(T_a - I_F'') \tag{5.46}$$

where there are three unknown numbers T_g, T_w, and a_F, which can be determined through experimentation.

In actuality, flame temperature in the furnace varies along the furnace height. Different researchers have proposed many different calculation methods for average flame temperature. The following equation is recommended for industrial boiler thermal calculation in China:

$$T_g = T_F''^n T_a^{(1-n)} \tag{5.47}$$

Supposing that the ratio of flame temperature and theoretical combustion temperature is nondimensional, denoted by θ:

$$\theta_g = \frac{T_g}{T_a}, \theta_F'' = \frac{T_F''}{T_a} \tag{5.48}$$

then:

$$\theta_g = \theta_F''^n \tag{5.49}$$

where index n reflects the influence of the combustion condition on the temperature field in the furnace. According to experimental data, for a spreader stoker boiler, n is 0.6; for other grate-firing boilers, n is 0.7. Eq. (5.49) is suitable for calculating burn-out chambers and cooling chambers, as well, but index n in these cases is 0.5.

Furnace surface temperature T_w denotes the surface temperature of the water wall tube external ash layer, and is expressed as follows:

$$T_w = \rho q + T_t \tag{5.50}$$

where ρ is the ash deposition coefficient (ie, the thermal resistance of the external deposition layer), which is determined according to combustion characteristics and the combustion condition in the furnace. Usually ρ is 2.6 $m^2 \cdot K/kW$. T_t is the metal wall temperature of the water wall tube (with unit K), which is usually the water saturation temperature at working pressure.

See Eq. (5.37a) for heat flow per area:

$$q + a_F \sigma T_w^4 = a_F \sigma T_g^4$$

where:

$$q = \frac{\sigma T_g^4}{\dfrac{1}{a_F} + \dfrac{\sigma}{q} T_w^4} \tag{5.51}$$

Set:

$$m = \frac{\sigma}{q} T_w^4 = \frac{\sigma}{q} (\rho q + T_t)^4 \tag{5.52}$$

then:

$$q = \frac{\sigma T_g^4}{\dfrac{1}{a_F} + m} \tag{5.53}$$

From the following thermal balance equation:

$$q = \frac{\varphi B_{cal} V\bar{C}(T_a - T_F'')}{A_r} \tag{5.54}$$

the furnace heat transfer basic formula can be obtained:

$$\frac{\sigma T_g^4}{\frac{1}{a_F} + m} = \frac{\varphi B_{cal} V\bar{C}(T_a - T_F'')}{A_r} \tag{5.55}$$

then:

$$Bo\left(\frac{1}{a_F} + m\right) = \frac{\theta_g^4}{1 - \theta_F''} = \frac{\theta_F''^{4n}}{1 - \theta_F''} \tag{5.56}$$

where Bo is the Boltzmann number:

$$Bo = \frac{\varphi B_{cal} V\bar{C}}{\sigma A_r T_a^3}$$

Coefficient m is used to consider the influence of the water wall deposition layer temperature T_w on furnace heat transfer. For a grate-firing furnace, when the furnace release rate is $50\sim120$ kW/m^2, m can be approximately regarded as a constant to simplify calculation under certain working pressures. The value of m has been tabulated and can be obtained from table or nomograph in the book "*General Methods of Calculation and Design for Industrial Boiler.*" The relationship between θ_F'' and $Bo\left(\dfrac{1}{a_F} + m\right)$ as calculated through Eq. (5.56) is shown in Tables 5.1 and 5.2.

Real-world heat transfer calculation in furnaces is very complicated. The grate-firing furnace thermal calculation standard in China uses the following equation:

$$\theta_F'' = k\left[Bo\left(\frac{1}{a_F} + m\right)\right]^p \tag{5.57}$$

where coefficients k and p are shown in Table 5.2. The function of θ_F'' is shown in Fig. 5.2.

The basic steps for calculation using the above method are as follows.

1. Calculate theoretical combustion temperature T_a.
2. Assuming furnace outlet gas temperature T_F'', calculate mean overall heat capacity of the combustion products $V\bar{C}$ and gas (flame) emissivity ε_g.

TABLE 5.1 Relationship of $Bo\left(\dfrac{1}{a_F}+m\right)$ and θ_F''

θ_F''	0.6	0.62	0.64	0.66	0.68	0.70	0.72	0.74	0.76	0.78	0.80
$n = 0.6$ Spreader stoker boiler	0.734	0.836	0.952	1.085	1.238	1.416	1.623	1.867	2.156	2.504	2.927
$n = 0.7$ Other grate-firing boiler	0.598	0.690	0.796	0.912	1.061	1.228	1.424	1.655	1.932	2.267	2.677

TABLE 5.2 Values of k and p

n	$Bo\left(\dfrac{1}{a_F}+m\right)$	k	p
0.6	0.6–1.4	0.6465	0.2345
Spreader stoker boiler	1.4–3.0	0.6383	0.1840
0.7	0.6–1.4	0.6711	0.2144
Other grate-firing boiler	1.4–3.0	0.6755	0.1714

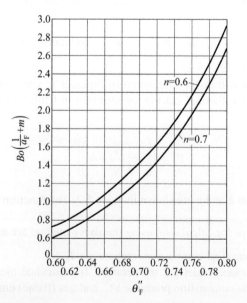

FIGURE 5.2 Function of θ_F''.

3. Calculate furnace system emissivity α_F.
4. Calculate Boltzmann number Bo.
5. Choose coefficient m and calculate furnace outlet gas temperature T_F'' accordingly. If the difference between the calculated and preassumed T_F'' is larger than 100 K, reiterate the calculation with the average of the previously assumed and calculated T_F'' as the newly assumed T_F'' (allowable error can be taken to be smaller than 100 K when using a computer).
6. Calculate heat transfer $q(Q)$.

Here, we would like to emphasize the calculation of α_F [see Eq. (3.46)]:

$$a_F = \frac{1}{\frac{1}{\varepsilon_w} + x\left(\frac{1}{M} - 1\right)}$$

$$M = \varepsilon_g + r(1 - \varepsilon_g)$$

where ε_w is the emissivity of the water wall (usually 0.8), ε_g is flame emissivity, x is the water-cooling ratio ($x = A_f/A$), and $r = R/A$ (where R is the grate area and A is the total furnace area except for the grate). The above equation can remove the intermediate variable M. See the following:

$$a_F = \frac{1}{\frac{1}{\varepsilon_w} + x\frac{(1 - \varepsilon_g)(1 - r)}{1 - (1 - \varepsilon_g)(1 - r)}} \tag{5.58}$$

5.3.2 Heat Transfer Calculation in Grate-Firing Furnaces

The following are known parameters of a grate-firing boiler with a double-drum (Type SHL10-13/350): output 10 t/h, steam pressure 1.3 MPaG, steam temperature 350°C, $t_{ha} = 150°C$; $t_{ca} = 30°C$. The system burns lean coal, $Q_{ar,net,p} = 18158$ kJ/kg, $[A]_{ar} = 33.12[\%]$, $a_{fa} = 0.2$. Total furnace area $A = 80.95$ m², grate area $R = 11.78$ m², enclosed area $A_F = 92.73$ m², furnace volume $V_F = 40.47$ m³, the radiation heating surface of front wall $L_1 = 4.86$ m (not including fire-resistant material), $L_2 = 3.54$ m (fire-resistant coating), the number of water wall tubes $n = 16$, tube spacing $s/d = 170/51 = 3.33$, and $e/d = 25.5/51 = 0.5$.

$$X_1 = 0.59$$

$$H_1 = (16 - 1) \times 0.17 \times 4.86 \times 0.59 = 7.31(m^2)$$

$$A_2 = (16 - 1) \times 0.17 \times 3.54 = 9.03(m^2)$$

$$H_2 = 0.3 \times A_2 = 2.71(m^2)$$

$$H_{front} = H_1 + H_2 = 10.02(m^2)$$

TABLE 5.3 Enthalpy-Temperature Table ($\alpha = 1.5$)

$\theta/°C$	900	1000	1500	1600
$I_g/(kJ/kg)$	10551.5	11848.9	18536.6	19906.7

The effective radiation heating surface (not including the front wall) is 27.95 m².

$$H_r = 10.02 + 27.95 = 37.97 (m^2)$$
$$x = H_r/(A_F - R) = 37.97/80.85 = 0.469$$
$$s = 3.6V_F/A_F = 3.6 \times 40.47/92.73 = 1.57 (m)$$
$$r = R/A = 11.78/80.95 = 0.146$$

From auxiliary calculation (not given herein):

$B = 0.633 (kg/s)$

Set $q_{ug} = 1\%$, $q_{uc} = 15\%$, $q_{rad} = 1.58\%$, $q_{ph} = 0.82\%$, $\eta = 73\%$

$a_F'' = 1.5$; $\Delta\alpha_1 = 0.1$

V^0 (Standard stadus) = 5.025 (m^3/kg)

$r_{H_2O} = 0.0394$; $r_{co_2} = 0.1.13$

$I_{ca}^0 = V^0 (C\theta)_{ca} = 5.025 \times 39.54 = 198.7 (kJ/kg)$

$I_{ha}^0 = V^0 (C\theta)_{ca} 5.025 \times 198.55 = 997.71 (kJ/kg)$

The enthalpy-temperature table related to the above conditions is shown in Table 5.3. V_{RO2}, V_{H2O}^0, and V_{N2}^0 are first calculated prior to making the enthalpy-temperature table. I_g^0 can be determined based on these known variables and their respective $C\theta$. I_a^0 can be obtained from $I_a^0 = V^0(C\theta)_a$. When $[A]_{ar}$ and a_{fa} are given, by $I_{fa} = (C\theta)_{fa} [A]_{ar}/100a_{fa}$, I_{fa} can be determined. The enthalpy of gas at different temperatures and α are: $I_g = I_g^0 + (\alpha - 1)I_a^0 + I_{fa}$.

See the following stepwise process for thermal calculation in grate-firing furnaces:

1. $B_{cal} = B(1 - q_{uc}/100) = 0.538 (kg/s)$

2. $\varphi = 1 - q_{rad}/(\eta + q_{rad}) = 0.979$

3. $Q_a = (a_F'' - \Delta a_1) I_{ha}^0 + \Delta a_1 I_{ca}^0 = 1416.7 (kJ/kg)$

4. $Q_a = \dfrac{Q_{ar,net,p}(100 - q_{ug} - q_{uc} - q_{ph})}{100 - q_{uc}} + Q_a = 19185.9 (kJ/kg)$

5. $\theta_a = 1547.4°C$ (refer to the above H–T table from Q_a)

6. Suppose $\theta_F'' = 960°C$, $T_F'' = 1233K$

7. $I_F'' = 11329.9 (kJ/kg)$ (from the H-T table)

8. $V\bar{C} = \dfrac{Q_a - I_F''}{T_a - T_F''} = \dfrac{19185.9 - 11329.9}{1820.6 - 1233} = 13.36\,(\text{kJ/kg}^\circ\text{C})$

9. $r_n = r_{H_2O} + r_{CO_2} = 0.1707$

10. $p_n = r_n p = 0.01707 \times 10^6\,(\text{Pa})$ $(p = 0.1 \times 10^6\,\text{Pa})$

11. $k_g = \left(\dfrac{0.78 + 1.6 r_{H2O}}{\sqrt{\dfrac{P_n}{P_0}}\,S} - 0.1 \right)\left(1 - 0.37\dfrac{T_F''}{1000} \right) \times 10^{-5}$

$= \left(\dfrac{0.78 + 1.6 \times 0.0394}{\sqrt{\dfrac{0.01707 \times 10^6}{9.81 \times 10^5}} \times 1.57} - 0.1 \right)\left(1 - 0.37\dfrac{1233}{1000} \right) \times 10^{-5}$

$= 8.227 \times 10^{-6}\left((\text{m} \cdot \text{Pa})^{-1} \right)$

12. $K_g = k_g p_n = 8.227 \times 10^{-6} \times 0.01707 \times 10^6 = 0.1404\,(\text{m}^{-1})$

13. $G_g = 1 - [A]_{ar} / 100 + 1.306 \times \alpha V^0$

$= 1 - 33.12 / 100 + 1.306 \times 1.5 \times 5.025$

$= 10.51\,(\text{kg/kg})$

14. $\mu_{fa} = \dfrac{[A]_{ar}\,a_{fa}}{100 G_g} = \dfrac{33.12 \times 0.2}{100 \times 10.51} = 0.0063\,(\text{kg/kg})$

15. $k_{fa} = \dfrac{4300 \rho_g \times 10^{-5}}{\left(T_F''^2 d_m^2 \right)^{1/3}} = \dfrac{4300 \times 1.30 \times 10^{-5}}{(1233 \times 20)^{2/3}} = 6.60 \times 10^{-5}\,[(1/\text{m}.\text{Pa})^{-1}]$

16. $K_{fa} = k_{fa}\mu_{fa} = 6.60 \times 10^{-5} \times 0.0063 \times 10^5 = 0.0416\,(\text{m}^{-1})$

17. $K = K_g + K_{fa} + x_1 x_2 p = 0.1404 + 0.0416 + 10^{-5} \times 0.03 \times 0.1 \times 10^6 = 0.212\,(\text{m}^{-1})$

(lean coal $x_1 = 10^{-5}$, grate-firing $x_2 = 0.03$)

18. $\varepsilon_g = 1 - e^{-Ks} = 1 - e^{-0.212 \times 1.57} = 0.283$

19. $\varepsilon_w = 0.8\,(\text{as artificially selected according to experience})$

20. $a_s = \dfrac{1}{\dfrac{1}{\varepsilon_w} + x\dfrac{(1-\varepsilon_g)(1-r)}{1-(1-\varepsilon_g)(1-r)}} = \dfrac{1}{\dfrac{1}{0.8} + 0.469\dfrac{(1-0.283)(1-0.146)}{1-(1-0.283)(1-0.146)}} = 0.502$

21. $B_O = \dfrac{\varphi B_{cal} V\bar{C}}{\sigma H_r T_a^3} = \dfrac{0.979 \times 0.538 \times 13.36}{5.67 \times 10^{-3} \times 37.97 \times 1820.4^3} = 0.542$

22. $m = 0.15$ (working pressure $= 1.3$ MPaG)

23. $n = 0.7$

Then θ_F'' can be obtained by substituting the variables above into the equation θ_F''

$$= k[B_o(1/a_s + m)]^p = 0.6711[0.542(1/0.504 + 0.150)]^{0.2144}$$
$$= 0.6711 \times 1.1567^{0.2144} = 0.692$$

24. $T_F'' = \theta_F'' T_a = 0.692 \times 1820.4 = 1260\,\text{K}$

$\theta_F'' = T_F'' - 273 = 987\,^\circ\text{C}$ (the error with supposed $\theta_F'' = 960\,^\circ\text{C}$ is acceptable, reiteration is not needed)

25. $I_F'' = 11680.2(\text{kJ/kg})$ (looked up from the enthalpy-temperature table)

26. $Q_r = \varphi(Q_a - I_F'') = 0.979(19185.9 - 11680.2) = 7348.1(\text{kJ/kg})$

$$Q = B_{cal}Q_r = 0.538 \times 7348.1 = 3953(\text{kW})$$

27. The furnace release rate is:

$$q = Q/H_r = 3953 \times 10^3/37.97 = 104.1 \times 10^3\,(\text{W/m}^2)$$

5.4 HEAT TRANSFER CALCULATION IN FLUIDIZED BED FURNACES [16,19,32,33,38,39,41]

5.4.1 Heat Transfer Calculation in Bubbling Fluidized Bed (BFB) Furnaces

The BFB boiler is the first-generation fluidized bed (FB) boiler. The furnace includes a dense-phase zone (also referred to as the bubbling bed) and suspension zone. The BFB boiler is typically used as an industrial boiler with 4~35 t/h capacity. Its structural characteristics involve heating surfaces with immersed tubes in the dense-phase zone (Fig. 5.3). Heat transfer calculation for this type

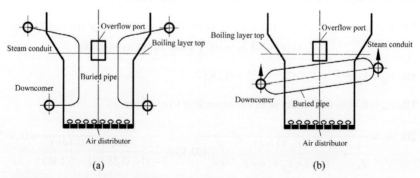

(a)   (b)

FIGURE 5.3 BFB bed dense-phase zone with immersed tubes. (a) Vertical buried pipe. (b) Transverse buried pipe.

of furnace therefore must include the heat transfer from the bed material to the immersed tubes, and is conducted in two parts: fluidized zone calculation, and suspension zone calculation.

Heat transfer calculation in the bubbling bed examines the heat transferred from the fluidized material to the immersed tubes. Design calculation and verification calculation can be applied to this area differently according to different aims. Design calculation is performed according to coal information and boiler parameters to determine fluidization velocity, bed material height (usually the height of the overflow port), and the air distributor area. The immersed tube area is calculated according to the chosen bed material temperature, then the immersed tubes are arranged accordingly. Verification calculation is performed according to an already existing immersed heating surface area and its physical design, as well as coal information and boiler parameters, to calculate bed material temperature.

The equations applied to the burn-out chamber of a grate-firing boiler can also be applied to heat transfer calculation in the suspension zone; the suspension system emissivity can also be calculated according to the system emissivity of a grate-firing furnace. The heat transfer calculation for a BFB boiler back-end heating surface also adopts that of a grate-firing furnace.

The BFB boiler is an outdated product that is rarely used in the current industry. Most commonly used are CFB boilers, so they are the focus of this book. Interested readers can refer to the former National Mechanical Industry Ministry Standard, "Thermal Calculation Method of Grate Combustion and Fluidize Bed Combustion" (JB/DQ 1060-82).

5.4.2   CFB Furnace Structure and Characteristics

The bed material circulation in a CFB furnace as it transfers heat involves energy balance, material balance, and loop pressure balance. Calculating heat transfer in a CFB, to this effect, is far more complex than calculating the same in a grate-firing furnace, BFB, or suspension-firing furnace, in which the material in the gas passes through the furnace only once. Because CFB technology is now developing, there is as yet no commonly accepted CFB furnace heat transfer calculation method.

Besides the basic principles and calculation methods of two-phase flow heat transfer introduced in chapter: Heat Transfer in Fluidized Beds , this section introduces a few basic processes and their relationships to heat transfer in CFB furnaces. In principle, the contents of this book can be combined with experimental and industrial data for any specific CFB boiler to calculate the heat transfer in CFB furnaces. There will be no detailed calculation method introduced here.

In a CFB boiler, the furnace not only undergoes chemical reaction (combustion and desulfurization), but also the heat exchanging of the gas-solid material with the working medium, as well as the gas-solid circulating in a closed loop. The following are the basic characteristics of a CFB furnace.

1. Combustion temperature
   The operation temperature of a typical CFB furnace is about 850°C, and the combustion temperature is maintained at 800~900°C.
2. Furnace cross section heat release rate
   The furnace cross section heat release rate is heat produced from the furnace unit's cross sectional area, which is a very important parameter in CFB boiler design. The furnace cross section heat release rate is the function of the rate of air flow through the furnace. Generally, for fossil fuels, furnace cross section heat release rate Q_F has the following approximate relation with superficial gas velocity U_0:

$$Q_F = \frac{3.3U_0}{\alpha_F''} \tag{5.59}$$

where U_0 is the furnace sectional superficial gas velocity at 300 K in m/s, α_F'' is the excess air coefficient at the furnace exit, and Q_F is in MW/m² units. The sectional superficial gas velocity is limited somewhat concerning its ability to maintain fast fluidization in the furnace. The excess air coefficient at the furnace exit of a typical CFB boiler is 1.2, and the sectional superficial gas velocity is 5~6 m/s at 850°C, so the furnace cross section heat release rate is 3.6~4.4 MW/m².

   The furnace cross-section heat release rate is not often used during CFB boiler design, because furnace height and cross section area are opposing factors that must be balanced. CFB boiler designers must also consider other factors.
3. Fuel
   For design and operation of any boiler, including a CFB boiler, the influence of fuel is crucial. The heating value, proximate analysis, and ultimate analysis data should be known during the preliminary design stage. The fuel heating value and boiler output and efficiency determine the input of the fuel. Proximate analysis of fuel affects the design of the cyclone separator and back-end surfaces, and determines air distribution to some extent. Better fuel reaction characteristics can improve combustion efficiency. Fuel particles that are large or hard to crush reduce combustion efficiency.

   In a boiler without a desulfurizer such as limestone powder, the bed material's average particle size is greatly affected by the fuel ash properties. The average particle size of the bed material determines the flow dynamic and heat transfer characteristics in the furnace. Further details regarding the influence of fuel properties on CFB boiler design and operation are listed in Table 5.4.
4. Thermal balance
   Fuel combusts in the furnace of a CFB boiler. Part of the heat from combustion is taken to the back-end heating surface by high-temperature gas, but because not all of the heat can be absorbed by the back-end heating surface, extra heating surface must be arranged within the solid circulating loop

TABLE 5.4 Influence of Coal Properties on CFB Boiler Design and Performance

Coal properties	Influenced design parameters	Performance specific to the coal properties
Fragility	Separator staged efficiency	Boiler efficiency and entrainment of carbon particles
Reactivity	Airflow distribution	Boiler efficiency, entrainment of carbon particles, and CO emission
Ratio of intrinsic ash and extraneous ash	Ash discharge, ash distribution, and heating surface design	Ash entrainment, emission in bed, and duty of fly ash separator
Ash chemical reactivity	Ash discharge, back-end flue flow area, and heating surface design	Bed agglomeration, slagging
Moisture	Absorbed heat, physical design of separator, and downstream equipment	Thermal efficiency and excess air coefficient
Heating value and sulfur content	Physical size and desulfurizer handling unit	Capacity, thermal efficiency, pollutant emission, and slag discharge equipment

system. Arrangements of the heating surface determine the heat distribution in the CFB boiler.

Let's examine the heat distribution in a specific type of CFB boiler [33] according to the arrangement of its heating surfaces. Fig. 5.4 shows the heat distribution of a CFB boiler designed with Lurgi technology. As shown in Fig. 5.4, evaporator heating surfaces in the combustion chamber absorb heat at 48 MW, the fluidized bed heat exchanger absorbs heat at 88 MW, and the back-end heating surface absorbs heat at 73 MW. The only cooling medium in the fluidized bed dense-phase zone is cold recycled bed material of about 400°C, for which about 90 MW heat is needed to heat it to 850°C.

Fig. 5.4 shows the heat distribution of the heating surface in a CFB boiler. Thermal balance of boilers can be calculated according to standard boiler thermal calculation methods, but for a CFB boiler that has desulfurization, the thermal effects of desulfurizer calcination and desulfurization reaction must be considered.

5. Mass balance

Solid material must be kept in balance during CFB boiler operation. Solid material sent into the CFB boiler is mainly fuel and desulfurizer (with added bed material for some furnaces). The combustible elements of C, H, O, N, and S and all moisture transform into gas. The remaining solid material

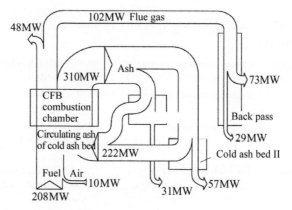

FIGURE 5.4   **Heat distribution in a Lurgi CFB boiler.**

(mainly ash and some unburned carbon) should be discharged through proper equipment and at a corresponding position to maintain the material balance in the furnace. The CFB boiler has two ash outlets positioned at the back-end pass and at the furnace bottom, respectively. Generally, a loop seal (or external fluidized bed heat exchanger) also discharges a portion of the ash; in addition, at a reversal chamber below the convection shaft some of the ash possibly leaves the system as well.

Previous researchers have provided an example of mass balance [33]. Fig. 5.5 shows the ash flow of a 95.8 MW CFB boiler in a German power

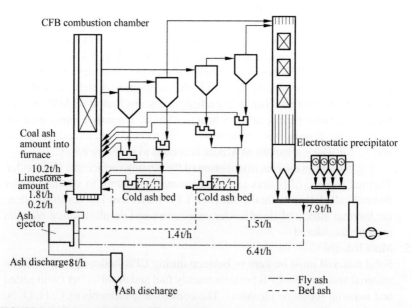

FIGURE 5.5   **Schematic diagram of ash balance.**

station. The ash balance data in the case of 100% load is shown in Fig. 5.5. There are four ash exits in the system: the furnace, external fluidized bed heat exchanger, convection shaft, and electrostatic precipitator.

The amount of ash and slag discharged from the system can be determined according to the material balance described above. The amount of circulated material is another very critical parameter in the CFB boiler concerning the design of the separator and loop seal, and should be examined in detail during material balance calculations.

5.4.3 Heat Transfer Calculation in CFB Furnaces

The heat transfer calculation methods and equations for CFB boiler furnaces were introduced in chapter: Heat Transfer in Fluidized Beds. Let's review the process step-by-step.

1. According to fuel and the requirements of boiler performance, calculate the air requirement, combustion products, the gas enthalpy-temperature table, and thermal balance to establish the necessary foundation data for thermal calculation in the furnace. This step is largely in accordance with other types of furnaces. The major difference lies in the mass and heat balance caused by material circulation.
2. Calculate the furnace structural dimensions, including all heating surfaces in the furnace.
3. Determine the furnace outlet gas temperature and conduct the heat transfer calculation. Heat transfer in a CFB boiler furnace is:

$$Q = \varphi B_{cal} \left(Q_{ar.net.p} \frac{100 - q_{ug} - q_{uc} - q_{ph}}{100 - q_{uc}} + Q_{air} + I_{cc} - I_F'' \right) \tag{5.60}$$

where Q_{air} is air enthalpy per kg of fuel:

$$Q_{air} = (\alpha_F'' - \Delta\alpha_F) \cdot I_{ha}^0 + \Delta\alpha_F \cdot I_{ca}^0 \tag{5.61}$$

where I_{cc} is heat brought back by air from the loop seal and ash cooler, and I_F'' is gas enthalpy at the furnace exit.

The lower part of a fluidized bed furnace is generally a dense-phase zone with refractory materials, the middle part is covered with a bare tube water wall, and in the upper part is a suspended heating surface (such as a platen superheater, platen reheater, evaporator panel, or even an economizer). Calculation should be divided into three parts at low loads of the boiler: suppose the outlet gas temperature at each zone, set the combustion fraction at each zone, calculate using trial and error, and verify whether the supposed gas temperature at each zone is within the allowable error using the thermal balance method.

At normal 100% load, due to the influence of the high internal and external bed material circulating ratio, the temperature profile of the whole furnace is almost uniform. At low loads, however, the material circulating rate in the furnace decreases so greatly that the CFB operates like a BFB, thus the dense-zone temperature is much higher than the temperature at the furnace exit. At this time, to obtain the bed temperature, zone calculations are needed and the heat transfer calculation in the dense-phase zone has to be performed separately.

4. Verify the thermal balance for all heating surfaces in the furnace. If the calculated absorbed heat through the heat transfer equation differs considerably (by over 5%) from that obtained through heat balance, start over.

Next, let's discuss heat transfer calculation in a CFB boiler furnace, specifically. As opposed to the BFB boiler, the transition from the dense-phase zone to the dilute-phase zone is not clearly found in the CFB boiler, so their definitions vary. Generally, the bed height at static state is in the range of 0.5~0.8 m, and the bed height at fluidized bed state is 2~2.8 times that of the static state. Thus, the bed height at the fluidized bed is between 1.0 and 2.24 m. This value varies considerably according to the state of fluidization, so it is best to suggest the elevation of secondary air as a definition boundary. The dense-phase zone is from the air distributor to the lower secondary air, the transitional zone is from the lower secondary air to the upper secondary air, and the dilute-phase zone is from the upper secondary air to the furnace exit.

Because there are no immersed tubes in the dense-phase zone and the water wall has fire/wear-resistant covers, the absorbed heat of the water wall in this zone is very low and the heat carried out by the material is high. The calculation of this area is fairly simple, mainly due to the thermal resistance of the fire-resistant layer. The material concentration in this area is very high. In addition, the concentrations of material leaving the dense zone from its upper part and falling into the dense zone along the wall are all high.

Calculating the fly ash concentration at the dense-phase zone outlet is a trial and error calculation, that is, the dense-phase outlet gas temperature must be known in order to calculate the heat transfer process. For this reason, the following parameters are required to be assumed beforehand: the combustion fraction at dense-phase zone δ, the ratio of the amount of falling material to rising material, and the temperatures of the rising and falling material, which are calculated as follows.

The heat input into the dense-phase zone is Q_{db}:

$$Q_{db} = \varphi B_{cal}\delta\left[Q_{ar,net,p}\frac{100 - q_{ug} - q_{uc} - q_{ph}}{100 - q_{uc}} + Q_{air} - I_s - I''_{db} - (I''_{ash} - I'_{ash}) \right] \quad (5.62)$$

where δ is the combustion fraction at the dense-phase zone; generally, for bituminous coal, $\delta = 0.5$ and for lean coal, $\delta = 0.45 \sim 0.47$. φ is the heat preservation coefficient (0.995) and B_{cal} is the design fuel supply rate.

The heat loss of the boiler should be determined according to the industrial operation data for specific coal and furnace types – if there are no such data available, they can be considered, $q_{ug} = 0.5\%$; $q_{uc} = 3\%$; $q_{ph} = 0.3\%$, where Q_{air} is the heat carried by hot air, I_s is the thermal enthalpy carried by hot slag, I''_{db} is the known gas enthalpy at the dense-phase zone outlet, I'_{ash} is inlet material enthalpy, and I''_{ash} is outlet material enthalpy.

Heat transferred to the water wall heating surface at dense-phase zone Q is:

$$Q = h_{db} \cdot A_{db} \cdot \Delta t \tag{5.63}$$

where h_{db} is the heat transfer coefficient in the dense-phase zone, which can be considered $h_{db} = 250$ W/(m·°C) if it cannot be calculated accurately. A_{db} is the heating surface area of the dense-phase zone with fire-resistant and wear-resistant material layers, and Δt is the temperature difference between the gas and working medium at the dense-phase zone (in °C).

The heat enthalpy carried by hot slag is:

$$I_s = \alpha_s \cdot T_s \cdot V_s \cdot c_s \tag{5.64}$$

where α_s is slag fraction, T_s is slag temperature, V_s is the slag amount per kg of fuel (in kg/kg), and c_s is the specific heat capacity of the slag.

Transform Eq. (5.62) into the following:

$$I''_{ash} - I'_{ash} = \frac{\varphi B_{cal} \delta \left(Q_{ar,net,p} \dfrac{100 - q_{ug} - q_{uc} - q_{ph}}{100 - q_{uc}} + Q_{air} - I_s - I''_{db} \right) - Q}{\varphi B_{cal} \delta} \tag{5.65}$$

The left-hand side of Eq. (5.65) is the difference between the outlet enthalpy and inlet enthalpy of the material, which can be calculated as follows:

$$I''_{ash} - I'_{ash} = V_{up} \cdot c \cdot T_1 - V_{down} \cdot c \cdot T_2 \tag{5.66}$$

where V_{up} and V_{down} are rising and falling material mass, respectively, c is material specific heat capacity, and T_1 and T_2 are rising and falling material temperature, respectively.

Defining m as the ratio of falling ash amount V_{down} and rising ash amount V_{up}, that is, $m = \dfrac{V_{down}}{V_{up}}$, then Eq. (5.66) can be rewritten as follows:

$$I''_{ash} - I'_{ash} = V_{up} \cdot c \cdot T_1 - V_{up} \cdot c \cdot m \cdot T_2 = V_{up} \cdot c(T_1 - mT_2) \tag{5.67}$$

Substituting Eq. (5.67) into Eq. (5.65) yields:

$$V_{up} = \frac{\varphi B_{cal} \delta \left(Q_{ar,net,p} \dfrac{100 - q_{ug} - q_{uc} - q_{ph}}{100 - q_{uc}} + Q_{air} - I''_{db} \right) - Q_1}{\varphi B_{cal} \cdot \delta \cdot c(T_1 - mT_2)} \tag{5.68}$$

According to experimental research, the ratio of falling ash and rising ash can be considered 0.97.

The ash carryover ratio is:

$$a_{ash} = \frac{V_{up}}{G_g} \tag{5.69}$$

So ash concentration can be calculated as follows:

$$C_{ash} = a_{fa} \cdot \rho_{TF} \tag{5.70}$$

where ρ_{TF} is the gas density at furnace temperature.

According to the above calculation, the material transfers most of the heat from fuel combustion at the dense-phase zone to the dilute-phase zone. In addition, the calculated ash enthalpy $(I''_{ash} - I'_{ash})$ is quite accurate. The main parameter that influences this enthalpy is the mass difference and temperature difference of the rising and falling material, as well as the fly ash amount V_{fa}. If the exact amount and temperature difference of the rising and falling material can be accurately determined, ash concentration C_{ash} can be calculated effectively.

The above is a verification calculation—the given gas temperature at dense-phase outlet T_1 is identified to calculate the material mass concentration C_{ash}. The known material mass concentration C_{ash} and other parameters can also be utilized to calculate the gas temperature at the dense-phase zone outlet.

5.5 HEAT TRANSFER CALCULATION IN BACK-END HEATING SURFACES

Heat transfer calculation in boiler furnaces was discussed systematically above. Calculating heat transfer in the back-end heating surface (convection heating surface) is also necessary. A "convection heating surface" is the heating surface positioned in the boiler gas pass directly heated by flowing gas, and "convective heat transfer" occurs between the boiler bundles or gas tubes, such as a superheater, economizer, and air preheater. These heating surfaces may be quite disparate in structure, arrangement, working medium, and gas properties, but their heat transfer process is similar, so heat transfer calculation can be conducted using the same method for all surfaces.

The primary goal of convective heat transfer calculation is either to determine the needed heating surface when the heat transfer rate is known, or to determine the rate of heat transfer when the heating surface is known. The heating surface is usually tentatively planned prior to actual calculation. The difference between the heat calculated by heat transfer equations and that calculated by thermal balance equations should not exceed ±2%.

5.5.1 Basic Heat Transfer Equations

The transferred heat of convection heating surface Q is in direct proportion to heating surface area A and the temperature difference between cold and hot fluid Δt, the heat transfer equation of which is:

$$Q = K\Delta tA \tag{5.71}$$

Proportionality coefficient K is the heat transfer coefficient, which can reflect the intensity of the heat transfer process, or the heat transfer per square meter of heating surface when the temperature difference is 1°C. The larger the heat transfer coefficient, the stronger the heat transfer process.

Per kg of fuel, the heat transfer equation is:

$$Q_c = \frac{K\Delta tA}{B_{cal}} \tag{5.71a}$$

where Q_c denotes the heat from the gas to the working medium per kg of fuel for the calculated convection heating surface.

The heat transfer of the unit's heating surface area is:

$$q = \frac{Q}{A} = K\Delta t \tag{5.72}$$

which is called "heating surface thermal load" or "heat flux."

When gas flows across the heating surface outside the tubes, the heating surface area is the superficial area of outside tubes on the gas side; when gas flows inside the tubes, the heating surface area is calculated according to the inner diameter of the tubes. The heating surface area of the tubular air preheater is calculated according to the average surface area of the gas side and air side.

In the thermal balance equation, the heat released by the gas equals the heat absorbed by water, steam, or air.

Heat from the gas to the working medium is:

$$Q_c = \varphi\left(I'_g - I''_g + \Delta\alpha I^0_{air}\right) \tag{5.73}$$

where φ is the heat preservation coefficient, I' and I''_g are gas enthalpy at the inlet and outlet of the heating surface, $\Delta\alpha$ is an air leakage coefficient (determined in the light of empirical data), and I^0_{air} is air enthalpy, for an air preheater. I^0_{air} should be evaluated according to average air temperature; for other heating surfaces, it should be calculated according to the ambient air temperature.

Working medium absorbed heat can be calculated using the following equation:

$$Q_c = \frac{D}{B_{cal}}(i'' - i') - Q_{r,F} \tag{5.74}$$

where D is the flow rate of the working medium, i' and i'' are inlet and outlet enthalpy of the working medium, and $Q_{r,F}$ is additionally absorbed heat from the furnace radiation.

When a slag screen or boiler bank is positioned at the exit of the furnace, if the tube rows are equal to or more than 5, all the radiation heat at the furnace exit is considered absorbed by the tube bundles. If the tube rows are fewer, part of the heat passing through the bundles is absorbed then by the downstream heating surface. At this time, the radiation heat absorbed by the bundles is:

$$Q_{r,F} = \frac{\varphi_{ct} y q_r A''_E}{B_{cal}} \tag{5.75}$$

where q_r is the average thermal load of the furnace heating surface as calculated, A''_E is the cross section area of the furnace exit window, and y is the uneven distribution coefficient of thermal load—when the furnace exit window is at the top of the furnace, $y = 0.6$, when the exit window is on one side of the furnace wall, $y = 0.8$. φ_{ct} is the configuration factor of the bundles.

The heat absorbed by air in the air preheater is:

$$Q_{aph} = \left(\beta''_{aph} + \frac{\Delta \alpha_{aph}}{2} \right) \left(I^{0''}_{aph} - I^{0'}_{aph} \right) \tag{5.76}$$

where β''_{aph} is the excess air coefficient at the air preheater outlet, which is calculated using the following equation:

$$\beta''_{aph} = \alpha''_F - \Delta \alpha_F \tag{5.77}$$

where $\Delta \alpha_{aph}$ is the air leakage coefficient (air side) of the air preheater, and $I^{0''}_{aph}$ and $I^{0'}_{aph}$ are theoretical air enthalpy at the exit and entrance of the air preheater.

5.5.2 Heat Transfer Coefficient

Within a boiler's convection heating surface the hot gas is used to heat water, steam, and air. The hot gas and heated working medium do not mix together at either side of the heating surface; heat moves from the hot gas through the tube wall to the working medium. To this effect, the heat transfer process is a combination of three separate processes: (1) heat release from hot gas to the outer tube surface, (2) heat conduction through the tube wall from the outer surface to the inner surface, and (3) heat release from the inner tube surface to the fluid.

We know that there are three basic modes of heat transfer: conduction, convection, and radiation. In actuality, the heat transfer process is usually a combination of these three modes, thus it is very complicated. In the convection heating surface of a boiler, heat released from the hot gas to the outer surface of tubes generally includes convection and radiation, heat transfer from the outer surface to the inner surface of tubes is a conduction process, and heat transfer from the inner surface of tubes to the working medium is a convection process. See the following diagram of the general serial heat transfer process:

| Hot gas | $\xrightarrow{\text{Convention+radiation}}$ | Outer surface | $\xrightarrow{\text{Conduction}}$ | Inner surface | $\xrightarrow{\text{Convection}}$ | Working medium |

Convection heat transfer can be expressed as follows:

$$Q = h_c A \Delta t \tag{5.78}$$

where h_c is the convective heat transfer coefficient.

Radiation heat transfer can be expressed as follows:

$$Q = h_r A \Delta t \tag{5.79}$$

where h_r is the radiative heat transfer coefficient.

Conduction heat transfer can be expressed as follows:

$$Q = \frac{\lambda}{\delta} A \Delta t \tag{5.80}$$

where λ is the thermal conductivity of the mental tube, and δ is tube wall thickness.

See the following three equations for expressions of heat transfer in series through the boiler heating surface discussed above.

1. Heat from hot gas to outer surface of tubes through convection and radiation:

$$Q_1 = (h_c + h_r)A(t_1 - t_{os}) = h_1 A(t_1 - t_{os}) \tag{5.81}$$

2. Heat from outer surface to inner surface of tubes through conduction:

$$Q_w = \frac{\lambda}{\delta} A(t_{os} - t_{is}) \tag{5.82}$$

3. Heat from inner surface of tubes to working medium through convection:

$$Q_2 = h_2 A(t_{is} - t_2) \tag{5.83}$$

In the above three equations, h_1 is the gas side convective heat transfer coefficient, h_2 is the working medium side heat transfer coefficient, t_1 and t_2 are gas temperature and working medium temperature, respectively, and t_{os} and t_{is} are outer surface temperature and inner surface temperature of the tubes, respectively.

According to the energy conservation principle, during the steady-state process, the heat transferred in a serial process is equal:

$$Q_1 = Q_w = Q_2 = Q \tag{5.84}$$

If a heated tube is simplified to a plate, that is, the areas of the outer and inner surfaces of the tube are considered equal, then the thermal loads of all three processes are equal:

$$q_1 = q_w = q_2 = q \tag{5.85}$$

Transform Eqs. (5.81–5.83) to equations of temperature difference as follows:

$$\left.\begin{aligned} t_1 - t_{os} &= q\frac{1}{h_1} \\ t_{os} - t_{is} &= q\frac{\delta}{\lambda} \\ t_{is} - t_2 &= q\frac{1}{h_2} \end{aligned}\right\} \tag{5.86}$$

Rearrange the above three equations to obtain:

$$t_1 - t_2 = q\left(\frac{1}{h_1} + \frac{\delta}{\lambda} + \frac{1}{h_2}\right) = qR \tag{5.87}$$

where R is the total thermal resistance of the heat transfer process, which equals the sum of thermal resistance of each serial process. In other words, the thermal resistance of all three processes can be expressed by linear superposition.

Transform the thermal load mode of Eq. (5.87) as follows:

$$q = \frac{t_1 - t_2}{\left(\frac{1}{h_1} + \frac{\delta}{\lambda} + \frac{1}{h_2}\right)} \tag{5.88}$$

The heat transfer coefficient can then be obtained by comparing Eq. (5.88) and the basic heat transfer equation (5.72):

$$K = \frac{1}{\left(\frac{1}{h_1} + \frac{\delta}{\lambda} + \frac{1}{h_2}\right)} \tag{5.89}$$

Then the total thermal resistance is:

$$R = \frac{1}{K} = \frac{1}{h_1} + \frac{\delta}{\lambda} + \frac{1}{h_2} \tag{5.90}$$

In an actual heat transfer process, there is ash outside the tube and scale inside the tube. According to the thermal resistance liner superposition principle, the total thermal resistance of the heat transfer process is:

$$R = \frac{1}{h_1} + \frac{\delta_a}{\lambda_a} + \frac{\delta_w}{\lambda_w} + \frac{\delta'}{\lambda'} + \frac{1}{h_2} \tag{5.91}$$

where $\dfrac{\delta_a}{\lambda_a}, \dfrac{\delta_w}{\lambda_w}$, and $\dfrac{\delta'}{\lambda'}$ are the thermal resistance of the ash layer, tube wall, and scale layer, respectively.

The following general equation for the heat transfer coefficient can then be obtained:

$$K = \frac{1}{\dfrac{1}{h_1} + \dfrac{\delta_a}{\lambda_a} + \dfrac{\delta_w}{\lambda_w} + \dfrac{\delta'}{\lambda'} + \dfrac{1}{h_2}} \tag{5.92}$$

Please note that the effects of ash deposition on heat transfer are introduced as a special topic in chapter: Effects of Ash Deposition and Slagging on Heat Transfer.

5.6 THERMAL CALCULATION OF THE BOILER

Whether we're considering a utility boiler or an industrial boiler, "heat transfer in the furnace" refers not only to the heat transfer in the combustion chamber but also to the flow and cooling process of gas after it leaves the combustion chamber. For boilers, complete thermal calculation is part of the overall heat transfer calculation. In this section, a suspension-firing furnace serves as an example to introduce the thermal calculation of boilers and illustrate the differences between the grate-firing furnace, fluidized bed, and suspension-firing furnace.

The basic definitions of boiler heating surfaces are first introduced briefly to inform the thermal calculation process provided in the Appendix C.

5.6.1 Basic Definitions of Boiler Heating Surfaces

1. The basic requirement of the boiler proper

   The function of a boiler is to transfer the heat from fuel combustion to working media through heating surfaces, and heat a low-temperature working medium to high temperatures (heating water to steam, for example). The boiler involves a series of processes including combustion, heat transfer, draft, water circulation, and steam–water separation. The design of a boiler proper is, in effect, the design of heating surfaces. The boiler proper usually contains a furnace, convection heating surface, drum (note that there is no drum in a once-through boiler), steel frame, and furnace wall.

   The design calculation of a boiler includes thermal calculation, hydrodynamic calculation, strength calculation, draft resistance calculation, and tube wall temperature calculation. Thermal calculation is the main calculation of boiler design, the task and aim of which is to determine the necessary physical design and size of each heating surface and provide initial data for other calculations according to given fuel characteristics, feedwater temperature, other technical conditions, and expected rated capacity, as well as steam conditions and other technical and economic constraints. Thermal calculation is the foundation of the entire boiler design and all other calculations, which has crucial effects on boiler performance.

Creating a reasonable arrangement and structure is the most important issue for both newly designed boilers and refitted boilers. Whether the designed or modified thermal parameters are appropriate directly influences the safety and economy of manufacturing, installation, operation, and maintenance of the boiler. Boiler structure and arrangement must be designed with the utmost care and elaborate detail.

Boiler surfaces can be divided according to heat transfer mode (eg, radiation heating surfaces and convection heating surfaces). Radiation heating surfaces, such as the water wall of a large boiler or the furnace flue and drum of a small boiler, are arranged in the boiler furnace to absorb radiation heat. Convection heating surfaces are contacted directly by heated gas and absorb convection heat; these include superheaters, reheaters, economizers, and air preheaters in utility boilers, and boiler banks, economizers, and air preheaters in industrial boilers.

2. Design and arrangement of the furnace and water wall

The water wall tubes of natural circulation boilers are arranged directly on the walls around the furnace – the lower part of the tubes is connected to the lower header outside the furnace wall, and the lower header is connected to the water space of the drum through a downcomer. The upper part of the tubes can be connected to the drum directly or through the upper header/steam conduit to form the water circulation loops of the water wall. The water wall tubes are usually seamless steel, and are plain or finned. As shown in Fig. 5.6, plain tubes are usually adopted for industrial boilers and finned tubes are usually adopted for utility boiler membrane walls. The advantage of a membrane wall is the favorable protection of the furnace wall, which significantly reduces furnace wall temperature, thus reducing the requirement for furnace wall thickness and weight. The furnace must be air/gas tight to reduce air/gas leakage into/from the furnace to enhance boiler thermal efficiency and keep the environment clean and safe.

The spacing of water wall tubes s should be considered carefully to protect the furnace wall from high temperatures and ensure metal resources are applied economically. From a heat transfer point of view, the water wall tubes also should not be too close to increase the utilization of the metal as a

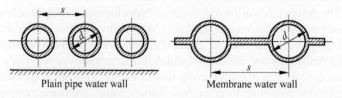

Plain pipe water wall   Membrane water wall

FIGURE 5.6 Water wall diagram.

heating surface. Expansion problems in the water wall should be considered. Large boilers usually adopt a suspension structure in which the water wall top is hung onto a steel frame and the lower part of the water wall can freely expand; small boilers usually adopt a braced structure in which the bottom is fixed and can expand upward.

3. The design and arrangement of back-end heating surfaces
There are two ways that gas flows across bundles: transverse flow and longitudinal flow. The heat transfer effect of transverse flow is better than that of longitudinal flow, so transverse flow should be adopted when possible during convection heating surface design. Longitudinal flow permits higher gas velocity, but its heat transfer effect is worse than that of transverse flow.

Tube arrangement configurations can be staggered or in-line. In general, the heat transfer coefficient of a staggered arrangement is larger than that of an in-line arrangement by 5~6%, so staggered arrangement is more beneficial for heat transfer. However, for boilers burning easily slagging coal, waste boilers, and biomass boilers, staggered arrangement is not used as far as possible to avoid slagging and deposition.

Selecting a reasonable gas velocity is very important when designing a convection heating surface. Enhancing gas velocity can improve heat transfer and protect the heating surface, but will increase gas resistance and operation expense. If gas velocity is too low, heating surface ash deposition will severely affect heat transfer; if gas velocity is too high, the heating surface will be severely abraded. For a water tube boiler, gas velocity should not be less than 6 m/s when burning coal, and is usually about 10 m/s. Oil- and gas-fired boilers can accommodate higher gas velocity. In addition, design and arrangement should ensure that heating surfaces are flushed well to avoid dead zones and enhance utilization of the heating surface.

Heat transfer is proportional to temperature difference. For small-scale boilers, convection banks of tubes are generally arranged near the furnace outlet. When gas flows through the heating surface, heat is absorbed gradually by the working medium as gas temperature decreases accordingly; the working medium temperature remains at saturation temperature. When gas temperature is below 350°C, the temperature difference between the working medium and the gas decreases considerably. At this time, increasing the heating surface does not significantly increase the heat transfer ability, but increases metal consumption. To reduce the gas temperature further, it is more reasonable to take advantage of an economizer or air preheater.

In boilers with high gas temperature at the furnace exit, to avoid slagging, several wide pitch slag screen bundles should be arranged at the furnace exit. Relative transverse pitch $s_1/d \geq 4.5$, relative longitudinal pitch $s_2/d \geq 3.5$, and in addition, transverse and diagonal spacing should be not less than 250 mm.

According to statistics, the average evaporation rate of a convection heating surface is $10\sim15$ kg/(m^2·h) for a coal-fired furnace, and $15\sim20$ kg/(m^2·h) for an oil- or gas-fired furnace. These values can be referred to during initial evaluation and heating surface arrangement design.

4. Overview of heat transfer calculation in furnaces

The furnace is the most important part of a steam boiler. In the furnace, combustion and heat transfer occur at the same time. All the factors involved in the combustion and heat transfer processes impact each other—as such, the furnace includes a complex series of physical and chemical processes including fuel combustion, heat transfer from the flame to the water wall, flow and mass transfer between the flame and gas, and fouling of the water wall. The goal of analyzing heat transfer in a furnace is to determine the absorbed heat of the furnace's heating surface and the gas temperature at the furnace exit.

For suspension-firing furnaces and grate-firing furnaces, a mixture of fuel and air combusts and produces high-temperature flame and gas. Heat is transferred to the water wall tubes through radiation heat transfer. At the furnace exit, the gas then cools to a certain temperature before entering the convection pass.

The heat transfer process in a furnace is subject to many factors. Under certain conditions, the greater the radiation heating surface in the furnace, the more heat is transferred and the lower the gas temperature at the furnace exit; conversely, the smaller the radiation heating surface, the lower the heat transfer and the higher the gas temperature at the exit. Furnaces are designed successfully by determining the heating surface area needed based on furnace outlet gas temperature; furnaces are verified by confirming the outlet gas temperature is reasonable after the radiation heating surfaces are arranged.

When determining the arrangement of radiation heating surfaces, an appropriate furnace size should be first determined. Generally, thermal load is selected according to coal type and combustion mode, then furnace volume is determined based on the selected thermal load.

For example, the thermal load of grate combustion q_R and the volumetric thermal load of the furnace q_v should be used for a grate-firing furnace, then the furnace volume and grate area can be evaluated according to q_R and q_v. Cross section thermal load q_A and volumetric thermal load q_v should be used for a suspension-firing furnace to calculate the furnace cross section area.

Gas temperature at the furnace exit is extremely important; it is a parameter that reflects the absorbed heat of the furnace, which determines the ratio of absorbed heat of radiation heating surfaces and convection heating surfaces in the boiler. The more heat a furnace absorbs, the lower the outlet gas temperature is; if the absorbed heat is low, the outlet gas temperature is high. If the gas temperature at the furnace exit is too low, average flame temperature in the furnace will also be too low, and the radiative heat transfer intensity will naturally be reduced. Such a situation is not economical and disadvantageous for combustion, causing incomplete combustion (thus, loss) of gas and solids, and reducing fuel ignition and combustion stability. If gas temperature at the furnace exit is too high, slagging occurs on the convection heating surface and impacts overall boiler reliability.

We would like to reiterate here that due to the complexity of the furnace heat transfer process, theory alone does not suffice for calculation; instead, empirical or semiempirical heat transfer equations are necessary, based on large amounts of experimental data. Though heating surface structures vary, heat transfer calculations of convection heating surfaces only involve heat transfer and mass transfer without considering chemical reaction. These processes happen in a flue pass, thus the heat transfer calculation is quite simple and the calculation is similar at all stages of the heating surface. This was discussed briefly in Section 5.4—to provide more detail, thermal calculation methods for boiler units are discussed at length below.

5.6.2   Thermal Calculation Methods for Boilers

Boiler design is the first step of a boiler's manufacture—as mentioned above, it determines overall product performance and quality. Boilers should be designed to be small in volume, lightweight, simple in structure, convenient, efficient, and high-quality. The boiler mode should be determined first for a new boiler, followed by the structure and size of every component. The design should aim for safety and reliability, advanced technology, low metal consumption, convenience for manufacture and installation, and high fuel-saving efficiency. Designers should conduct exhaustive research and integrate all related theories and practical knowledge of operation processes, then calculate and compare all available technology. One of the most important calculations during boiler design is thermal calculation of the entire boiler (including determining the structure and size of all heating surfaces according to given technology and expected thermal characteristics parameters)—called "design calculation" for a new boiler, as discussed above.

Design calculation is carried out at rated load of the boiler, and is calculated in the following stepwise process:

1. Determine the initial data.
2. Calculate the air amount, gas amount, and enthalpy of combustion.
3. Calculate the thermal balance to determine each item of heat loss, and calculate boiler efficiency and fuel consumption accordingly.
4. Along the gas flow direction, calculate each stage of the heating surface from the furnace to the back-end in turn.
5. Summarize and tabulate the necessary data for the entire boiler unit.

Once the structure/size of the boiler is given, thermal characteristics can be calculated under other off-design conditions (eg, change of load, change of combustion, change of feedwater temperature), for verification calculation. Identify the exhaust gas temperature θ_{ex} and hot air temperature (when using an air preheater), then calculate the exhaust gas heat loss, boiler efficiency, and fuel consumption for every heating surface in turn. When calculating by hand, if the calculated exhaust gas temperature differs from the supposed value by less than ±10°C and the hot air temperature differs from the supposed value by less than ±40°C, the calculation can be considered acceptable. When using a computer for calculation, the acceptable difference can be much smaller (±1°C). Verify the exhaust gas heat loss, boiler efficiency, and fuel consumption using the last calculated exhaust gas temperature, then verify the absorbed heat of the radiation heating surface using the obtained hot air temperature.

Calculation error can be determined using the following equation:

$$\Delta Q = Q_a \eta - \left(\sum Q \right) \left(1 - \frac{q_{uc}}{100} \right) \tag{5.93}$$

where ΣQ is the sum of the absorbed heat of the water wall, superheater, boiler bundles, and economizer, calculated using the heat obtained from the thermal balance equation for each. Calculation error ΔQ should not exceed the heat entering into furnace Q_a by more than ±0.5%.

If the last calculated exhaust gas temperature or hot air temperature does not meet the above requirement, it must be recalculated. At this time, if the change in fuel consumption does not exceed that of the last calculation by 2%, the heat transfer coefficients of convection heating surfaces do not need to be recalculated; instead, only the temperature and temperature difference of each stage of the heating surface need to be corrected.

Case 5.2: for thermal calculation of a 410 t/h pulverized coal boiler, refer to Appendix D.

5.6.3 Thermal Calculation According to Different Furnace Types

First, let's examine a few similarities. Boilers with three types of furnace burning fossil fuels (the specific fuel in the boiler is not discussed here), including grate-firing, suspension-firing and fluidized-firing furnaces, have similar corresponding calculations and structures. Due to differences in fuel preparation, combustion types, and structures, the practical operation and calculation processes show a little bit of difference. Since oil-fired boilers and gas-fired boilers are also suspension-firing, the following discussion uses the suspension-firing coal-fired boiler as an example.

1. Identical requirements
   Thermal calculations for these three types of furnaces share the same aim and requirements – either designing a new boiler or verifying an existing boiler structure. The results all serve as basic guarantees of boiler performance and form the basis for subsequent work.
2. Similar working processes
   The working process is similar for the different furnaces: fuel enters combustion, burns in a combustion chamber given air and absorbs some amount of heat, enters a convective back-end heating surface, then leaves the boiler.

3. Similar calculation processes

The similar working process of the different types of furnace implies that the basic process for thermal calculation is also similar across the different furnace types, ie, air balance calculation, gas composition calculation, enthalpy-temperature table calculation, thermal balance calculation, heat transfer calculation in the combustion chamber (furnace), and heat transfer calculation of heating surfaces at each stage.

Next, let's consider some notable differences among these different types of furnace.

1. Combustion chamber calculation

Due to different boiler structures and combustion modes, heat transfer calculation in the combustion chamber can vary quite widely. (You've noticed, certainly, that this is the main idea of this book.)

2. Material balance

Fuel particle size is largest in a grate-firing furnace. After combustion, most of the ash is discharged from the bottom of the combustion chamber as slag, and a small fraction of ash leaves the boiler through the back-end heating surface as fly ash. Because the coal particle size in a suspension-firing (pulverized coal) furnace is small, most of the ash leaves the boiler through the back-end heating surface as fly ash, and only a small fraction (less than 10%) is discharged from the furnace hopper; the ash distribution of a CFB boiler falls somewhere in between. The CFB boiler has material circulation among the combustion chamber, separator, and loop seal, so material mass concentration in the combustion chamber is much higher than that in the grate-firing furnace or suspension-firing furnace. This characteristic affects the combustion process and energy balance in the combustion chamber, and requires quite different heat transfer calculation and thermal calculation in the furnace.

3. Energy balance

Fuel stays in the combustion facility (grate) for a very long time in a grate-firing furnace where the temperature is low and uneven, and the combustion center is low. For a suspension-firing furnace the fuel and formed fly ash pass through the combustion chamber very rapidly (generally just a few seconds) and the combustion center is high. For a CFB boiler fuel burns within the entire combustion chamber, that is, there is combustion fraction along the furnace height. In a CFB furnace material mass concentration is high due to material circulation, causing uniform temperature distribution in the furnace. Since the combustion process, heat release process, and energy balance are different, so the calculation method is likewise considerably different.

4. Ash deposition conditions

Due to differences in the material circulation and combustion process (especially regarding temperature), the fly ash of different boilers is different in terms of both amount and property, so their influence on heat transfer is different. This is actually a most likely neglected problem during the course of calculation.

5. Capacity and complexity

Before the CFB boiler was developed, grate-firing boilers were small or medium-scale boilers (with capacity less than 130 t/h) and suspension-firing boilers were large or medium-scale boilers (220 t/h or so). At present, boiler capacity can be roughly ranked from small to large as follows: grate-firing, CFB boiler, and suspension-firing, with some capacity overlap between corresponding furnace types. In general, grate-firing furnaces are small in capacity, simple in structure, and with low calculation complexity and precision demand. Suspension-firing furnaces have generally large capacity (2000~3000 t/h) and complex structure, with high calculation precision demand. Because the energy balance is coupled with the material balance in a CFB boiler furnace (and because the technology is still relatively under-researched and still under development), it is more difficult to analyze.

Chapter 6

Effects of Ash Deposition and Slagging on Heat Transfer

Chapter Outline

6.1 ASH DEPOSITION AND SLAGGING PROCESSES AND CHARACTERISTICS [15]

6.1.1 Deposition and Slagging

For boilers that use solid fuel, deposition on the heating surface, such as in pulverized coal furnaces, in grate-fired furnaces, or on the suspended heating surface of CFB boilers, is inevitable. Deposition and slagging cause the thermal resistance from the flame to the working medium to increase, which decreases the absorbed heat of the heating surface and reduces the boiler's thermal efficiency.

Deposition is characterized by loose ash particles adhering to the heating surface. Only when suspended ash particles have already solidified before touching the heating surface, can loose deposition form. In contrast to deposition,

Theory and Calculation of Heat Transfer in Furnaces. http://dx.doi.org/10.1016/B978-0-12-800966-6.00001-6

slagging is characterized by a tight ash layer adhering to the heating surface. Generally, melting or viscous ash particles attacking the heating surface will form slagging rather than deposition.

Ash layers form through some of the following ways.

1. The finest particles in the vicinity of the water wall surface move to the sublayer of the gas boundary through molecular diffusion, turbulence diffusion, or Brownian movement.
2. Alkali sulfates, chlorides, and hydroxide in gas condense on the surface of the water wall.
3. Large particles move, being carried by the gas flow.
4. High-temperature electrophoresis.
5. Static electricity between fly ash particles and the water wall.
6. Softening and melting particles form sedimentary layers on the surface of the water wall.

Deposition and slagging have clear differences, but are closely related. Deposition forms when ash accumulates on the heating surface at temperatures lower than the ash fusion point, and slagging forms when melting ash accumulates on the heating surface. Slagging is related to the composition, fusion temperature, and surface temperature of ash particles that transfer to the surface under various types of force, usually occurring on the radiation heating surface of the high-temperature zone. Severe deposition or slagging worsens the heat transfer conditions, influencing output and hydrodynamic performance to the point where an accident may occur. So it is very important to avoid deposition and slagging.

Slagging in a pulverized coal furnace usually occurs due to fusion deposition, which is usually related to the transfer conditions of melting or viscous ash particles in the flue gas. When the gas cools down, vaporized matter in the high-temperature area condenses, potentially making these elements accumulate as fusion ash on the furnace wall and forming a tight ash layer.

Slagging and deposition on the boiler heating surface are related to not only the fusion point and composition of the fuel and its ash, but also the design parameters of the boiler such as combustor layout, furnace thermal load, furnace outlet temperature, superheater location, gas velocity/temperature, steam temperature, tube wall temperature, and pitch and arrangement of the heating surface. Slagging and deposition are affected by operating conditions such as load changes, fuel changes, the flow field in the furnace, the multiphase properties of fly ash, the heat transfer process in the furnace, installation of the soot blower and its blowing frequency, the excess air coefficient, and the combustion adjustment process. To this effect, slagging and deposition on the boiler heating surface is a multidisciplinary practical problem that involves the boiler, fuel chemistry, multiphase fluid dynamics, heat and mass transfer, combustion theory, and technology and material science principles. This chapter only introduces content related to heat transfer in furnaces. Interested readers are urged to refer to the literature.

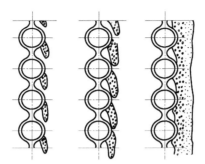

FIGURE 6.1   Slagging on a water wall.

6.1.2   Formation and Characteristics of Deposition and Slagging

Let's first define some different types of deposition and slagging, first by examining the morphology of deposition and slagging on a boiler's heating furnace. During boiler operation, slagging happens only when a portion of the ash particles are viscous enough to adhere to the wall's surface. An ash layer first forms on the surface, then it extends from the heating surface to the furnace interior until reaching the fusion state. Fig. 6.1 shows the process of slagging on a furnace heating surface.

The slagging process is closely related to the flow and temperature field inside the furnace. When part of the gas in the furnace stagnates or changes flow direction, ash particles entrained by the gas may deposit on the furnace wall due to inertia settlement. If the furnace wall temperature is high enough, or the ash surface is molten, ash accumulates to a certain degree on the furnace wall then flows downward to cooler places such as manholes, sight holes, or any measuring holes due to gravity, then condenses into a slag lump looking like a stalactite as shown in Fig. 6.2.

The typical deposition and slagging morphology of convective heating surface tube bundles is shown in Fig. 6.3. Deposition is most likely on the leeward side of the tubes; when gas flows across the tube bundles, a stagnant area forms on this leeward side. Ash particles deposit, again, due to inertia, forming slag arches and bridges. When the temperature is low, the deposited ash is present in the form of fine particle powder.

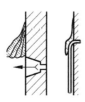

FIGURE 6.2   Sight hole and slag lumps on a leading tube.

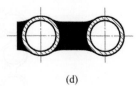

(a)   (b)   (c)   (d)

FIGURE 6.3 **Deposition morphology.** (a) Bilateral wedge deposition. (b) Unilateral wedge deposition. (c) Unilateral melt transformative deposition. (d) Deposition bridge.

The processes of deposition and slagging are complicated, and are affected by alternating physical and chemical factors. As such, there are various types of deposition and slagging that are sometimes difficult to distinguish, but can be roughly classified according to certain properties.

According to ash particle temperature, deposition and slagging can be classified as fusion slag, high-temperature deposition, or low-temperature deposition. When the gas temperature is higher than 800°C, the ash particle temperature reaches the fusion point and the slag is fusion slag. When the gas temperature is 600~800°C, the ash particles are solid and a high-temperature deposition is formed. When the gas temperature is lower than 600°C, fine ash particles on the heating surface of the tubes form a low-temperature deposition.

Deposition can be divided by strength into loose deposition and viscous deposition. Loose deposition is primarily unilateral wedge deposition formed on the leeward of the tubes. Viscous deposition forms on the windward of the tubes when velocity is very low and ash particles are very small, and grows toward the gas flow. Unlike loose deposition, viscous deposition will not stop growing upon reaching a certain size, but instead will continue to grow. As a result, the bundle flow resistance increases until the gas passage is completely blocked.

The properties and differences of loose deposition and viscous deposition are shown in Table 6.1.

Let's look at some specific processes next, starting with the heating surface deposition process. SEM (scanning electron microscope) analysis of heating surface deposition has shown that deposition formation can be divided into three stages.

1. Initial depositions form the substrate. One type of deposition, iron-rich fusion slag, hits and sticks to the tube wall in a tight, iron-rich, spherical glass ball; the other type of deposition is mainly formed by the sublimation and agglomeration of sintering, or the high-temperature effects of aluminum-rich andalusite with silica and aluminum minerals.
2. Fly ash particles then adhere to the top of the substrate. The strength of the deposition layer is increased due to mutual particle adherence. A layer of viscous clusters is formed as the deposition grows. The thermal insulation of the deposition layer raises the temperature of the deposition layer, causing stronger particle adhesion.

TABLE 6.1 Differences Between Loose and Viscous Deposition

Factors considered	Loose deposition	Viscous deposition
Forming location when flowing across bundles	Mainly on leeward side, casually on windward side	Mainly formed on windward side
Growing characteristics	Growth until force balance of fine ash deposition and deposited layer collapse caused by large particles	Infinite growth tendency
Flow resistance	Does not increase bundle resistance	Increases bundle resistance obviously
Ash content in fuel	No influence basically	Deposition becomes severe with large amount of ash
Gas velocity	Increasing velocity decreases deposition size	Increasing velocity increases deposition
Mechanical strength	Loose without mechanical strength	With various mechanical strength

3. As furnace temperature increases, the depositing rate also increases, even causing deposition sintering, so that forms the slag layer.

Now, let's take a look at the slagging process on a heating surface. For pulverized furnaces, the fusible matter in the ash first melts to liquid during combustion, then shrinks into a spherical shape under the effects of surface tension. The fused spherical ash particles have high density, and readily separate from the gas into the ash hopper or slag pool, or adhere to the furnace heating surface. The nonmelted ash particles maintain their original irregular shape. Low-density porous particles form when the combustible matter burns out, then escape the furnace to form fly ash or build up on the cooler heating surface.

In the high-temperature zone in a furnace, the ash in the coal is in a fused or quasi-fused state. If the ash is not cold enough to become solid before reaching the heating surface, the ash still has strong adhesive capacity and will readily adhere to the heating surface or furnace wall facing the high-temperature gas or flame. Fusible or easily gasified substances (such as alkali metal compounds) volatilize rapidly and enter into the flue gas. When the temperature decreases, these substances condense partially on the heating surface to form an initial ash layer. As the adhesive force of the ash layer increases, the thickness of the ash layer increases, and the surface temperature of the ash layer increases. If the fuel ash fusion point is low, the ash forms a plastic slag film with high viscosity when it reaches the ash deformation temperature, then sinters gradually.

Research has shown that the process involves three stages.

1. Diffusion: where a thin ash deposition layer is formed around the tubes that is not affected by flue gas velocity.
2. Internal sintering: where the ash layer is formed on the windward of the tubes because of the hitting of ash particles (about several millimeters in thickness). Particles in this layer bond with each other due to surface viscosity, then gradually sinter into solid form.
3. External sintering: where as the internal sintering layer grows, the ash surface temperature increases, even up to the gas temperature. When the gas temperature and alkali metal proportion in the ash are high enough, a fusion layer is formed on the windward of the ash layer. These fusion materials can further capture oncoming particles and bond with them to form high-strength ash deposition.

Naturally, if the flow field in a furnace causes fuel jet stream deflection, a high-temperature flame blows directly onto the heating surface and directly causes slagging without the above stages.

6.1.3 Damage of Deposition and Slagging

According to the aforementioned analysis of the mechanisms and processes of deposition and slagging in the boiler, the following problems deserve attention:

1. Deposition and slagging reduce the heat transfer ability of the furnace's heating surface; after ash particles deposit on the heating surface, the thermal conductivity is small and the thermal resistance becomes greater. The heat transfer ability decreases by 30~60% after being covered by ash, causing upward movement of the flame center in the furnace, decreased absorption of heat, and increased furnace outlet temperature.
2. Due to increased furnace outlet temperature, fly ash readily adheres to the pendant superheater and convective superheater of the back-end high-temperature zone, which causes deposition, slagging, and corrosion in the superheater. This not only influences heat transfer, but also can lead to corrosion and bursting of the tubes.
3. Severe deposition will cause blockage in the economizer and air preheater gas passage, deteriorating the heat transfer and increasing the exhaust gas temperature, which decrease boiler thermal efficiency and operational economy.
4. Increased total thermal resistance (namely, decreased effective heating area) leaves the boiler unable to keep working at rated load in relation to the designed coal consumption, thus, coal feed has to be increased, causing increased furnace outlet temperature, in turn making ash more readily adhere to the heating surface, and forming a vicious circle. Dangerous accidents such as superheater and economizer blockage, cracks, air preheater blockage, and/or air leaking, then result. Increased furnace outlet temperature

may raise the superheated steam temperature, causing the superheater tubes to work above allowable temperatures.

5. Under the effects of high-temperature gas, ash on the water wall or high-temperature superheater will react with the metal on the tube wall to form high-temperature corrosion. Research has shown that the average water wall corrosion rate is 0.8~2.6 mm/a with high-temperature corrosion, therefore, deposition and slagging are the beginning of high-temperature corrosion.

The above problems caused by deposition and slagging directly result in economic losses including decreased average boiler thermal efficiency, decreased boiler output, decreased available operation time, and increased maintenance time and cost. Avoiding deposition and slagging should be given a great deal of attention during the design, operation, and maintenance of the boiler. It is also important to note that designers should avoid calculating blindly according to ideal, purely theoretical situations, or the results will not reflect actual situations in practice.

6.1.4 Ash Composition

For convenience, we'll use a coal and a pulverized coal-fired boiler as examples to illustrate the influence of fuel composition and ash properties on heating surface deposition and slagging. The ash composition is very complex, and includes mainly Si, Al, Fe, Ti, Ca, Mg, V, Mn, K, Na, S, P, and O. During chemical analysis, these elements are usually denoted by oxide forms (such as SiO_2, Al_2O_3, Fe_2O_3 or FeO, CaO, MgO, V_2O_5, Mn_3O_4 or MnO, K_2O, Na_2O, SO_3, and P_2O_5.) In addition to the oxides, these elements are present in the form of silicates, aluminosilicates, and sulfates. The oxide concentration in ash from high to low is SiO_2, Al_2O_3, Fe_2O_3, FeO, CaO, MgO, Na_2O, and K_2O.

The ash composition combines with the following factors to influence the deposition and slagging.

1. Thermodynamic conditions: gas velocity, gas flow direction, gas temperature, and wall temperature.
2. Operation conditions: coal particle size, fly ash concentration, excess air coefficient, boiler local heat flux, and boiler output.
3. Geometric conditions: tube diameter, tube bundle pitch, number of tube rows, and the approach taken for enhancing heat transfer such as spiral pipes or a bundle with fins.

6.2 EFFECTS OF ASH DEPOSITION AND SLAGGING ON HEAT TRANSFER IN FURNACES

Let's start this section with a look into changes in the thermal efficiency coefficient ψ during deposition and slagging processes. According to the information provided earlier, the furnace wall can be considered a graybody, and the relationship among its own radiation, reflected radiation, projected radiation, and radiosity can be expressed as follows:

$$q_R = \varepsilon \sigma T^4 + (1-\varepsilon)q_I \tag{6.1}$$

The radiative heat transfer flux q is the amount of heat transferred between this surface and the external system. For energy balance, calculate the following:

$$q = q_I - q_R \quad \text{(It is positive if heat is received.)} \tag{6.2a}$$

or

$$q = \varepsilon q_I - \varepsilon \sigma T^4 \tag{6.2b}$$

The effect of fuel characteristics, combustion mode, and boiler thermal load on the in-furnace heat transfer can be lump-sum represented by the thermal efficiency coefficient ψ, which can be measured from experiments. Projected radiation and radiosity can be measured with a radiation heat flowmeter. If the heat flowmeter faces the flame, the projected radiation q_I from flame to wall can be measured; if the heat flowmeter faces the water wall, the radiosity q_R from wall to flame can be obtained. The thermal efficiency coefficient of the water wall is:

$$\psi = \frac{q_I - q_R}{q_I} = \frac{q}{q_I} \tag{6.3}$$

where ψ denotes the ratio of heat absorbed by the water wall to the projected radiation. The larger ψ is, the stronger the radiative heat transfer between the water wall and flame.

The factors influencing the water wall thermal efficiency coefficient ψ include the following.

1. The emissivity of the water wall surface, namely, the ash surface emissivity after deposition and slagging, is related to the slagging structure and temperature. The surface emissivity of loose slag is lower than that of tight slag. The emissivity of the water wall is usually around 0.8.

2. ψ is proportional to the difference of projected radiation and radiosity. If projected radiation is large but reflected radiation is also large, then the thermal efficiency coefficient will not necessarily increase. Fig. 6.4 shows where the measured thermal efficiency coefficient varies along the furnace height of a 230 t/h wet bottom boiler. Projected radiation is high at the slag melting zone, but the radiosity is also high, so ψ is low. At the upper stage of the furnace, projected radiation is low because the gas temperature is low, but radiosity is also low, so the thermal efficiency coefficient ψ is high.

3. ψ is related to the temperature of the water wall, namely the temperature of the ash surface. Surface temperature is related to the thickness of deposition and slagging. If the ash layer is thin, the thermal conduction resistance of

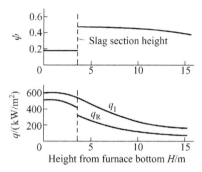

FIGURE 6.4   Changes in thermal efficiency coefficient and heat flux of a 230 t/h wet bottom furnace.

ash is low, and the slag surface temperature is low. If the ash layer is thick, the thermal conduction resistance increases and the slag surface temperature grows rapidly. A previous study provided the example depicted in Fig. 6.5 [15]. For the case of the same flame temperature, when slag layer thickness is 5 mm, the difference between slag surface temperature and flame temperature is about 350°C, whereas at slag layer thickness of 50 mm, the slag surface temperature is only about 40°C lower than flame temperature. Meanwhile, the radiative heat transfer of the flame and water wall decreases from 200×10^3 W/m² to 30×10^3 W/m². In brief, the influence of surface temperature (slag layer temperature) is the vital one.

Finally, based on Eq. (6.3) and the conception of radiative heat transfer:

$$\psi = \frac{q_1 - q_R}{q_1} = \frac{q_1 - \left[\varepsilon\sigma T^4 + (1-\varepsilon)q_1\right]}{q_1} = \varepsilon\left(1 - \frac{\sigma T^4}{q_1}\right) \tag{6.4}$$

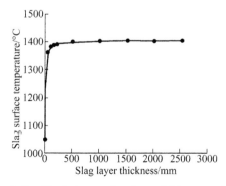

FIGURE 6.5   The relationship between slag layer thickness and slag layer surface temperature.

Evidenced by this equation, the thermal efficiency coefficient ψ is related to water wall surface emissivity ε, surface temperature T, and projected radiative heat flow to water wall surface q_1, where q_R is the effective radiative heat flow of the water wall surface.

In the absence of deposition, the water wall surface temperature equals the metal tube surface temperature, which is close to the temperature of the working medium in the tube without considering the thermal resistance of scales deposited on the inner side of the heating tubes. With the same projected radiative heat flow q_1, $\dfrac{\sigma T^4}{q_1}$ is smallest at this time while the thermal efficiency coefficient ψ reaches its maximum. After deposition, the water wall surface temperature T is equivalent to the ash surface temperature and varies with ash layer thickness/compactness (the thicker the ash layer, the higher the ash layer surface temperature). When the ash is thick enough that the ash layer surface temperature rises close to the flame temperature, the difference is nearly zero, and the heat transfer tends toward zero. At this time, the thermal efficiency coefficient ψ reaches its minimum (also zero).

Let's consider the influence of the fouling factor ζ on heat transfer. According to the above definition, the fouling factor ζ is:

$$\zeta = \psi/x \tag{6.5}$$

where x is the effective configuration factor of the water wall tube, and is dependent on the structure of the water wall. Large coal boilers usually adopt a membrane wall with fins ($x = 1$), thus the fouling factor is equal to the thermal efficiency coefficient. During boiler design and analysis, the fouling factor is usually used instead of the thermal efficiency coefficient. The fouling factor of a general boiler is detailed in Table 3.1.

For a water wall with refractory material, its fouling factor is determined by the following equation:

$$\zeta = 0.53 - 0.25\frac{T_{FT}}{1000} \tag{6.6}$$

where T_{FT} is the fluid temperature of ash.

When fouling factor ζ decreases, the ability of radiative heat transfer weakens and the heat absorbed by the furnace decreases, causing a rise in the furnace outlet gas temperature.

6.2.1 Heat Transfer Characteristics and Ash Layer Calculation with Slagging

The ash deposition coefficient ρ is characterized by the thermal conduction resistance of the ash layer on the water wall, which is related to fuel type and water wall mode. Table 6.2 shows the recommended values of ash deposition coefficients for the water wall. The ash deposition coefficients for other types

TABLE 6.2 Recommended Ash Deposition Coefficient Values

Water wall type	Fuel/combustion type	$\rho/(m^2 \cdot {}^{\circ}C/W)$
Bare water wall	Gas	0
	Heavy oil	0.0017
	Pulverized coal	0.0034
	Coal powder R_{90} = 12–15%	0.0052
	Oil shale	0.0060
	Grate-firing furnace	0.0026
With refractory coating		0.0067
Water wall with refractory brick		0.0086

of heating surface can be found in D6.31, D6.33, D6.37, and D6.38 of the Appendix.

Thermal conduction resistance is related to ash structure and thickness. The following balance equation describes the relationship between heat transfer flow and slag layer thickness (or ash deposition coefficient):

Measuring the heat transfer process provides the sum of radiative heat transfer and convective heat transfer absorbed by the heating surface, which is equal to the amount of heat both conducted through the ash layer and the convective heat transferred into the working medium in the water wall tubes, and can be expressed as follows:

$$\sigma a_F \left(T_g^4 - T_s^4 \right) + h_2 \left(T_g - T_w \right) = \frac{T_s - T_t}{\rho} \tag{6.7}$$

$$\frac{T_s - T_t}{\rho} = h_1 \left(T_t - T_1 \right) \tag{6.8}$$

where T_g is the flame temperature in the furnace, T_s is the ash layer surface temperature, T_t is the average temperature of the water wall tube (ignore any temperature difference caused by thermal resistance of the tube itself and internal deposition), and a_F is furnace emissivity, which is calculated by flame emissivity ε_g and slag surface emissivity ε_s (see the following):

$$a_F = \frac{1}{\dfrac{1}{\varepsilon_g} + \dfrac{1}{\varepsilon_s} - 1} \tag{6.9}$$

h_2 is the convective heat transfer coefficient from the flame to the ash surface, h_1 is the heat transfer coefficient between the water wall tube inner surface and the working medium in the tube, T_1 is the working medium temperature, σ is the radiation constant, and ρ is the ash deposition coefficient (ash thermal conduction

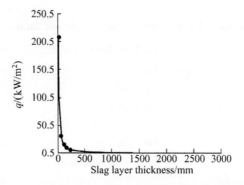

FIGURE 6.6 **The relationship between slag layer thickness and heat flux.**

resistance δ/λ, where δ is the ash layer thickness and λ is the ash thermal conductivity). From Eqs. (6.7) and (6.8), the relationship between heat flux q and ash thickness can be calculated (supposing λ is constant), as shown in Fig. 6.6.

The following is a practical example.

Example 6.1

A pulverized coal boiler has capacity 130 t/h, furnace water wall area $A = 479.5$ m², calculated fuel consumption $B_{cal} = 4.437$ kg/s, and amount of heat transfer in furnace for 1 kg fuel burnt $Q_r = 12246$ kJ/kg. Known deposition thickness is 0.5 mm and deposition thermal conductivity is 0.1 W/(m·°C). Calculate the temperature inside and outside the deposition layer.

The total heat absorbed by the water wall is:

$$Q_r B_{cal} = 12246 \times 4.437 = 54335.5 \, kW$$

The heat flux of the heating surface area is:

$$q = \frac{Q_r B_{cal}}{A} = 113.3 \, \left(kW/m^2 \right)$$

The temperature difference is:

$$\Delta t = \frac{\delta}{\lambda} q = \frac{0.5 \times 10^{-3}}{0.1} \times 113.3 = 566.5 °C$$

If the water wall surface temperature is 450°C, then the external ash temperature is 1016.5°C.

6.2.2 Heat Transfer Calculation with Deposition and Slagging

Furnace heat transfer calculation equations are obtained based on radiative heat transfer. After comprehensive boiler tests, the integrated influence of boiler

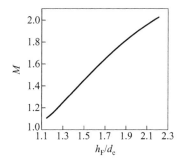

FIGURE 6.7 Furnace geometric shape coefficient.

structure, combustion mode, fuel characteristics, and deposition and slagging can be reflected by equations using coefficients M and ψ.

The furnace outlet gas temperature T_F'' can be determined using Eq. (5.27):

$$T_F'' = \frac{T_a}{M\left(\dfrac{\sigma\psi A\tilde{a}_F T_a^3}{\varphi B_{cal} V\overline{C}}\right) + 1}$$

When furnace structure, fuel characteristics, and boiler parameters are given, the furnace outlet gas temperature can be calculated from the above equation. The influence of deposition and slagging on the furnace thermal efficiency coefficient ψ has been introduced before in this section.

Coefficient M is related to furnace geometric shape, and is a function of the ratio of furnace height h_F to furnace cross section equivalent diameter d_e. As a previous study pointed out, according to a large number of tests on a dry bottom pulverized coal boiler, the relationship of M and h_F/d_e can be obtained as shown in Fig. 6.7 [15]. The furnace cross section equivalent diameter can be calculated using the following equation:

$$d_e = \frac{4F}{U} \tag{6.10}$$

where F is the furnace cross section area, and U is the perimeter of the furnace cross section.

6.3 EFFECTS OF ASH DEPOSITION AND SLAGGING ON HEAT TRANSFER IN CONVECTIVE HEATING SURFACES

6.3.1 Effects of Severe Ash Deposition and Slagging

Radiative heat transfer in a furnace worsens owing to ash deposition and slagging, leading to increases in the furnace outlet gas temperature, and increases in the heat load of the convective heating surface positioned immediately after the furnace exit. Previous research has provided a good example of this [15]. Calculation of a 600 MWe boiler water wall showed that as deposition grew severe, the fouling factor ζ decreased from 0.45 to 0.25 and the furnace outlet gas temperature increased by 65°C. The more serious the deposition was, the higher the furnace outlet gas temperature. As furnace gas flows through the platen superheater, reheater, superheater, economizer, and air preheater, the inlet gas temperature of all heating surfaces increases. The closer the heating surface is located downstream from the furnace exit, the higher the magnitude of the temperature rise. The platen superheater inlet gas temperature increases by 145°C, the last-stage reheater inlet gas temperature increases by 80°C, the economizer inlet gas temperature increases by 12°C, and the air preheater inlet gas temperature increases by 5°C when deposition is severe. To reiterate, deposition and slagging in the furnace have a considerable impact on the high-temperature heating surface, and to this effect, the security issue attributable to overheating of the tube wall of the reheater and superheater needs special attention.

6.3.2   Basic Heat Transfer Equation for Convective Heating Surfaces

For heating surfaces in the boiler back-end ductwork, the working medium absorbs heat from the gas by convective heat transfer. These heating surfaces are therefore called "convective heating surfaces." Convective heating surfaces in large boilers include slag screens, convective superheaters, reheaters, economizers, and air preheaters, all of which belong to recuperative heat exchangers, follow the same heat transfer principle, and are calculated in almost the same way.

The thermal balance equation of the convective heating surface gas side is:

$$Q = \varphi\left(I' - I'' + \Delta\alpha I_{air}^0\right) \tag{6.11}$$

The thermal balance equation of the working medium side is:

$$Q = \frac{D(i'' - i')}{B_{cal}} \tag{6.12}$$

The heat transfer equation is as follows:

$$Q = \frac{K \Delta t A}{B_{cal}} \tag{6.13}$$

where Q is the convective heat transfer of the heating surface, I' is the gas enthalpy at the heating surface inlet, I'' is the gas enthalpy at the heating surface outlet, $\Delta\alpha$ is the air leakage coefficient within the calculated heating surface, I_{air}^0 is the theoretical leaked air enthalpy (usually the enthalpy of cold air), D is the flow rate of the working medium, i', i'' is the working medium enthalpy at the heating surface inlet and outlet, respectively, B_{cal} is the designed fuel supply rate (really burnt), A is the convective heating surface area (usually the external area of the heating surface tube whereas the mean diameter between the inner and external diameters should be used for a tubular air preheater area calculation), Δt is the average temperature difference between the gas and the working medium, and K is the convective heat transfer coefficient.

The effects of deposition and slagging on heat transfer are integrated by convective heat transfer K in the following heat transfer equation:

$$K = \frac{1}{\dfrac{1}{h_1} + \dfrac{\delta_a}{\lambda_a} + \dfrac{\delta_w}{\lambda_w} + \dfrac{\delta'}{\lambda'} + \dfrac{1}{h_2}}$$

(The meaning of this equation was discussed in detail in Section 5.5.)

The metal walls of tubes in the boiler are very thin and their thermal conductivity is large. Therefore, thermal resistance δ_w/λ_w is small enough to be ignored. Because scale deposition under normal operating conditions does not occur, the resistance of scale deposition δ'/λ' is not considered. Bear in mind that if there is deposition, its thermal conductivity is very small, so the heat transfer coefficient will drop sharply.

The thermal resistance of the ash layer is related to many factors including fuel type, gas velocity, tube diameter and arrangement, ash particle size, and others. At present, the ash deposition coefficient $\rho = \delta_a/\lambda_a$ or the effectiveness factor ψ is adopted to account for this. To distinguish this factor from the furnace thermal efficiency coefficient, ψ_a is used here. ψ_a is defined as the ratio of the heat transfer coefficient of a dirty (fouled) tube K and the heat transfer coefficient of a clean tube K_0:

$$\psi_a = K/K_0 \tag{6.14}$$

The heat transfer coefficient can be expressed by the following equation:

$$K = \frac{1}{\dfrac{1}{h_1} + \rho + \dfrac{1}{h_2}} \tag{6.15a}$$

or by:

$$K = \psi_a \frac{1}{\dfrac{1}{h_1} + \dfrac{1}{h_2}} \tag{6.15b}$$

The heat transfer coefficient from the gas to the tube wall can be calculated by the following equation:

$$h_1 = \xi(h_c + h_r) \tag{6.16}$$

where ξ is a utilization coefficient that makes a correction for reduced absorption of heat due to nonuniform flow of gas passing across the heating surface. Note that for the evaporating heating surface and economizer, the heat transfer coefficient from the tube wall to the working medium $h_2 >> h_1$, thus thermal resistance $1/h_2$ can be ignored.

Based on experimental data for cross-flow staggered tube bundles in solid fuel fired boilers, the effect of deposition on the heat transfer coefficient can be observed according to the ash deposition coefficient. For a superheater:

$$K = \frac{1}{\dfrac{1}{h_1} + \rho + \dfrac{1}{h_2}} \tag{6.17}$$

For an economizer and boiler convection bank:

$$K = \frac{1}{\dfrac{1}{h_1} + \rho} \tag{6.18}$$

The ash deposition coefficient is determined by experimental data, as discussed in Section 6.3.3.

For cross-flow in-line bundles in solid fuel fired boilers, heat transfer coefficients are calculated according to the effectiveness coefficient. For a superheater:

$$K = \psi_a \frac{1}{\dfrac{1}{h_1} + \dfrac{1}{h_2}} \tag{6.19}$$

For an economizer and boiler convection bank:

$$K = \psi_a h_1 \tag{6.20}$$

where for anthracite and mean coal, $\psi_a = 0.6$; for bituminous coal and lignite, $\psi_a = 0.65$; for oil shale, $\psi_a = 0.5$.

For tubular air preheaters, the influence of ash deposition and the uneven gas flow field is considered using the utilization coefficient ξ as follows:

$$K = \xi \frac{1}{\dfrac{1}{h_1} + \dfrac{1}{h_2}} \tag{6.21}$$

6.3.3 Coefficients Evaluating the Ash Deposition Effect

"Convective heating surface deposition" refers to the deposition process on convective heating surfaces by gas laden with ash. Heating surface deposition severely influences heat transfer; heat transfer capacity generally decreases by about 30%, and potentially even up to 50%, after deposition. Fouling effects are observed using the ash deposition coefficient ρ, the effectiveness coefficient ψ_a, and the utilization coefficient ξ during a convective heat transfer calculation. Because the deposition process is related to fuel type, heating surface arrangement, and operating conditions, the above coefficients can only be obtained from model experiments and measurements taken during practical operation. We'll discuss them separately below.

1. Ash deposition coefficient ρ

The ash deposition coefficient is the thermal resistance caused by deposition on the convective heating surface, $\dfrac{\delta_a}{\lambda_a}$. Because deposition layer thickness δ_a and thermal conductivity λ_a are difficult to measure, the recommended ρ is determined based on the difference of the reciprocal of the dirty tube wall heat transfer coefficient K and the clean tube wall heat transfer coefficient K_0: $\rho = \dfrac{1}{K} - \dfrac{1}{K_0}$.

For loose deposition, field tests have shown that ρ is related to gas velocity, heating surface arrangement, fuel characteristics, and ash particle size. The higher the gas velocity is, the smaller the ρ value. Gas easily washes the back side of a tube if the tubes are in a staggered arrangement, so there is less deposition on the leeward of the tubes and the value of ρ is small. If the arrangement of the tubes is dense (the vertical spacing is small), the gas more readily washes the front of the tubes, which decreases the ρ value. If large-diameter tubes are in an in-line arrangement, deposition tends to be severe and the ρ value tends to be high. As for the ash particle size, the bigger the particle is, the greater the self-cleaning effect is, and as a result, the ash deposition becomes slow and the ρ value decreases.

For staggered bundles arrangements including a single row of tubes when burning solid fuel, ρ can be calculated using the following equation:

$$\rho = C_d C_a \rho_0 + \Delta\rho \tag{6.22}$$

where ρ_0 is the basic ash deposition coefficient related to gas velocity ω_g, which is obtained as shown in Fig. 6.8. S_2 is the vertical pitch of tube rows, d is the tube external diameter, and C_d is the correction coefficient for tube diameter (Fig. 6.8). C_a is the correction coefficient for ash particle size, $C_a = 1 - 1.18\lg\dfrac{R_{30}}{33.7}$, and R_{30} is the mass percentage of ash particles larger than 30 μm. Generally, for coal and oil shale, $C_a = 1.0$; for peat, $C_a = 0.7$. $\Delta\rho$ is an additional correction value (See Table D19).

For a platen heating surface, the ash deposition coefficient is related to fuel characteristics, average gas temperature $\bar{T}_g$, and soot blowing condition, which can be determined as shown in Fig. D9 of the Appendix.

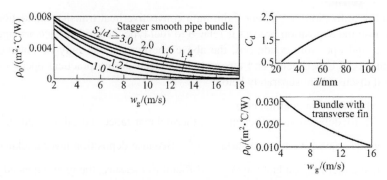

FIGURE 6.8 Basic ash deposition coefficient and correction coefficient.

2. Effectiveness factor ψ_a

Eq. (6.14) indicates that the effectiveness factor ψ_a reflects the heat transfer coefficient of a dirty tube wall and a clean tube wall. The effectiveness factor is adopted regardless of fuel type for the slag screen of a large boiler or tube bank of a small boiler, and also for the convective heating surface of a boiler burning liquid or gas fuel.

The effectiveness factor ψ_a, as well as the ash deposition coefficient ρ, depend on fuel characteristics, gas velocity, heating surface structure, and working systems. The value of ψ_a should be determined according to several different conditions. For heating surfaces such as convective in-line superheaters, slag screens, boiler banks, reheaters, and the transition areas of once-through boilers burning lean coal and anthracite, $\psi_a = 0.6$; burning bituminous coal, lignite, and middling, $\psi_a = 0.65$; burning oil shale, $\psi_a = 0.5$. For heavy oil fired boilers, $\psi_a = 0.5 \sim 0.7$ for all the convective heating surfaces except for the air preheater, which is dependent on the gas velocity and soot blowing condition.

For gas fuel fired boiler the effectiveness factor is taken to consider the influence of fouling on heat transfer in all types of convective heating surface. For single-stage economizers with inlet gas temperatures below 400°C, first stage economizers with a two-stage arrangement, and reheaters, $\psi_a = 0.85$.

3. Utilization coefficient ξ

The utilization coefficient ξ is a correction that reflects any uneven flow onto the heating surface and the resultant reduction in utilization efficiency of the heating surface. During calculation of the convective heating surface, the ξ value is generally considered. For convective heating surfaces with mixed flow, ξ is 0.95. For cross flow heating surfaces of modern boilers, due to their simple structure and effective sweeping, ξ is 1. For boiler evaporating banks with complex gas flow, ξ is 0.9.

To calculate the gas side heat transfer coefficient of a platen superheater, ξ should be adopted to consider the influence of uneven flow from the gas to the platen surface, because gas velocity in the equation of the heat transfer coefficient must be determined by the average flow velocity assuming gas flowing uniformly across the heating surface. Because most platen heating surfaces are arranged at the position where flue gas in the furnace top enters the horizontal flue gas passage, gas velocity in the furnace exit section is uneven (despite the arrangement of the furnace nose or the fact that the combustor is arranged according to the furnace air dynamic field). The convective heat transfer coefficient has to be corrected, to this effect, and the utilization coefficient should be measured from experiments. When average gas velocity $\omega_g \geq 4$ m/s, $\xi = 0.85$; as gas velocity decreases, the influence of uneven flow increases and the ξ value decreases.

For tubular air preheaters, the utilization coefficient ζ is the influence of the air side where the air cannot flow across bundles uniformly due to direction turning of air flow induced by the baffling plate. The correction method of this influence involves multiplying the heat transfer coefficient under average velocity by a factor less than 1, which is similar to the correction method for the gas side of a dirty tube. During heat transfer calculation the combined effect of the air side and flue gas side is considered and denoted by utilization coefficients.

Under all the conditions introduced above, the values of ρ, ψ_a, and ξ are suitable for the situation of a single-fuel boiler. If the boiler uses blended fuel, such as coal–oil blends or oil–gas blends, the values of these coefficients should be calculated according to the degree of fouling. If boilers burn gas fuel after solid fuel, they should be calculated in accordance with the solid fuel burned.

The ash deposition coefficient ρ, effectiveness factor ψ_a, and utilization coefficient ξ are empirical corrections for the additional thermal resistance caused by ash deposition on the heating surface, which is obtained from experiments. Generally, for solid fuel and cross flow staggered bundles, the ash deposition coefficient ρ is applied to correct the influence of heating surface fouling on heat transfer. For in-line arranged bundles and various fuels, ash deposition is usually accounted for by the effectiveness factor ψ_a. For air preheaters, besides the thermal resistance of ash deposition, the effects of an uneven flow field are quite significant, so the utilization coefficient ξ is adopted to reflect both.

Chapter 7

Measuring Heat Transfer in the Furnace

Chapter Outline

Any formula established theoretically must be verified through real-world test results and experiments in a laboratory and/or industry setting. Occasionally, a coefficient is needed to correct the original formula. When studying the characteristics of heat transfer in furnaces according to scaling laws, sufficient and reliable data gathered from experiments are necessary to verify (and amend, if necessary) the theoretical equations empirically.

Development of testing and experimental technologies also assists researchers to better understand the fundamentals of furnace heat transfer phenomena. Results obtained and observations made in the field strengthen theoretical analysis and improve the accuracy of calculation formulae. Indeed, the entire body of knowledge regarding heat transfer in furnaces has been built by combining theory and practice, gradually forming a complete and credible system from empirical subject matter.

Skilled and methodical measurement of flame emissivity allows comprehensive and accurate understanding of flame radiation characteristics, and the ability to predict (and therefore, to manage) necessary related factors. The distribution of radiative heat flow and the effects of ash on the radiative heating surface in the furnace become clear once irradiation and reflective heat flow are accurately measured—this then allows us to precisely and clearly illustrate heat

Theory and Calculation of Heat Transfer in Furnaces. http://dx.doi.org/10.1016/B978-0-12-800966-6.00007-7

transfer in the furnace, improve theoretical calculation methods, and enhance the reliability of empirical formulae.

There are many methods available for experimenting on heat transfer in furnaces, but we will only discuss those for flame emissivity and radiant heat flux here. We will also provide an example of local heat transfer coefficient measurement in the heating surface of back-end ductwork of a boiler with combined radiation and convection, plus another example of local heat transfer coefficient measurement in the furnace of a CFB boiler.

7.1 FLAME EMISSIVITY MEASUREMENT [42]

Based on Planck's law:

$$E_{b\lambda} = \frac{2\pi hc^2}{\lambda^5(e^{\frac{hc}{\lambda kT}} - 1)} \tag{7.1}$$

$$I_{b\lambda} = \frac{E_{b\lambda}}{\pi} \tag{7.2}$$

we know that the monochromatic radiation intensity of a black body in one wavelength is the single-valued function of temperature T. Therefore, once the monochromatic radiation intensity is measured, we can calculate T. At the same time, the optical radiance emitted from a black body is proportional to its radiation intensity, so we can obtain the temperature of the black body based on optical radiance.

We can measure the emissivity of luminous flames with optical pyrometers (eg, a bichromatic optical pyrometer) or with auxiliary radiative resources.

7.1.1 Bichromatic Optical Pyrometer

As described earlier, the monochromatic radiation intensity of a black body is a function of temperature and wavelength. If wavelength is fixed to one value, for example, $\lambda = 0.65\,\mu m$, the monochromatic radiation intensity of the black body is only dependent on temperature; on the other hand, the light luminance emitted by an object is proportional to its radiation intensity, so the temperature of a measured black body can be calculated according to its optical radiance.

Eqs. (7.1) and (7.2) calculate the radiation intensity of a black body with temperature T_b. See the following:

$$I_{b\lambda} = \frac{2hc^2}{\lambda^5(e^{\frac{hc}{\lambda kT_b}} - 1)} \tag{7.3}$$

As we know, when $\dfrac{hc}{\lambda kT} \gg 1$, the Planck formula can be simplified into the Wien formula, therefore Eq. (7.3) can be transformed into the following:

$$I_{b\lambda} = \frac{2hc^2}{\lambda^5} e^{-\frac{hc}{\lambda kT_b}} \tag{7.4}$$

If the measured body with monochromatic emissivity ε_λ is not a black body, then:

$$I_\lambda = \frac{2hc^2 \varepsilon_\lambda}{\lambda^5} e^{-\frac{hc}{\lambda kT}} \tag{7.5}$$

Because an optical pyrometer is calibrated according to a black body, and the radiation intensity of black bodies is larger than that of real bodies at the same temperature, the radiance temperature T_b measured by optical pyrometer will be lower than the real temperature T in Eqs. (7.4) and (7.5), so:

$$\frac{1}{T} - \frac{1}{T_b} = \frac{\lambda k}{hc} \ln \varepsilon_\lambda \tag{7.6}$$

where T is the absolute temperature of the measured body, and T_b is the radiance temperature measured by pyrometer.

The body being measured is usually not a black body, so the result must be corrected.

We can measure the emissivity of a luminous flame with a pyrometer very conveniently. Two types of optical filter, red ($\lambda_1 = 0.6651\,\mu m$) and green ($\lambda_2 = 0.5553\,\mu m$), are available for measuring two radiance temperatures. From Eq. (7.6), we can obtain:

$$\left.\begin{array}{l} \dfrac{1}{T} - \dfrac{1}{T_{b1}} = \dfrac{\lambda_1 k}{hc} \ln \varepsilon_{\lambda_1} \\[2mm] \dfrac{1}{T} - \dfrac{1}{T_{b2}} = \dfrac{\lambda_2 k}{hc} \ln \varepsilon_{\lambda_2} \end{array}\right\} \tag{7.7}$$

Research has shown that the monochromatic extinction coefficient of carbon black is similar to that of ash particles in a luminous flame, represented by the following formula:

$$K_\lambda = \frac{c\mu_c}{\lambda^n} \tag{7.8}$$

where μ_c is the concentration of carbon black in the flue gas. Remember that $M = c\mu_c$, then:

$$K_\lambda = M \lambda^{-n} \tag{7.8a}$$

Based on $\varepsilon = 1 - e^{-K_\lambda s}$, the monochromatic emissivity of the flame can be calculated as follows:

$$\left.\begin{aligned}\varepsilon_{\lambda_1} &= 1 - e^{-\frac{Ms}{\lambda_1^n}} \\ \varepsilon_{\lambda_2} &= 1 - e^{-\frac{Ms}{\lambda_2^n}}\end{aligned}\right\} \tag{7.9}$$

Hottel suggested that the power n can be 1.39 in Eq. (7.9), so:

$$\left.\begin{aligned}\frac{1}{T} - \frac{1}{T_{b1}} &= \frac{\lambda_1 k}{hc} \ln(1 - e^{-\frac{Ms}{\lambda_1^{1.39}}}) \\ \frac{1}{T} - \frac{1}{T_{b2}} &= \frac{\lambda_2 k}{hc} \ln(1 - e^{-\frac{Ms}{\lambda_2^{1.39}}})\end{aligned}\right\} \tag{7.10}$$

Once the radiance temperatures T_{b1} and T_{b2} of the flame have been measured, the real temperature and coefficient Ms can be calculated according to Eq. (7.10) and flame emissivity can be calculated using Eq. (7.9).

In the above method, the power n in Eq. (7.9) was assumed to be constant, but it is different in real boilers based on different fuels, boiler structures, and operation conditions. Sometimes, n is related to the distribution of carbon black particles, which is dependent on fuel and combustion conditions. Basically, there are errors inherent to applying this method to measure flame emissivity.

7.1.2 Auxiliary Radiative Resources

There are two types of auxiliary radiative resource: radiance adjustment, and radiance nonadjustment.

When applying radiance adjustment methods to measure flame emissivity, an auxiliary radiative resource is typically set on the opposite side of the optical pyrometer face toward the flame, then its temperature is adjusted to ensure the radiance temperature is constant with or without flame. Kirchhoff's law asserts that the radiance temperature $T_{b\lambda}$ of the auxiliary radiative resource equals the real temperature of the flame, that is, the radiation intensity emitted by the auxiliary radiative resource and absorbed by the flame equals the radiation intensity emitted by the flame. See the following:

$$I'_\lambda (1 - \alpha_\lambda) + I_\lambda = I'_\lambda \tag{7.11}$$

where α_λ is the monochromatic absorptivity of the flame, I_λ is the radiation intensity of the flame, and I'_λ is the radiation intensity of the auxiliary radiative resource.

The above formula can be rewritten as:

$$\alpha_\lambda I'_\lambda = I_\lambda \tag{7.12}$$

The monochromatic absorptivity of the flame can be obtained as such, which is equal to monochromatic emissivity $\varepsilon_\lambda = \alpha_\lambda$:

$$\varepsilon_\lambda = \frac{I_\lambda}{I'_\lambda} \tag{7.13}$$

If we apply a black body as the auxiliary radiative resource with radiation intensity $I'_{b\lambda}$, we can obtain the flame emissivity as follows:

$$\varepsilon = \frac{I_\lambda}{I'_{b\lambda}} \tag{7.14}$$

When applying radiance nonadjustment methods to measure flame emissivity, three separate radiation intensities must be measured: radiation intensity I_λ of the flame only, radiation intensity I'_λ of the auxiliary radiative resource without the flame, and radiation intensity I''_λ of the auxiliary radiative resource penetrating the flame. Based on the energy conservation relationship:

$$I''_\lambda = I_\lambda + I'_\lambda \, (1-\alpha_\lambda) \tag{7.15}$$

Then flame emissivity (if $\varepsilon_\lambda = \alpha_\lambda$) is:

$$\varepsilon_\lambda = \frac{I_\lambda + I'_\lambda - I''_\lambda}{I'_\lambda} \tag{7.16}$$

Eq. (7.16) can then be substituted into (7.6) after measuring the radiance temperature $T_{b\lambda}$ of the flame to calculate the real temperature of the flame:

$$T = \left[\frac{1}{T_{b\lambda}} + \frac{\lambda k}{hc} \ln \frac{I_\lambda + I'_\lambda - I''_\lambda}{I'_\lambda} \right]^{-1} \tag{7.17}$$

Of course, Eq. (7.2) can also be substituted into (7.16) to obtain the following, given the temperature of $T_{b\lambda}$, $T'_{b\lambda}$, T''_λ is read using instruments:

$$\varepsilon_\lambda = 1 - e^{\frac{hc}{\lambda k}\left(\frac{1}{T'_{b\lambda}} - \frac{1}{T''_{b\lambda}}\right)} + e^{\frac{hc}{\lambda k}\left(\frac{1}{T'_{b\lambda}} - \frac{1}{T_{b\lambda}}\right)} \tag{7.18}$$

The real temperature of the flame can then be calculated as follows:

$$T = \left\{ \frac{1}{T_{b\lambda}} + \frac{\lambda k}{hc} \ln \left[1 - e^{\frac{hc}{\lambda k}\left(\frac{1}{T_{b\lambda}} - \frac{1}{T''_\lambda}\right)} + e^{\frac{hc}{\lambda k}\left(\frac{1}{T'_\lambda} - \frac{1}{T_{b\lambda}}\right)} \right] \right\}^{-1} \tag{7.19}$$

7.2   RADIATIVE FLUX MEASUREMENT

It is very important to accurately measure radiative flux in the radiative heating surface for the design and operation of both grate-firing boilers and suspension-firing boilers, as it is the basis of reasonable scientific design, diagnosis, and operation, as well as a necessary condition for combustion examination. There are many types of radiation heat flux meter (heat conductive, heat capacitive, calorimetric, and others) that are employed according to related measurement principles.

Radiation flux is the radiative energy emitted by media in the furnace as it irradiates to the heating surface in unit time. In large-scale, pulverized coal fired boilers used in power stations, technicians typically employ radiation pyrometry to measure the distribution of radiative heat flux, thermal deviation between heating surfaces, thermal efficiency coefficient, radiative heat transfer coefficient of the panel heating surface, flame center position, temperature distribution in the furnace, and other necessary factors. In medium-scale or small-scale grate-firing boilers, radiation pyrometry is usually applied to test the radiative ability of the fired bed, absorption and radiation of the furnace surface, and central flame temperature. In CFB boilers, because the combustion temperature is very low (usually below 950°C) and there is gas–solid flow with high bed material concentration in the furnace, the share is almost the same for both convective and radiative heat transfer—for this reason, we will discuss the characteristics of local heat transfer instead of radiative heat transfer in the following section.

7.2.1 Conductive Radiation Heat Flux Meter

Temperature differences between metal surfaces from hot to cool are characterized by thermal resistance controlled by thermal flux flow over the metal body; the greater the heat flux, the steeper the temperature gradient is. This temperature difference can, as such, be measured in order to calculate the radiative heat flux projected to the metal surface. This is the basic principle of the conductive radiation heat flux meter.

A schematic diagram of the conductive radiation heat flux meter is shown in Fig. 7.1. The hot end of the cylinder absorbs heat flux q from the flame area, where the temperature is t_1 and distance from the hot end of the cylinder is x_1; the cold part is cooled by water, where the temperature is t_2 and distance from the hot end of the cylinder is x_2. Heat is transferred from the hot end to the cold

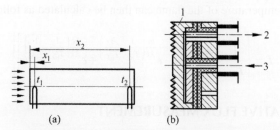

FIGURE 7.1 Schematic diagram of conductive radiation heat flux meter. (a) Cylinder, (b) Cooling structure (1) Heating surface, (2) Water inlet, (3) Water outlet.

end by conduction. Because the cylindrical surface is isothermal, heat conduction occurs only in the axial direction of the cylinder in a one-dimensional manner. Under steady-state conditions, Fourier's law can be employed to calculate one-dimensional heat flux conduction as follows:

$$q = -\lambda \frac{dt}{dx} \tag{7.20}$$

where λ is the thermal conductivity of the cylinder material, and $\dfrac{dt}{dx}$ is the temperature gradient along the heat flow.

If we know the temperatures t_1 and t_2 of the cylinder with the spacing along the axial of x_1 and x_2, we can determine heat flux q. Eq. (7.20) can be integrated into the following:

$$q \int_{x_1}^{x_2} dx = -\int_{t_1}^{t_2} \lambda\, dt$$

or:

$$q = -\frac{1}{x_2 - x_1} \int_{t_1}^{t_2} \lambda\, dt \tag{7.21}$$

When the heat flux meter was designed, spacing between x_1 and x_2 was typically small. To this effect, the following average thermal conductivity can be defined as

$$\bar{\lambda} = \frac{\int_{t_1}^{t_2} \lambda\, dt}{t_2 - t_1} \tag{7.22}$$

The above formula can be substituted into Eq. (7.21) to form the following formula for heat flux:

$$q = \bar{\lambda}\, \frac{(t_1 - t_2)}{(x_2 - x_1)} \tag{7.23}$$

7.2.2 Capacitive Radiation Heat Flux Meter

The rate of temperature increase in an area of heat measurement (such as the plate of a heat flux meter) can be determined during the heating process to determine the heat flux of irradiation. The heat flux of a heat flux meter can be expressed as follows:

$$q = \delta \rho C \frac{\Delta t}{\Delta \tau} \tag{7.24}$$

where δ is the thickness of the heat flux meter plate, ρ is the mass density of the plate, C is the specific heat capacity of the plate, Δt is the increment of the plate temperature during time $\Delta\tau$, and $\dfrac{\Delta t}{\Delta\tau}$ is the rate at which the plate temperature rises.

When a millivoltmeter is employed to measure temperature, its heat flux can be measured as follows:

1. Preset two temperatures in the millivoltmeter $\Delta t = t_2 - t_1$.
2. Immerse the heat flux meter in boiled water (100°C).
3. Insert the heat flux meter immediately into the furnace.
4. Run a stopwatch, and record the time interval as the millivoltmeter passes through the two temperature readings preset before.
5. Calculate heat flux q by plugging the data into Eq. (7.24).

7.2.3 Calorimetric Radiation Heat Flux Meter

Calorimetric radiation heat flux meters work by inserting a measuring unit into a furnace with cooling water, then calculating the heat that the unit absorbs from the furnace according to the increased enthalpy of the cooling water. Because the heat flux is calculated according to the heat that the water flowing through the measuring unit absorbs from the furnace, this type of heat flux meter is also called a "water circulating heat flux meter." Please refer to Fig. 7.2 for a schematic diagram.

When measuring water flow rate $\dot{m}$, inlet temperature t_i, and outlet temperature t_o, the heat flux of the heating surface is:

$$q = C\dot{m}(t_o - t_i) \tag{7.25}$$

where C is the specific heat capacity of water.

The response time of this type of heat flux meter is very short, usually about 10 s. Water temperature (t_i and t_o) cannot respond as quickly, because it needs much more time to become steady (ie, to reach heat balance).

7.3 TWO OTHER TYPES OF HEAT FLUX METER [34]

Flame emissivity and radiative heat flux measurement methods are described earlier—these are suitable for measuring flue gas radiation in the furnace. The heat pipe flux meter, a relatively novel type of heat flux meter, is also worth discussing. As we know, the local heat transfer coefficient in a CFB furnace

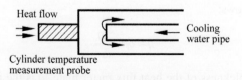

Heat flow

Cooling water pipe

Cylinder temperature measurement probe

FIGURE 7.2 Diagram of calorimetric radiation heat flux meter.

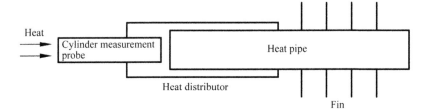

FIGURE 7.3   Structure of heat pipe heat flux meter.

reflects two types of heat transfer: radiation and convection. An example that illustrates local heat transfer coefficient measurement in a CFB furnace using a heat pipe heat flux meter is mentioned later.

7.3.1   Heat Pipe Heat Flux Meter

Because calorimetric radiation heat flux meters need to connect input and output water tubes onsite, they are not convenient for use in boilers that are over a few meters tall. Researchers from Xi'an Jiaotong University (China) developed a heat flux meter, the heat pipe heat flux meter. See Fig. 7.3 for a diagram.

Based on vacuum phase-change heat transfer, the heat transfer ability of heat pipes is 100 times that of copper pipes, and so can overcome the issue of heat dissipation over a long distance. Therefore, the heat pipe heat flux meter is applied with fins instead of water cooling tubes. Heat absorbed on the front face of the probe is transferred to the distributer by conduction, then the distributer transfers the heat to the vaporizing section of the heat pipe. The heat then dissipates to the surrounding air from the fins located on the condensation section of the heat pipe. The heat distributor is made of copper or aluminum, which have high thermal conductivity, with two threaded ends connected to the cylinder measurement probe and heat pipe, respectively. The heat the probe receives from the furnace can be expressed as follows:

$$q = q_H(\frac{\pi}{4}d^2)$$ (7.26)

where q_H is the heat load of the furnace heating surface, and d is the diameter of the probe.

The maximum heat load of the heating surface in a boiler furnace is 6 MW/m^2, and the diameter of the probe is known, so it is possible to calculate the absorbed heat q of the probe using Eq. (7.26).

After we calculate the heat measured by the heat pipe heat flux meter using Eq. (7.26), the number of fins on the heat pipe must be determined to ensure the probe dissipates heat appropriately from the furnace to the surrounding air. Because the diameter of the heat pipe is usually very small, and the diameter of the fin is about twice that of the heat pipe, we can assume that the temperature is the same on any point of the fin and in the media in the tube. The heat dissipating to the air can be determined as follows:

$$q_1 = hN \frac{\pi(D^2 - d^2)}{2}(t_1 - t_2) \tag{7.27}$$

where q_1 is the heat dissipating to the air, h is the convective coefficient between the fin and surrounding air, D is fin diameter, d is heat pipe diameter, N is the number of fins, t_1 is the temperature of media in the tube, and t_2 is the surrounding air temperature.

The convective coefficient between the fin and surrounding air can be assumed to be 10 W/m^2 K. Then the number of fins can be calculated from Eq. (7.27) so that the fins can ensure $q_1 = 1.2q$. More detail regarding the design and calibration of heat pipe heat flux meters is available, but not discussed in this book due to space limitations.

7.3.2 Measuring Local Heat Transfer Coefficient in CFB Furnaces

There are immersed tubes in the density area of bubble fluidization bed boilers (BFB boilers), so the heat transfer between the gas–solid flow and immersed tubes poses a problem directly related to the safety of the structure and its hydrodynamic cycle. Although it is very important, we will not discuss this topic here because BFB boilers are rarely applied in the industry currently. Interested readers may consult our references.

Refractory materials cover the wall of the density area in a circulating fluidization bed boiler (CFB boiler) without immersed tubes; most of the heating surfaces are located in the diluted area. Measuring the local heat transfer coefficient of heating surfaces in the diluted area is the basis of engineering design and verification for CFB boilers.

According to the two-phase flow heat transfer calculation introduced in chapter: Heat Transfer in Fluidized Beds, the appropriate method for measuring the local heat transfer coefficient of the heating surface in the diluted area is as follows.

In the CFB boiler furnace, the heat transfer coefficient on the flue gas side consists of two parts (with convection and radiation) expressed as follows:

$$h = h_c + h_r$$

where the convective heat transfer coefficient h_c consists of a gas convection component and a particle convection component:

$$h_c = h_{gc} + h_{pc}$$

The radiative heat transfer coefficient h_r also consists of two parts:

$$h_r = h_{gr} + h_{pr}$$

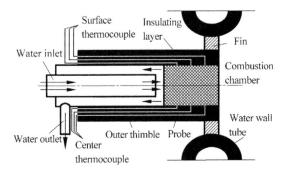

FIGURE 7.4   Schematic diagram of heat flux meter structure.

In practice, radiative/convective heat cannot be identified clearly due to its complexity, so a radiative heat flux meter is typically applied to measure these two values directly.

The heat transfer coefficient between the bed and wall can be calculated according to the conductive heat, bed temperature, and wall temperature measured by conductive heat flux meter. The heat flux meter depicted in Fig. 7.4 can be applied to measure the local heat transfer coefficient in the circulating fluidization bed. The probe in the heat flux meter is a carbon steel cylinder, in which there are three thermocouples melted along the axis to measure axial heat flux, and three thermocouples at a certain cross section along the outer radial to measure radial heat flux. In practice, the heat flux meter is inserted along the measuring hole to keep its end face aligning to the inner surface of the water wall fin and receive the heat transferred from the bed. The other end of the conductive unit in the heat flux meter is cooled by water. When the detector in the heat flux meter reaches heat balance, the temperature field along the axis is steady and axial heat flux q can be calculated. The detector is covered by insulating materials.

After measuring the temperatures of the detector end surface and bed, the heat exchange coefficient can be calculated as follows:

$$h = q/(t_b - t_s) \tag{7.28}$$

Assuming the cylinder is a one-dimensional conduction heat transferer, the temperature distribution along the axis of the heat flux meter is steady and the heat flux is:

$$q = \lambda \cdot \Delta t / \Delta l \tag{7.29}$$

where λ is the thermal conductivity of the probe, Δl is the space between the thermocouples along the axis, and Δt is the temperature difference.

The thermal conductivity of carbon steel changes with temperature, so the thermal conductivity of the probe should be calibrated in the laboratory before application.

Appendix A

Common Physical Constants of Heat Radiation

Light velocity in vacuum $\quad c_0 = 2.9979 \times 10^8 \, \text{m/s}$
Planck constant $\qquad\qquad\quad h = 6.6262 \times 10^{-34} \, \text{J s}$
Boltzmann constant $\qquad\quad k = 1.3806 \times 10^{-23} \, \text{J/K}$

First radiation constant of Planck spectrum radiation formula

$$c_1 = 0.59553 \times 10^8 \, \text{W} \cdot \mu\text{m}^4/\text{m}^2$$

$$c_1 = 0.59553 \times 10^{-16} \, \text{W} \cdot \text{m}^2$$

Second radiation constant of Planck spectrum radiation formula

$$c_2 = 14388 \, \mu\text{m} \cdot \text{K}$$

Wien displacement law constant $\quad c_3 = 2897.8 \, \mu\text{m} \cdot \text{K}$
Stefan–Boltzmann constant $\qquad\quad \sigma = 5.670 \times 10^{-8} \, \text{W}/(\text{m}^2 \cdot \text{K}^4)$

Theory and Calculation of Heat Transfer in Furnaces. http://dx.doi.org/10.1016/B978-0-12-800966-6.00014-4

Appendix A

Common Physical Constants of Heat Radiation

Appendix B

Common Configuration Factor Calculation Formulas [18,20]

The configuration factor from cell surface to cell surface is depicted below.

For two finite-length strips with parallel generatix (Fig. B1), the configuration factor from ds_1 to ds_2 φ_{ds_1,ds_2} is:

$$\varphi_{ds_1,ds_2} = \frac{1}{\pi} d(\sin\theta)\arctan\left(\frac{z}{l}\right) \tag{B1}$$

where θ is the included angle between the plane of the strip ds_1 and the line of the two strips, z is the length of the strips, and l is the distance between the two strips. For infinite-length strips ($z \to \infty$), this equation changes to:

$$d\varphi_{ds_1,ds_2}(\infty) = \frac{1}{2} d(\sin\varphi)$$

The following curve is given in the form of $d\varphi_{ds_1,ds_2}/d\varphi_{ds_1,ds_2}(\infty)$ (Fig. B2).

The configuration factor $\varphi_{dj,dk}$ between two cell surfaces dA_j and dA_k in the surface of a hollow sphere is depicted in Fig. B3 and expressed as follows:

$$\varphi_{dj,dk} = \frac{dA_k}{4\pi R^2} \tag{B2}$$

where R is the sphere radius.

The configuration factor φ_{d_1,d_2} between the cell rings of two parallel circular flanges dA_1 and dA_2 (Fig. B4) in a cylinder is:

$$d\varphi_{d_1,d_2} = -\frac{1}{2\pi}\frac{\partial I}{\partial \rho_2}d\rho_2 \tag{B3}$$

where:

$$I(\rho_1,\rho_2) = -\phi_m + \left[2(\rho_1^2 - \rho_2^2 + N_L^2)/K\right]$$
$$\times \arctan\left\{\left[K\tan\left(\frac{\phi_m}{2}\right)\right](\rho_1^2 + \rho_2^2 - 2\rho_1\rho_2 + N_L^2)\right\};$$

Theory and Calculation of Heat Transfer in Furnaces. http://dx.doi.org/10.1016/B978-0-12-800966-6.00015-6

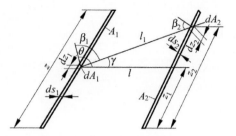

FIGURE B1   Two finite-length strips with parallel generatix.

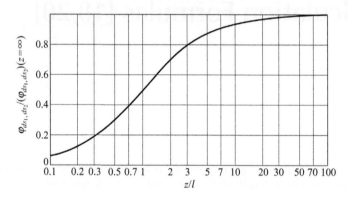

FIGURE B2   Configuration factor between two strips with parallel generatix.

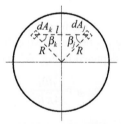

FIGURE B3   Two cell surfaces in the surface of a hollow sphere.

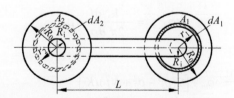

FIGURE B4   Cell rings in two parallel circular flanges in a cylinder.

$$K = \left[\left(\rho_1^2 + \rho_2^2 + N_L^2\right)^2 - 4\rho_1^2\rho_2^2\right]^{1/2};$$

$$\phi_m = \arccos\left(N_R/\rho_1\right) + \arccos\left(N_R/\rho_2\right);$$

$$\rho_1 = \frac{r_1}{R_0};$$

$$\rho_2 = \frac{r_2}{R_0};$$

$$N_L = \frac{L}{R_0};$$

$$N_R = \frac{R_i}{R_0};$$

where R_0 is the radius of the circular flange, R_i is the radius of the cylinder, L is the distance between the two flanges, r_1 is the radius of cell ring dA_1, and r_2 is the radius of cell ring dA_2.

The configuration factor $\phi_{dr,d\xi}$ from a cell ring surface at the top of a cylinder, concentric with the cylindrical surface to a cell ring surface in the cylindrical surface and parallel to the top of the cylinder (Fig. B5) is:

$$\phi_{dr,d\xi} = 2\xi R^2 \frac{r^2 - \xi^2 - R^2}{\left[\left(\xi^2 + r^2 + R^2\right)^2 - 4r^2R^2\right]^{3/2}} d\xi \qquad (B4)$$

where R is the cylinder's radius, r is the radius of the cell ring in the top surface, and ξ is the distance from the cell ring in the cylindrical surface to the top surface.

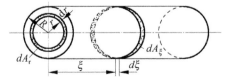

FIGURE B5 Configuration factor from a cell ring surface at a top concentric cylinder surface to a cell ring on the cylindrical surface parallel to the top of the cylinder.

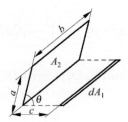

FIGURE B6 Configuration factor from cell strip dA_1 to the rectangle with the same length A_2.

The configuration factor $\varphi_{ds_1,2}$ from cell strip dA_1 to a rectangle with the same length A_2 (Fig. B6) is:

$$
\begin{aligned}
\phi_{ds_1,2} = \frac{1}{\pi}\Bigg\{ &\arctan\left(\frac{1}{L}\right) + L\frac{\sin^2\theta}{2}\ln\left[\frac{L^2\left(L^2 - 2NL\cos\theta + 1 + N^2\right)}{\left(1+L^2\right)\left(L^2 - 2NL\cos\theta + N^2\right)}\right] \\
&- L\sin\theta\cos\theta\left[\frac{\pi}{2} - \theta + \arctan\left(\frac{N - L\cos\theta}{L\sin\theta}\right)\right] \\
&+ \cos\theta\sqrt{1 + L^2\sin^2\theta}\left[\arctan\left(\frac{N - L\cos\theta}{\sqrt{1+L^2\sin^2\theta}}\right) + \arctan\left(\frac{L\cos\theta}{\sqrt{1+L^2\sin^2\theta}}\right)\right] \\
&+ \frac{N\cos\theta - L}{\sqrt{L^2 - 2NL\cos\theta + N^2}}\arctan\left(\frac{1}{\sqrt{L^2 - 2NL\cos\theta + N^2}}\right)\Bigg\}
\end{aligned}
\tag{B5}
$$

where $L = \dfrac{c}{b}$, $N = \dfrac{a}{b}$, a and b are the width and length of the rectangle, respectively, c is the distance from dA_1 to one edge of A_2, and θ is the angle between dA_1 and the rectangle.

The configuration factor $\phi_{d_1,2}$ from cell surface dA_1 to infinite plane A_2 (Fig. B7) is:

$$
\phi_{d_1,2} = \frac{1}{2}(1 + \cos\theta)
\tag{B6}
$$

FIGURE B7 Configuration factor from cell surface dA_1 to the infinite plane A_2.

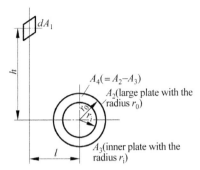

FIGURE B8   Configuration factor from cell surface dA_1 to the ring surface A_4 vertical to it.

where θ is the angle between cell surface dA_1 and plane A_2.

The configuration factor $\varphi_{d_1,4}$ from cell surface dA_1 to the ring surface A_4 (Fig. B8) vertical to it is:

$$\phi_{d_1,4} = \phi_{d_1,2} - \phi_{d_1,3}$$

$$= \frac{H}{2}\left\{\frac{H^2 + R_0^2 + 1}{\left[\left(H^2 + R_0^2 + 1\right) - 4R_0^2\right]^{1/2}} - \frac{H^2 + R_i^2 + 1}{\left[\left(H^2 + R_i^2 + 1\right)^2 - 4R_i^2\right]^{1/2}}\right\} \quad (B7)$$

where $H = \dfrac{h}{l}, R_0 = \dfrac{r_0}{l}, R_i = \dfrac{r_i}{l}$; r_0 is the radius of the large ring plate, r_i is the radius of the small ring plate, h is the vertical distance from the cell surface to the ring, and l is the transverse distance from the cell surface to the ring center.

The following is the configuration factor $\phi_{d_1,2}$ from cell surface dA_1 to the rectangle A_2 parallel to dA_1 (Fig. B9) and the normal line of dA_1 passing through an angular of rectangle A_2:

$$\phi_{d_1,2} = \frac{1}{2\pi}\left[\frac{X}{\sqrt{1+X^2}}\arctan\frac{Y}{\sqrt{1+X^2}} + \frac{Y}{\sqrt{1+Y^2}}\arctan\frac{X}{\sqrt{1+Y^2}}\right] \quad (B8)$$

where $X = \dfrac{a}{c}, \; Y = \dfrac{b}{c}$, a and b are the edge lengths of rectangle A_2, and c is the distance from dA_1 to A_2.

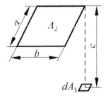

FIGURE B9   Configuration factor from cell surface dA_1 to the rectangle A_2 parallel to dA_1.

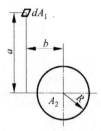

FIGURE B10   Configuration factor from cell surface dA_1 to the finite disc A_2 parallel to dA_1.

The configuration factor $\phi_{d_1,2}$ from a cell surface dA_1 to a finite disc A_2 parallel to dA_1 (Fig. B10) is:

$$\phi_{d_1,2} = \frac{1}{2} - \frac{1+C^2-B^2}{\sqrt{C^4+2C^2(1-B^2)+(1+B^2)^2}} \tag{B9}$$

where $B = \dfrac{R}{a}, C = \dfrac{b}{a}$, R is the radius of the disc, a is the vertical distance from dA_1 to A_2, and b is the transverse distance from dA_1 to A_2.

The configuration factor $\phi_{d_1,2}$ from cell surface dA_1 to rectangle A_2 (Fig. B11), where the angle between dA_1 and A_2 is θ, 0 degree$<\theta<$180 degree, is:

$$\phi_{d_1,2} = \frac{1}{2\pi}\left\{ \arctan\left(\frac{l}{L}\right) + V(N\cos\theta - L)\arctan V \right.$$

$$\left. + \frac{\cos\theta}{W}\left[\arctan\left(\frac{N-L\cos\theta}{W}\right) + \arctan\left(\frac{L\cos\theta}{W}\right) \right] \right\} \tag{B10}$$

where $V = \dfrac{1}{\sqrt{N^2+L^2-2NL\cos\theta}}$, $W = \sqrt{1+L^2\sin^2\theta}$, $L = \dfrac{c}{b}$, $N = \dfrac{a}{b}$, a and b are the lengths of two edges of rectangle A_2, c is the distance from dA_1 to any edge of A_2, and θ is the angle between dA_1 and A_2.

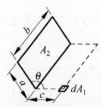

FIGURE B11   Configuration from cell surface dA_1 to the rectangle A_2.

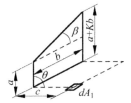

FIGURE B12   Configuration factor from cell surface dA_1 to the surface A_2 formed by the rectangle and right triangle A_2.

The configuration factor $\phi_{d_1,2}$ from cell surface dA_1 to the surface A_2 being formed by a rectangle and right triangle (Fig. B12), where dA_1 is near the short edge of A_2, is as follows:

$$\phi_{d_1,2} = \frac{1}{2\pi}\left(\arctan\left(\frac{1}{L}\right) + \frac{N\cos\beta - L}{V}\left\{\arctan\left[\frac{(K^2+1) + K(N - L\cos\beta)}{V}\right]\right.\right.$$
$$\left.\left. - \arctan\left[\frac{K(N - L\cos\beta)}{V}\right]\right\} + \frac{\cos\beta}{W}\left[\arctan\left(\frac{L\cos\beta}{W}\right)\right.\right.$$
$$\left.\left. + \arctan\left(\frac{N - L\cos\beta + K}{W}\right)\right]\right) \tag{B11}$$

where V, W, L, and N are the same symbols labeled earlier, and $K = \tan\beta$.

For two infinite-length planes with the same width A_1 and A_2 (Fig. B13), the configuration factor from A_1 to A_2 can be calculated as follows:

$$\varphi_{12} = \sqrt{1 + H^2} - H \tag{B12}$$

where $H = h/d$.

For two infinite-length planes with different widths A_1 and A_2 (Fig. B14), the bisector of plane A_1 faces the bisector of plane A_2 directly, so the configuration factor from A_1 to A_2 is calculated as follows:

$$\varphi_{12} = \frac{1}{2D_1}\left[\sqrt{4 + (D_1 + D_2)^2} - \sqrt{4 + (D_2 - D_1)^2}\right] \tag{B13}$$

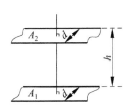

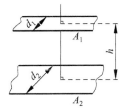

FIGURE B13   Configuration factor between two infinite-length planes with the same width A_1 and A_2.

FIGURE B14   Two infinite-length parallel planes A_1 and A_2 with different widths.

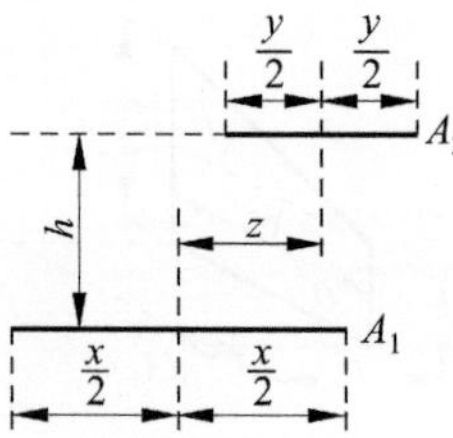

FIGURE B15   **Two infinite-length parallel planes with different widths.**

$$\varphi_{21} = \frac{1}{2D_2}\left[\sqrt{4+(D_1+D_2)^2} - \sqrt{4+(D_1-D_2)^2}\right] \tag{B14}$$

where $D_1 = \dfrac{d_1}{h}$ and $D_2 = \dfrac{d_2}{h}$.

The configuration factor between two infinite-length parallel planes A_1 and A_2 with different widths (Fig. B15) is:

$$\varphi_{12} = \frac{Y}{X}\varphi_{21} = \frac{1}{2X}\left[\sqrt{1+\left(\frac{Y+X-2Z}{2}\right)^2} + \sqrt{1+\left(\frac{Y+X+2Z}{2}\right)^2}\right.$$

$$\left. -\sqrt{1+\left(\frac{X-Y-2Z}{2}\right)^2} - \sqrt{1+\left(\frac{X-Y+2Z}{2}\right)^2}\right] \tag{B15}$$

where $X = \dfrac{x}{h}$, $Y = \dfrac{y}{h}$, and $Z = \dfrac{z}{h}$.

The configuration factor between two parallel, identical rectangles A_1 and A_2 (Fig. B16) is:

$$\phi_{12} = \frac{2}{\pi XY}\left\{\ln\left[\frac{(1+X^2)(1+Y^2)}{1+X^2+Y^2}\right]^{1/2} + X\sqrt{1+Y^2}\,\arctan\frac{X}{\sqrt{1+Y^2}} + \right.$$

$$\left. Y\sqrt{1+X^2}\,\arctan\frac{Y}{\sqrt{1+X^2}} - X\arctan X - Y\arctan Y\right\} \tag{B16}$$

where $X = \dfrac{a}{h}$ and $Y = \dfrac{b}{h}$.

Consider two parallel rectangles (Fig. B17), where a rectangle A_1 faces a part of another rectangle $A_{2,4}$. Calculate the configuration factor from A_1 to $A_{2,4}$ as follows:

$$\varphi_{1(2,4)} = \frac{1}{2A_1}\left[A_{(1,3)}\varphi_{(1,3)(2,4)} + A_1\varphi_{12} - A_3\varphi_{34}\right] \tag{B17}$$

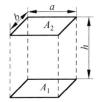

FIGURE B16   Two identical parallel rectangles.

FIGURE B17   Two parallel rectangles.

The configuration factor between two rectangles with arbitrary length and parallel edges A_1 and A_2 in two parallel planes (Fig. B18) is:

$$\varphi_{12} = \frac{1}{\pi A_1}[f(\beta - b, \delta - c) - f(\beta - a, \delta - c)$$

$$+ f(a - a, \delta - c) - f(a - b, \delta - c) + f(a - b, \gamma - c)$$

$$- f(a - a, \gamma - c) + f(\beta - a, \gamma - c) - f(\beta - b, \gamma - c)$$

$$+ f(\beta - b, \gamma - d) - f(\beta - a, \gamma - d) + f(a - a, \gamma - d)$$

$$- f(a - b, \gamma - d) + f(a - b, \delta - d) - f(a - a, \delta - d)$$

$$+ f(\beta - a, \delta - d) - f(\beta - b, \delta - d) \tag{B18}$$

where function f is defined by the following equation:

$$f(v, \xi) = \frac{1}{2}\left(hv \arctan\frac{v}{h} - \xi\sqrt{h^2 + v} t \arctan\frac{\xi}{\sqrt{h^2 + v^2}}\right.$$

$$\left. -v\sqrt{h^2 + \xi^2}\arctan\frac{v}{\sqrt{h^2 + \xi^2}} + \frac{h^2}{2}\ln\frac{h^2 + v^2 + \xi^2}{h^2 + \xi^2}\right) \tag{B18a}$$

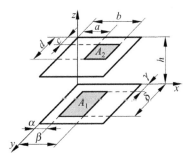

FIGURE B18   Two rectangles with arbitrary length and parallel edges within two parallel planes.

FIGURE B19 Two infinite-length planes with one common edge, equal width, and the included angle θ.

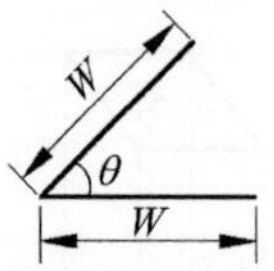

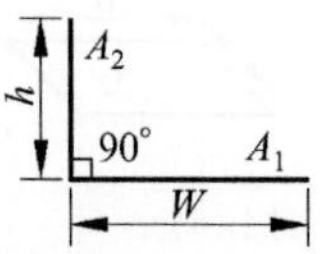

FIGURE B20 Two mutually perpendicular infinite-length planes with one common edge and different widths.

The configuration factor between two infinite-length planes with one common edge, equal width, and the included angle θ (Fig. B19) is:

$$\phi_{12} = \phi_{21} = 1 - \sin\frac{\theta}{2} \tag{B19}$$

The configuration factor between two mutually perpendicular, infinite-length planes with one common edge and different widths (Fig. B20) is:

$$\phi_{12} = H\varphi_{21} = \frac{1}{2}\left[1 + H - \sqrt{1 + H^2}\right] \tag{B20}$$

where $H = \dfrac{h}{W}$.

The configuration factor between two mutually perpendicular, infinite-length planes with one common edge, different widths, and the included angle θ (Fig. B21) is:

$$\varphi_{12} = \frac{l_1 + l_2 - l_3}{2l_1} \tag{B21}$$

where l_1 and l_2 are the widths of the two infinite-length planes, and l_3 is the length between the external ends of A_1 and A_2 in the plane perpendicular to the common edge.

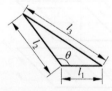

FIGURE B21 Two mutually perpendicular, infinite-length planes with one common edge, different widths, and the included angle θ.

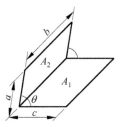

FIGURE B22   Two rectangles A_1 and A_2 with one common edge and the included angle θ.

The configuration factor between two rectangles A_1 and A_2 with one common edge and the included angle θ (Fig. B22) is:

$$
\phi_{12} = \frac{1}{\pi L}\left(-\frac{\sin 2\theta}{4}\left[NL\sin\theta + \left(\frac{\pi}{2}-\theta\right)(N^2+L^2) + L^2\arctan\left(\frac{N-L\cos\theta}{L\sin\theta}\right) \right.\right.
$$

$$
\left. +N^2\arctan\left(\frac{L-N\cos\theta}{N\sin\theta}\right)\right] + \frac{\sin^2\theta}{4}\ln\left\{\left[\frac{(1+N^2)(1+L^2)}{1+N^2+L^2-2NL\cos\theta}\right]^{\cos 2\theta + \cot 2\theta}\right.
$$

$$
\left.\times\left[\frac{L^2(1+N^2+L^2-2NL\cos\theta)}{(1+L^2)(N^2+L^2-2NL\cos\theta)}\right]^{L^2}\right\}
$$

$$
+\frac{N^2\sin^2\theta}{4}\ln\left[\left(\frac{N^2}{N^2+L^2-2NL\cos\theta}\right)\left(\frac{1+N^2}{1+N^2+L^2-2NL\cos\theta}\right)^{\cos 2\theta}\right]
$$

$$
+L\arctan\left(\frac{1}{L}\right)+N\arctan\left(\frac{1}{N}\right)
$$

$$
-\sqrt{N^2+L^2-2NL\cos\theta}\,\arctan\sqrt{N^2+L^2-2NL\cos\theta}
$$

$$
+\frac{N\sin\theta\sin 2\theta}{2}\sqrt{1+N^2\sin^2\theta}\left[\begin{array}{l}\arctan\left(\dfrac{N\cos\theta}{\sqrt{1+N^2\sin^2\theta}}\right)\\[2mm] +\arctan\left(\dfrac{L-N\cos\theta}{\sqrt{1+N^2\sin^2\theta}}\right)\end{array}\right]
$$

$$
+\cos\theta\int_0^L \sqrt{1+z^2\sin^2\theta}\left[tg^{-1}\left(\frac{N-z\cos\theta}{\sqrt{1+z^2\sin^2\theta}}\right)\right.
$$

$$
\left.+\arctan\left(\frac{z\cos\theta}{\sqrt{1+z^2\sin^2\theta}}\right)\right]dz \right) \tag{B22}
$$

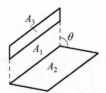

FIGURE B23 Two rectangles A_2 and A_3 in two planes with the included angle θ.

where $N = \dfrac{a}{b}$ and $L = \dfrac{c}{b}$.

Consider two rectangles A_2 and A_3 in two planes with the included angle θ (Fig. B23) where one group of edges is parallel to the arris, while one edge of A_2 is the arris itself. The other pairs of edges of A_2 and A_3 are in two planes perpendicular to the arris. See the following:

$$A_3\varphi_{32} = A_2\varphi_{23} = A_{(1,3)}\varphi_{(1,3)2} - A_1\varphi_{12} \tag{B23}$$

where $\varphi_{(1,3)2}$, φ_{12} are calculated from Eq. (B22).

Appendix C

Example of Thermal Calculation of 113.89 kg/s (410 t/h) Ultra-High-Pressure, Coal-Fired Boiler [17,21]

C1 DESIGN REQUIREMENT

1. Boiler rated capacity D_1 = 113.89 kg/s (410 t/h).
2. Steam conditions:
   Superheated steam outlet pressure p_{ss} = 13.7 MPa.
   Superheated steam outlet temperature t_{ss} = 540°C.
   Steam pressure in drum t_s = 15.07 MPa.
3. Feed water temperature t_{fw} = 235°C.
4. Feed water pressure p_{fw} = 15.6 MPa.
5. Blowdown percentage δ_{bd} = 1%.
6. Exhaust gas temperature θ_{ex} = 135°C.
7. Hot air temperature t_{ha} = 320°C.
8. Ambient (cold) air temperature t_{ca} = 20°C.

C2 FUEL CHARACTERISTICS

1. Coal ultimate and proximate analysis on as-received basis:
   Carbon C_{ar} = 70.8%.
   Hydrogen H_{ar} = 4.5%.
   Oxygen O_{ar} = 7.13%.
   Nitrogen N_{ar} = 0.72%.
   Sulfur S_{ar} = 2.21%.
   Ash A_{ar} = 11.67%.
   Moisture M_{ar} = 2.97%.
2. Volatile matter on dry ash-free basis V_{daf} = 24.96%.

Theory and Calculation of Heat Transfer in Furnaces. http://dx.doi.org/10.1016/B978-0-12-800966-6.00016-8

3. Ash fusion point properties:
   $t_{DT} > 1500°C$.
   $t_{ST} > 1500°C$.
   $t_{FT} > 1500°C$.
4. Lower heating value on as-received basis $Q_{ar,net,p} = 27797$ kJ/kg.

C3 BASIC BOILER STRUCTURE

The boiler is characteristic of a single-drum, natural circulation Π shape, and dry bottom furnace. The furnace is in the front of the boiler, surrounded by a membrane wall. A platen superheater is positioned in the furnace exit, a two-stage convection superheater is located in the horizontal flue gas passage, and a two-stage economizer and two-stage air preheater are arranged in the back-end shaft. The horizontal flue gas passage and reversing chamber are covered by a membrane wall. The system is illustrated in Fig. C1.

The furnace section is square, surrounded by a membrane wall. The lower tubes of the front and rear membrane wall are bent 50 degrees to the horizontal line to form the dry bottom hopper. The rear membrane wall bends at the furnace exit to form the furnace nose with a 35 degree up-dip angle and 30 degree elevation angle. The membrane wall then separates into two parts: one rises vertically upward through the horizontal flue gas passage and enters the upper heater at the rear membrane wall, and the other goes upward with the same up-dip angle to form slope wall tubes at the bottom of the horizontal flue gas passage, then enters the upper header of the enclosure wall tubes. This system is depicted in Fig. C2.

The boiler employs a radiation superheater, convection superheater, cross connected piping, and a two-stage attemperator. The superheater includes roof

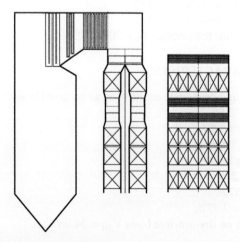

FIGURE C1 Diagram of boiler structure.

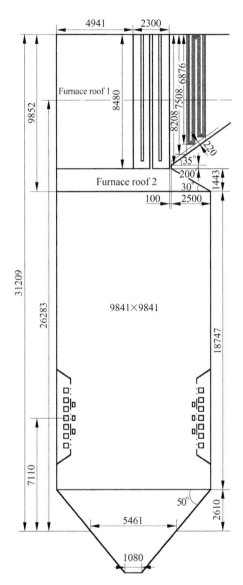

FIGURE C2   **Size of furnace structure.**

tubes, wall tubes, a platen superheater, and a two-stage convection superheater. Saturated vapor leaving the drum is introduced into the inlet header of the roof tubes through connecting piping, then enters the middle header of the roof tubes at the horizontal flue gas passage outlet through the roof tubes, where the saturated vapor is then divided into two paralleled routes.

One route is formed by 98 $\phi51\times5.5$ tubes with 100 mm spacing, moving first backward and then downward, and then entering the lower header of the rear enclosure wall tubes to serve as the roof tubes and rear enclosure wall tubes of the reversing chamber. The steam then enters the lower header of the rear side enclosure wall tubes through the wrought square bent at both sides of the header, then flows through the two rear side walls of the tube-covered reversing chamber, and finally arrives at the upper header of the side wall tubes.

Through another route, the steam is introduced to the lower header of the front enclosure wall tubes through connecting pipes, then introduced into the lower header of the front side enclosure wall tubes through branch connecting pipes. The steam then enters the upper header of the side wall tubes through the side wall horizontal flue gas passage, then enters the inlet header of the low-temperature superheater from the upper header of the side wall tubes through connecting pipes. The steam then enters the outlet header of the low-temperature superheater, countercurrent flowing along the flue gas passage. The steam then comes out of the header and is cooled by the first-stage attemperator. Then the steam is introduced into the 14-part platen superheater through the mixing heater and connecting pipes. The horizontal transverse distance of the panels is 700 mm, and the steam flows downstream in the panels. After leaving the panels, the steam enters the mixing header and two high-temperature inlet headers after which the steam separately countercurrent passes the left and right quarter areas of the flue gas passage, to form the cooler stage of the high-temperature convection superheater. The steam then enters the respective two second stage desuperheaters; after cross-connection the steam concurrently pass through the remaining 1/2 area in the middle of the flue gas passage to form the hotter stage of the high-temperature superheater. Finally, the steam enters the outlet header of the high-temperature superheater and is introduced into the steam header via connecting pipes, and the superheated steam enters the turbine from either side of the vapor header. The superheater system is illustrated in Fig. C3.

The economizers are located in the back-end shaft by upper and lower stages. The working medium flows countercurrent from top to bottom. The upper economizer is divided into left and right parts, as is the lower economizer. To reduce any abrasion of the heating surfaces by ash in the flue gas, guard tiles are attached to the upper two rows and side two rows of the heating surfaces of the upper and lower economizers.

The air preheater is a vertical, tubular structure that is arranged in two stages. The upper stage has one pass and the lower stage has three passes. Considering the level of corrosion that occurs at low temperatures, the lowest pass possible is designed for the tube box to minimize maintenance (or replacement). The second and third passes are in one tube box but partitioned by a middle tube sheet. Due to structure and system requirements, the flue gas is divided into four parts (two parts for the upper stage) in the horizontal section when the air flows countercurrent with flue gas from the front/rear wall of the lower stage to the front/rear walls of the upper stage.

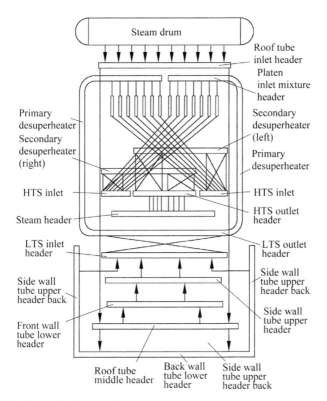

FIGURE C3   Systematic diagram of superheater.

C4   AUXILIARY CALCULATION

Thermal calculation can be performed using a computer in the following stepwise process. Due to the impact of the significant figures (rounded), computed results may have slight differences from those obtained by hand calculation.

1. Calculate the volume of combustion products.

    The theoretical air volumes and combustion product volumes from the complete combustion of coal are listed in Table C1.

2. Construct an air balance and enthalpy–temperature table.

    The excess air coefficient in the flue gas passage, air leakage factor of the heating surfaces, and combustion products volume of different excess air coefficients are shown in Table C2, the enthalpy–temperature table of different excess air coefficients is shown in Table C3, and boiler thermal balance and fuel consumption calculation processes are shown in Table C4.

TABLE C1 Coal-Combustion Product Volumes and Theoretical Air Volumes

No.	Parameter	Symbol	Unit	Formulae and data source	Result
1	Theoretical volume of air	V^0	Nm³/kg	$0.0889(C_{ar} + 0.375S_{ar}) + 0.265H_{ar} - 0.0333O_{ar}$	7.3229
2	Theoretical volume of N_2	$V^0_{N_2}$	Nm³/kg	$0.79V^0 + 0.8N_{ar}/100$	5.7908
3	Theoretical volume of water vapor	$V^0_{H_2O}$	Nm³/kg	$0.111H_{ar} + 0.0124M_{ar} + 0.0161V^0$	0.6542
4	Theoretical volume of CO_2	$V^0_{RO_2}$	Nm³/kg	$1.866(C_{ar} + 0.375S_{ar})/100$	1.3366
5	Theoretical flue gas volume	V^0_g	Nm³/kg	$V^0_{RO_2} + V^0_{N_2} + V^0_{H_2O}$	7.7816
6	Fly ash fraction	a_{fa}	–	Refer to Appendix D6.1	0.95
7	Fly ash concentration	μ_{fa}	kg/kg	$A_{ar} \times a_{fa}/(100G_{fa})$	0.1109
8	Converted ash content	$A_{ar,c}$	g/MJ	$10000A_{ar}/Q_{ar,net,p}$	4.1983

C5   COMBUSTION CHAMBER AND HEAT TRANSFER CALCULATION

The structure design of the combustion chamber (furnace) and platen super-heater are shown in Table C5.

The structure/size of the furnace is shown in Table C6.

The thermal calculation of the furnace according to the above structure/size is shown in Table C7.

Refer to the basic furnace design procedure.

The following describes the basic furnace design procedure step-by-step.

1. Determine furnace volume heat release rate and sectional thermal load, and calculate furnace volume and sectional area via equations.
2. Determine the width and depth of the furnace. When burners are corner-arranged, the aspect ratio should not be larger than 1.2.
3. Determine the structure/size of the furnace's dry bottom hopper and top. The inclination of the dry bottom hopper should not be less than 50 degrees to ensure ash falls smoothly. The exit window height of the furnace exit is determined by outlet gas velocity, which is approximately 6 m/s. The length

TABLE C2 Flue Gas Characteristics

Parameter	Symbol	Unit	Formulae and data source	Platen super-heater	High-temper-ature super-heater	Low-temper-ature super-heater	Reversing chamber	Upper stage econo-mizer	Upper stage air preheater	Lower stage econo-mizer	Lower stage air preheater
Excess air coefficient of flue inlet	α'	–		1.20	1.20	1.23	1.26	1.26	1.28	1.31	1.33
Excess air coefficient of flue outlet	α''	–			1.23	1.26	1.26	1.28	1.31	1.33	1.36
Average excess air coefficient of flue	α_{ave}	–	$(\alpha' + \alpha'')/2$	1.2	1.2150	1.2450	1.26	1.27	1.295	1.32	1.345
Excess air volume	ΔV	Nm³/kg	$(\alpha_{ave}-1)V^0$	1.465	1.574	1.794	1.904	1.977	2.160	2.343	2.526
Water vapor volume	V_{H_2O}	Nm³/kg	$V^0_{H_2O} + 0.0161(\alpha_{ave}-1)V^0$	0.678	0.680	0.683	0.685	0.686	0.689	0.692	0.695
Flue gas total volume	V_g	Nm³/kg	$V^0_g + (\alpha_{ave}-1)V^0 + 0.0161(\alpha_{ave}-1)V^0$	9.27	9.38	9.60	9.72	9.79	9.98	10.16	10.35
Total flue gas volume at exit	V_g''	Nm³/kg	$V^0_g + (\alpha''-1)V^0 + 0.0161(\alpha''-1)V^0$	9.27	9.49	9.72	9.72	9.87	10.09	10.24	10.46

(Continued)

TABLE C2 Flue Gas Characteristics (cont.)

Parameter	Symbol	Unit	Formulae and data source	Platen superheater	High-temperature superheater	Low-temperature superheater	Reversing chamber	Upper stage economizer	Upper stage air preheater	Lower stage economizer	Lower stage air preheater
Volume fraction of RO_2	r_{RO_2}	–	$V_{RO_2}^0/V_g$	0.144	0.142	0.139	0.138	0.137	0.134	0.132	0.129
Volume fraction of water vapor	r_{H_2O}	–	V_{H_2O}/V_g	0.0731	0.0724	0.0711	0.0705	0.0701	0.0691	0.0681	0.0671
Volume fraction of triatomic gas	r_n	–	$(V_{H_2O} + V_{CO_2}^0 + V_{SO_2}^0)/V_g$	0.217	0.215	0.210	0.208	0.207	0.203	0.200	0.196
Flue gas mass of 1 kg fuel	G_g	kg/kg	$1 - A_{ar}/100 + 1.306\alpha_{ave}V^0$	12.36	12.50	12.79	12.93	13.03	13.27	13.51	13.75
Flue gas density	ρ_g	kg/Nm^3	G_g/V_g	1.333	1.333	1.332	1.331	1.331	1.330	1.329	1.328
Dimensionless concentration of fly ash	μ_{fa}	kg/kg	$\mu_{fa} = A_{ar} \times a_{fa}/(100G_g)$	0.00897	0.00887	0.00867	0.00857	0.00851	0.00836	0.00821	0.00807

Note: the excess air coefficient of the platen superheater exit was selected from Appendix D6.2, and the excess air coefficients of other heating surface outlets were selected from Appendix D6.3.

TABLE C3 Flue Gas/Air Enthalpy–Temperature Table

Temperature θ (°C)	$V_{RO_2} = 1.3366$ (m³/kg)		$V_{N_2} = 5.7908$ (m³/kg)		$V_{H_2O} = 0.6542$ (m³/kg)		$G_{fa} = 0.1109$ (kg/kg)		$I_g^0 = I_{RO_2} + I_{N_2} + I_{H_2O} + I_{fa}$ (kJ/kg)	$V_a^0 = 7.3229$ (m³/kg)		$I_g = I_g^0 + I_{fa} + (\alpha'' - 1)I_a^0$							
	$C_{CO_2}\theta$ (kJ/m³)	$I_{RO_2} = V_{RO_2}C_{CO_2}\theta$ (kJ/kg)	$C_{N_2}\theta$ (kJ/m³)	$I_{N_2} = V_{N_2}C_{N_2}\theta$ (kJ/kg)	$C_{H_2O}\theta$ (kJ/m³)	$I_{H_2O} = V_{H_2O}C_{H_2O}\theta$ (kJ/kg)	$C_{fa}\theta$ (kJ/m³)	$I_{fa} = C_{fa}\theta G_{fa}$ (kJ/kg)		$C_a t$ (kJ/m³)	$I_a^0 = V_a^0 C_a t$ (kJ/kg)	α''_{psh}	α''_{pcsh}	α''_{scsh}	α''_{rc}	α''_{seco}	α''_{spah}	α''_{peco}	α''_{ppah}
												1.2	1.23	1.26	1.26	1.28	1.31	1.33	1.36
100	169.7	226.8	129.6	750.5	150.5	98.5	80.7	8.9	1075.8	132.0	966.6	1269.1	1298.1	1327.1	1327.1	1346.4	1375.4	1394.8	1423.8
200	357.0	477.2	259.6	1503.3	303.9	198.8	168.9	18.7	2179.3	265.9	1947.2	2568.7	2627.1	2685.5	2685.5	2724.5	2782.9	2821.8	2880.3
300	558.0	745.8	391.3	2265.9	461.9	302.2	263.3	29.2	3314.0	402.1	2944.5	3902.9	3991.2	4079.5	4079.5	4167.6	4226.8	4285.6	4374.0
400	770.8	1030.2	525.8	3044.8	625.3	409.1	359.5	39.9	4484.1	540.9	3960.9	5276.3	5395.2	5514.0	5514.0	5633.1	5712.0	5791.3	5910.1
500	994.8	1329.6	663.0	3839.3	793.4	519.1	457.7	50.7	5688.0	683.0	5001.5	6688.3	6838.4	6988.4	6988.4	7088.4	7238.5	7338.5	7488.6
600	1220.6	1631.4	802.6	4647.7	965.6	631.7	559.3	62.0	6910.9	828.5	6067.0	8124.3	8306.3	8488.3	8488.3	8609.6	8791.7	8913.0	9095.0
700	1458.8	1949.8	944.7	5470.6	1145.3	749.3	661.3	73.3	8169.7	978.1	7162.5	9602.2	9817.1	10031.9	10031.9	10175.2	10390.1	10533.3	10748.2
800	1701.3	2273.9	1091.0	6317.8	1333.4	872.3	765.8	84.9	9464.1	1128.6	8264.6	11117.0	11364.9	11612.9	11612.9	11778.2	12026.1	12191.4	12439.3
900	1947.9	2603.5	1241.5	7189.3	1521.5	995.4	873.6	96.9	10788.3	1279.1	9366.7	12661.6	12942.6	13223.6	13223.6	13410.9	13691.9	13879.3	14160.3
1000	2198.7	2938.7	1391.9	8060.2	1722.2	1126.7	982.3	108.9	12125.7	1433.7	10498.8	14225.7	14540.4	14855.4	14855.4	15065.4	15380.3	15590.3	15905.3
1100	2453.7	3279.6	1542.4	8931.8	1922.8	1257.9	1095.2	121.4	13469.3	1592.6	11662.4	15801.8	16151.7	16501.5	16501.5	16734.8	17084.7	17317.9	17667.8
1200	2712.8	3625.0	1692.9	9803.3	2127.6	1391.9	1203.8	133.5	14821.1	1751.4	12825.3	17386.2	17770.9	18155.7	18155.7	18412.2	18797.0	19053.5	19438.2
1300	2972.0	3972.4	1847.6	10699.1	2340.8	1531.4	1358.5	150.6	16202.9	1910.3	13988.9	19000.7	19420.3	19840.0	19840.0	20119.8	20539.4	20819.2	21238.9
1400	3235.3	4324.3	2006.4	11618.7	2554.0	1670.9	1580.0	175.2	17613.9	2073.3	15182.5	20650.4	21105.9	21561.3	21561.3	21865.0	22320.5	22624.1	23079.6
1500	3498.7	4676.3	2161.1	12514.6	2775.5	1815.8	1755.6	194.6	19006.7	2236.3	16376.1	22281.9	22773.2	23264.5	23264.5	23592.0	24083.3	24410.8	24902.1

(Continued)

TABLE C3 Flue Gas/Air Enthalpy–Temperature Table (cont.)

Temperature θ (°C)	$V_{RO_2}=1.3366$ (m³/kg)		$V_{N_2}=5.7908$ (m³/kg)		$V_{H_2O}=0.6542$ (m³/kg)		$G_{fa}=0.1109$ (kg/kg)		$I_g^0=I_{RO_2}+I_{N_2}+I_{H_2O}+I_{fa}$ (kJ/kg)	$V_a^0=7.3229$ (m³/kg)		$I_g=I_g^0+I_{fa}+(\alpha''-1)I_a^0$							
	$C_{CO_2}\theta$ (kJ/m³)	$I_{RO_2}=V_{RO_2}C_{CO_2}\theta$ (kJ/kg)	$C_{N_2}\theta$ (kJ/m³)	$I_{N_2}=V_{N_2}C_{N_2}\theta$ (kJ/kg)	$C_{H_2O}\theta$ (kJ/m³)	$I_{H_2O}=V_{H_2O}C_{H_2O}\theta$ (kJ/kg)	$C_{fa}\theta$ (kJ/m³)	$I_{fa}=C_{fa}\theta G_{fa}$ (kJ/kg)		$C_a t$ (kJ/m³)	$I_a^0=V_a^0C_a t$ (kJ/kg)	α''_{psh} 1.2	α''_{pcsh} 1.23	α''_{scsh} 1.26	α''_{rc} 1.26	α''_{seco} 1.28	α''_{spah} 1.31	α''_{peco} 1.33	α''_{ppah} 1.36
1600	3762.0	5028.3	2319.9	13434.1	2997.1	1960.8	1872.6	207.6	20423.2	2399.3	17569.8	23937.1	24464.2	24991.3	24991.3	25342.7	25869.8	26221.2	26748.3
1700	4029.5	5385.8	2478.7	14353.7	3222.8	2108.4	2060.7	228.5	21848.0	2562.3	18763.4	25000.6	26163.5	26726.4	26726.4	27101.7	27664.6	28039.9	28602.8
1800	4297.0	5743.3	2637.6	15273.9	3452.7	2258.8	2182.0	241.9	23276.1	2725.4	19957.7	27267.6	27866.3	28465.1	28465.1	28864.2	29463.0	29862.1	30460.9
1900	4564.6	6101.0	2800.6	16217.8	3682.6	2409.3	2382.6	264.1	24728.0	2892.6	21182.1	28964.5	29599.9	30235.4	30235.4	30659.0	31294.5	31718.1	32353.6
2000	4836.3	6464.2	2959.4	17137.4	3920.8	2565.1	2508.0	278.0	26166.6	3059.8	22406.5	30647.9	31320.1	31992.3	31992.3	32440.4	33112.6	33560.8	34233.0
2100	5108.0	6827.3	3122.5	18081.9	4154.9	2718.2			27627.4	3227.0	23630.9	32353.6	33062.5	33771.4	33771.4	34244.1	34953.0	35425.6	36134.5
2200	5379.7	7190.5	3285.5	19025.8	4393.2	2874.1			29090.4	3394.2	24855.3	34061.4	34807.1	35552.7	35552.7	36049.8	36795.5	37292.6	38038.3

Note: Because $1000 \times A_{ar} \times a_{fa}/Q_{ar,net,p} = 1000 \times 11.67 \times 0.95/27797 = 0.399 < 1.43^*$, the enthalpy of fly ash is negligible during flue gas enthalpy calculation. Refer to Appendix D6.4. Flue gas enthalpy is calculated by linear interpolation when this table is used.

TABLE C4 Thermal Balance and Fuel Consumption

No.	Parameter	Symbol	Unit	Formulae and data source	Result
1	Heat input by fuel	Q_{in}	kJ/kg	*Similar to lower heating value	27797
2	Exhaust gas temperature	θ_{ex}	°C	Design specification	135
3	Exhaust gas enthalpy	I_{ex}	kJ/kg	See Table C3	1933.5
4	Cold air temperature	t_{ca}	°C	Design specification	20
5	Cold air enthalpy	I_{ca}	kJ/kg	See Table C3	193.3
6	Heat loss due to unburnt carbon	q_{uc}	%	Refer to Appendix D6.1	1
7	Heat loss due to gas incomplete combustion	q_{ug}	%	Refer to Appendix D6.1	0
8	Heat loss due to exhaust gas	q_{ex}	%	$(I_{ex}-\alpha_{ex}I_{ca})(1-q_{uc}/100)100/Q_{ar,net,p}$	5.95
9	Heat loss due to furnace wall radiation and convection	q_{rad}	%	Refer to Appendix D6.5	0.40
10	Physical heat loss of ash	q_{ph}	%	Refer to Appendix D6.6	0
11	Heat preservation coefficient	φ	%	$1-q_{rad}/100$	0.996
12	Sum of heat loss	Σq	%	$q_{ex} + q_{ug} + q_{uc} + q_{ph} + q_{rad}$	7.35
13	Thermal efficiency by energy balance method	η_b	%	$100-\Sigma q$	92.65
14	Superheated steam enthalpy	i''_{ss}	kJ/kg	See Standard Steam Table	3437.5
15	Feed waster enthalpy	i_{fw}	kJ/kg	See Standard Steam Table	1016.1
16	Saturated temperature in drum	t_s	°C	See Standard Steam Table	342.53
17	Saturated steam enthalpy in drum (x = 1)	i''_s	kJ/kg	See Standard Steam Table	2608.9
18	Saturated water enthalpy in drum (x = 0)	i'_s	kJ/kg	See Standard Steam Table	1612.9
19	Blowdown percentage	δ_{bd}	%	Determine from design	1
20	Flow rate of superheated steam	D_{ss}	kg/h	Determine from design	410000
21	Heat absorbed by steam	Q_b	kJ/s	$[D_{ss}(i''_{ss}-i_{fw}) + D_{ss}\delta_{bd}(i_s'-i_{fw})]/3600$	276445.7
22	Actual fuel consumption	B	kg/s	$Q_b/(\eta_b Q_{ar,net,p})$	10.73
23	Design fuel consumption	B_{cal}	kg/s	$B(1-q_{uc}/100)$	10.63

*Similar to lower heating value, as a calculation reference.

TABLE C5 Structure Design of Furnace and Platen Superheater

No.	Parameter	Symbol	Unit	Formulae and data source	Result
1	Furance volume thermal load	q_v	W/m^3	*Refer to Appendix D6.7	120000
2	Calculated furnace volume	$V_{f,cal}$	m^3	$1000B_{cal}Q_{ar,net,p}/q_v$	2486.47
3	Furnace sectional thermal load	q_a	W/m^2	*Refer to Appendix D6.8	3050000
4	Calculated furnace cross section area	$A_{f,cal}$	m^2	$1000B_{cal}Q_{ar,net,p}/q_a$	96.85
5	Aspect ratio of furnace section	a/b	–	Determine according to $a/b = 1\sim1.2$, square is the best	1
6	Furnace width	a	m	Determine the value of a to make $a/b = 1$	9.841
7	Furnace depth	b	m	$A_{f,cal}/a$	9.841
8	Dry bottom hopper inclination	θ_{wch}	°	Geometric structure calculation can be completed by the reader	50
9	Dry bottom hopper outlet depth	b_{wch}	m	Geometric structure calculation can be completed by the reader	1.08
10	Outlet size of membrane wall at dry bottom hopper	l_{wch}	m	Geometric structure calculation can be completed by the reader	5.461
11	1/2 Dry bottom hopper height	h_{wch}	m	Geometric structure calculation can be completed by the reader	2.610
12	Dry bottom hopper volume	V_{wch}	m^3	Calculate frustum volume from $0.5h_{wch}$	156.97
13	Furnace nose length	l_{fa}	m	*Refer to Appendix D6.9	2.5
14	Furnace nose up-dip angle	θ_{up}	°	*Refer to Appendix D6.9	35
15	Furnace nose elevation angle	θ_{dw}	°	*Refer to Appendix D6.9	30
16	Platen superheater tube diameter	d	mm	Determine from design	42

TABLE C5 Structure Design of Furnace and Platen Superheater (*cont.*)

No.	Parameter	Symbol	Unit	Formulae and data source	Result
17	Platen superheater tube wall thickness	δ	mm	Determine from design	5
18	Working medium mass flux in platen superheater tubes	$\rho\omega$	kg/ (m$^2\cdot$s)	*Refer to Appendix D6.10	1000
19	Total flow area of platen superheater tubes	A	m^2	$(D_l-D_{dw2})/(3600$ $\rho\omega)$ (Suppose $D_{dw2} = 0.02D_1$)	0.1116
20	Area of single tube of platen superheater	A_i	m^2	$\pi d^2_i/4$ (d_i is inner diameter)	0.00080
21	Calculated platen superheater tube number	n_{cal}	–	A/A_i	138.78
22	Platen superheater tube number	n	–	Round from n_{cal}	139.00
23	Platen superheater transverse spacing	s_l	mm	*Refer to Appendix D6.11	700
24	Calculation platen superheater sheet number	$Z_{l,cal}$	–	Determine from a/s_l-1	13.1
25	Platen superheater sheet number	Z_l	–	Round from $Z_{l,cal}$	14
26	Calculated tube number of single sheet of platen superheater	$n_{l,cal}$	–	Determine from n/Z_l	9.9
27	Number of concurrent bent panel tubes	n_l	–	Determine from $n_{l,cal}$	10
28	Pass number of single sheet of platen superheater	n	–	Determine from design	4
29	Platen superheater longitudinal spacing	s_2	mm	*Refer to Appendix D6.12	50
30	Platen superheater minimum bending radius	R	mm	*Refer to Appendix D6.13	80

(Continued)

TABLE C5 Structure Design of Furnace and Platen Superheater (*cont.*)

No.	Parameter	Symbol	Unit	Formulae and data source	Result
31	Vertical spacing of platen superheater bottom	s_3	mm	*Refer to Appendix D6.12	60
32	Row numbers of longitudinal bypass	–	–	Determine from design	2
33	Platen superheater depth	b_p	m	See Fig. C4 (two tubes are directly joined to prevent overheating)	2.3
34	Distance from platen superheater back side to furnace nose	δ_{fa}	m	Determine from design	0.1
35	Gas velocity at furnace exit	ω_g	m/s	*Refer to Appendix D6.14	6.2
36	Gas temperature at furnace exit	θ_f''	°C	Trial and error	1115
37	Flow area of furnace exit	A_e	m²	$B_{cal}V_g/\omega_g \times (\theta_f'' + 273)/273$	80.78
38	Furnace nose outlet height	$h_{e,fa}$	m	A_e/a	8.208
39	Vertical part height of furnace nose	h_{fa}	m	*Refer to Appendix D6.9	0.2
40	Height of platen superheater	h_p	m	$R + h_{fa}$	8.408
41	Distance from high-temperature superheater to platen superheater	l_{p-ssh}	m	Determine from design	1.1
42	Calculated height of flue duct of high-temperature superheater	$h_{scs,c}$	m	See Fig. C2	7.508
43	Furnace nose downward inclination angle height	$h_{fa, b}$	m	$l_{fa}\tan30°$	1.443
44	Height of furnace top	h_{fr}	m	Since furnace top is designed, $h_p + h_{fa,b}$ is calculated according to the graph	9.852

TABLE C5 Structure Design of Furnace and Platen Superheater (*cont.*)

No.	Parameter	Symbol	Unit	Formulae and data source	Result
45	Furnace top volume 1	V_{fr1}	m^3	Furnace top is designed, calculate according to the graph	408.88
46	Furnace top volume 2	V_{fr2}	m^3	Furnace top is designed, calculate according to the graph	104.99
47	Furnace top volume	V_{fr}	m^3	Furnace top is designed, calculate according to the graph	513.87
48	Main body volume	V_{fr}	m^3	$V_{f,c}-V_{fr}-V_{wch}$	1815.62
49	Furnace main body height	h_F	m	$V_{f,c}-V_{fr}-V_{wch}/A_{f,cal}$	18.747
50	Loop number of front and rear wall water wall	$Z_{1,ww}$	–	Determine according to the heating length of each loop ≤2.5m	4
51	Loop number of left and right side wall water wall	$Z_{2,ww}$	–	Determine according to the heating length of each loop ≤2.5m	4
52	Tube diameter of water wall	d	mm	Determine from design	60
53	Tube thickness of water wall	δ	mm	Determine from design	5
54	Tube spacing	s	mm	Determine from design	80
55	Tube number of front and rear walls	n_1	–	Determine from a/s, must reasonably cover the furnace	123
56	Tuber number of left and right walls	n_2	–	Determine from b/s, must reasonably cover the furnace	123
57	Roof tube diameter	d	mm	Determine from design	51
58	Roof tube thickness	δ	mm	Determine from design	5
59	Roof tube spacing	s_{roof}	mm	Determine from design	100
60	Roof tube row number	Z_{roof}	–	Determine from a/s_{roof}, need to cover the furnace top reasonably	98

TABLE C6 Structure/Size of Furnace

No.	Parameter	Symbol	Unit	Formulae and data source	Result
1	Water wall tube diameter	d	mm	See Table C5	60
2	Water wall tube thickness	δ	mm	See Table C5	5
3	Water wall tube spacing	s	mm	See Table C5	80
4	Furnace width	a	m	See Table C5	9.841
5	Furnace depth	b	m	See Table C5	9.841
6	Furnace height	H	m	See Fig. C2 (from dry bottom hopper center line to furnace roof center line)	31.209
7	Dry bottom hopper area	H_{wch}	m²	$4(a + l_{wch})/2 \times h_{wch} = 4 \times (9.841 + 5.461) \times 2.610 \div 2$	79.89
8	Single side wall area	H_{sw}	m²	$(V_{f,cal} - V_{wch})/a = (2486.47 - 156.97) \div 9.841$	236.71
9	Front wall area	H_{fw}	m²	$a(h - h_{wch}) = 9.841 \times (31.209 - 2.610)$	281.44
10	Rear wall area	H_{rw}	m²	$a[l_{fa}/\cos(\theta_{dw}) + h_F] = 9.841 \times [2.5 \div \cos 30° + 18.747]$	212.90
11	Area of furnace exit window	H_{fe}	m²	$a(h_p + b_p + \delta_{fa}) = 9.841 \times (8.408 + 2.3 + 0.1)$	106.37
12	Covered area of furnace top	H_{roof}	m²	$a(b - b_p + \delta_{fa}) = 9.841 \times (9.841 - 2.3 - 0.1)$	73.23
13	Refractory belt area	H_{rb}	m²	Approximate value from design	0
14	Furnace total area	H_F	m²	$2H_{sw} + H_{fw} + H_{bq} + H_{fe} + H_{roof} + H_{rb} + H_{wch}$	1227.24
15	Area of doors and holes	H_{mh}	m²	Empirical value from design	12.00
16	Furnace configuration factor	X	–	Determine from design	1
17	Radiation heating area of furnace	H_r	m²	$x(F_F - F_{mh})$	1215.24
18	Thickness of effective radiation layer	s	m	$3.6 \times V_{f,cal}/H_F$	7.29

TABLE C7 Furnace Thermal Calculation

No.	Parameter	Symbol	Unit	Formulae and data source	Result
1	Hot air temperature	t_{ha}	°C	Determine from design	320
2	Hot air theoretical enthalpy	I°_{ha}	kJ/kg	See Table C3	3147.8
3	Furnace air leakage factor	$\Delta\alpha_F$	–	Refer to Appendix D6.3	0.05
4	Air leakage factor of coal pulverizing system	$\Delta\alpha_{zf}$	–	Refer to Appendix D6.15 (medium-speed mill)	0.04
5	Cold air temperature	t_{ca}	°C	Determine from design	20
6	Theoretical cold air enthalpy	I°_{ca}	kJ/kg	See Table C3	193.3
7	Excess air coefficient on air side	β''_{ah}	–	$\alpha'' - (\Delta\alpha_F + \Delta\alpha_{zf})$	1.11
8	Heat input by air sensible heat	Q_{air}	kJ/kg	$\beta''_{ah}I^0_{ha} + (\Delta\alpha_F + \Delta\alpha_{ms})I^0_{ca}$	3511.5
9	Heat input by 1 kg fuel	Q_l	kJ/kg	$Q_{in} \times (100 - q_{ug} - q_{rad} - q_{uc})/(100 - q_{uc}) + Q_{air}$	31308.5
10	Theoretical combustion temperature	θ_a	°C	See Table C3 according to Q_l	2038.7
		T_a	K	$273 + \theta_a$	2311.7
11	Furnace outlet gas temperature	θ''	°C	Assume/trial and error	1115
12	Furnace outlet gas enthalpy	I''	kJ/kg	See Table C3	16041.5
13	Mean diameter of ash particle	d_{fa}	μm	Refer to Appendix D6.16 (medium-speed mill)	16
14	Mean overall heat capacity of the combustion products	$V\overline{C}$	kJ/(kg·°C)	$(Q_l - I'')/(\theta_a - \theta'')$	16.53
15	Volume fraction of water vapor	r_{H_2O}	–	See Table C2	0.0731
16	Voume fraction of triatomic gases	r_n	–	See Table C2	0.2173
17	Gas density	ρ_g	kg/Nm^3	See Table C2	1.333
18	Furnace pressure	p	MPaA	Determine from design	0.1

(Continued)

TABLE C7 Furnace Thermal Calculation (*cont.*)

No.	Parameter	Symbol	Unit	Formulae and data source	Result
19	Partial pressure of triatomic gases	–	MPaA	pr_n	0.0217
20	Product of p_n and s	$p_n s$	m·MPaA	$pr_n s$	0.159
21	Radiant absorption coefficient of gas	k_g	l/(m·MPa)	$10.2[(0.78 + 1.6r_{H_2O})/(10.2p_n s)^{0.5} - 0.1](1 - 0.37T''/1000)$	3.00
22	Radiant absorption coefficient of fly ash	k_{fa}	l/(m·MPa)	$48350\rho_g/(T''^2 d_{fa}^2)^{1/3}$	74.00
23	Radiant absorption coefficient of coke particles	k_{co}	l/(m·MPa)	Refer to Appendix D6.17	10.2
24	Demensionless number	x_1	–	Refer to Appendix D6.18	0.5
25	Demensionless number	x_2	–	Refer to Appendix D6.19	0.1
26	Radiant absorption coefficient of flame radiation	K	l/(m·MPa)	$k_g r_n + k_{fa}\mu_{fa} + k_{co}x_1 x_2$	1.827
27	Exponent of Eq. (2.73)	kps	–	Kps	1.332
28	Furnace flame emissivity	a_{fl}	–	$1 - e^{-kPs}$	0.736
29	Water wall fouling factor	ζ	–	Refer to Appendix D6.20	0.45
30	Average thermal efficiency coefficient	ψ_{ave}	–	ζx	0.45
31	Furnace emissivity	a_F	–	$a_{fl}/[\psi_{ave}(1 - a_{fl}) + a_{fl}]$	0.861
32	Burner height	h_b	m	Empirical data determined from design	7.110
33	Height from dry bottom hopper to center of furnace exit	h_F	m	See Fig. C2	26.283
34	Burner relative height	x_b	–	h_b/h_F	0.271
35	Parameter	Δx	–	Refer to Appendix D6.21	0
36	Relative height of flame center	x_{fl}	–	$x_b + \Delta x$	0.271
37	Flame center modification factor	M	–	$B - Cx_{fl}$, refer to Appendix D6.21, $B = 0.59$, $C = 0.5$	0.455

TABLE C7 Furnace Thermal Calculation (*cont.*)

No.	Parameter	Symbol	Unit	Formulae and data source	Result
38	Furnace outlet gas temperature	θ''_{cal}	°C	$T_a/\{M \times [\sigma_0 T^3_a$ $a_F H_r \Psi_{ave}/(\varphi B_{cal} V\bar{C})]^{0.6}$ $+1\}-273$	1115.2
39	Furnace outlet gas enthalpy	I''	kJ/kg	See Table C3	16042.0
40	Radiative absorbed heat of furnace	Q_r	kJ/kg	$\varphi(Q_1 - I'')$	15205.4
41	Thermal load of radiation heating surface	–	kW/m²	$B_{cal}Q_r/H_r$	133.0
42	Furnace outlet gas temperature error	$\Delta\theta$	°C	$\theta''-\theta''_{cal}$	–0.16

of the furnace exit nose is generally determined as 1/4–1/3 of the furnace depth. The up-dip angle is 20–40 degrees, and downward inclination angle (elevation angle) is 20–30 degrees.

4. Calculate the height of the main body of the furnace.
5. According to fuel charateristics and combustion modes, determine burner type and design the burner arrangement.
6. Determine the water wall structure and recalibrate structure/size, furnace volume heat release rate, and sectional thermal load. Draw the furnace structure sketching.
7. According to the above furnace structure, perform the verified thermal calculation and determine furnace outlet gas temperature and other related thermal furnace parameters to verify their reasonability.

The following are stepwise instructions for platen superheater design.

1. Determine the total tube number of platen superheaters. The mass flux of a platen superheater with high thermal load is 700~1200 kg/(m²·s), thus the total number of platen superheater tubes can be calculated according to single tube diameter.
2. Determine transverse tube spacing of the platen superheater—s_1 is usually 700–900 mm. Flue gas velocity should be kept at approximately 6 m/s.
3. Determine longditudinal spacing and tube bending radius. The longitudinal tube spacing of a vertically suspended superheater should be small under

appropriate tube bending radius (50 mm is the empirical value). The minimum tube bending radius is usually two times its diameter; for a tube with 42 mm diameter, usually 80 mm or 85 mm.

Furnace outlet gas temperature is usually the gas temperature before entering the platen superheater. Furnace outlet gas temperature determines the radiative and convective fraction of the boiler unit. If the furnace outlet gas temperature is low, the radiative fraction of the furnace is high and total metal consumption and investment in heating surfaces decrease. However, if the furnace outlet gas temperature is too low, the average furnace tempature is low and radiative heat intensity decreases. In addition, the average temperature difference in the convection superheater decreases, creating a need for more expensive heating surfaces in the convection superheater. Furnace outlet gas temperature should be controlled to ensure that there is no slagging at the furnace exit, as well. Furnace outlet gas tempature must be lower than the softening temperature of the fuel (generally below 100°C.) Item 11 in Table C7 is the reference value according to experience.

The water wall fouling factor and average thermal efficiency coefficient are other important considerations. The fouling factor reflects the influence of reradiation of heating surfaces on heat exchange—its physical significance is the radiative heat fraction from the flame to the heating surface absorbed by the heating surface. If deposition on the water wall is severe, the temperature of the fouling layer increases, reradiation increases, and the heat absorbed by the water wall decreases. The thermal efficiency coefficient reflects the ratio of the effective heat flow and projected radiation heat—when a membrane wall is adopted, the thermal efficiency coefficient and fouling factor are almost equal. Item 28 in Table C7 is a reference value from experience.

The furnace's flame center is the zone with the highest working medium temperature. The influence of flame center position on the furnace outlet gas tempature is accounted for by coefficient M during thermal calculation. Item 31 in Table C7 is the reference value.

Hot air tempature is determined according to combustion type. Fast ignition and steady combustion of fuel are the designer's primary goals. If hot air temperature is too high, however, the air preheater will be too large and it will be difficult to arrange the back-end heating surface effectively.

C6 SUPERHEATER DESIGN AND HEAT EXCHANGE CALCULATION

A structural diagram of the superheater is shown in Figs. C4 and C5.

Structural design and thermal calculation of the platen superheater are shown in Tables C8 and C9.

The platen superheater at the furnace exit absorbs both the convection heat from flue gas and the radiation heat from the furnace. In addition, when the gas

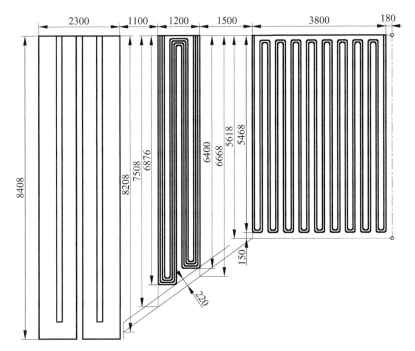

FIGURE C4   Structure diagram of superheater.

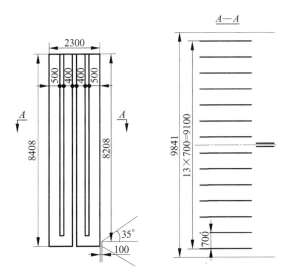

FIGURE C5   Structure diagram of platen superheater.

TABLE C8 Structure/Size of Platen Superheater

No.	Parameter	Symbol	Unit	Formulae and data source	Result
1	Tube diameter of platen superheater	d	mm	See Table C5	42
2	Platen superheater tube thickness	δ	mm	See Table C5	5
3	Platen superheater transverse rows	Z_1	–	See Table C5	14
4	Platen superheater longitudinal bypass rows	–	–	See Table C5	2
5	Platen superheater longitudinal rows	Z_2	–	Determine from design	40
6	Platen superheater height	h_p	m	See Table C5	8.408
7	Platen superheater depth	b_p	m	See Table C5	2.3
8	Platen superheater average transverse spacing	s_1	mm	$a/(Z_1 + 1)$	656.1
9	Platen superheater average longtitudinal spacing	s_2	mm	$b_p/(Z_2-1)$	59.0
10	Relative transverse pitch	σ_1	–	s_1/d	15.6
11	Relative vertical pitch	σ_2	–	s_2/d	1.4
12	Inlet radiation area	H'_r	m^2	$(h_p + b_p)aZ_1/(Z_1 + 1)$	98.36
13	Outlet radiation area	H''_r	m^2	$h_p aZ_1/(Z_1 + 1)$	77.23
14	Configuration factor	x	–	Refer to Appendix D6.22	0.96
15	Platen superheater total heating area	A_{psh}	m^2	$2h_p b_p Z_1 x$	519.84
16	Configuration factor from inlet to outlet	φ_h	–	$((b_p/s_1)^2 + 1)^{1/2} - (b_p/s_1)$	0.140
17	Exit window area	H_{fe}	m^2	See Table C6	106.37
18	Side water wall heating area 1 within panel	A_{ww1}	m^2	$(0.5\pi \times d + (s-d))/s \times 2 \times (h_p \times (b_p + \delta_{fa}))$	57.64
19	Side water wall heating area 2 within panel	A_{ww2}	m^2	$(0.5\pi \times d + (s-d))/s \times 2 \times 0.5 l_{p-ssh} \times (h_p + h_{scs,c})$	25.00
20	Side water wall heating area witin panel	A_{ww}	m^2	$A_{ww1} + A_{ww2}$	82.64
21	Roof tube diameter	d	mm	See Table C5	51
22	Roof tube rows	Z_{roof}	–	See Table C5	98

TABLE C8 Structure/Size of Platen Superheater (*cont.*)

No.	Parameter	Symbol	Unit	Formulae and data source	Result
23	Roof tube spacing	S	mm	See Table C5	100
24	Roof tube length within platen superheater zone	l_{roof}	m	$(b_p + l_{p\text{-ssh}})$	3.4
25	Heating area of roof tubes	A_{roof}	m²	$a(0.5 \times \pi \times d + (s-d))/s \times l_{roof}$	43.20
26	Effective radiation layer thickness	s	m	$1.8/(1/b_p + 1/h_p + 1/s_1)$	0.866
27	Gas inlet flow area	A'_g	m²	$(h_p + b_p)(a - dZ_1)$	99.09
28	Gas outlet flow area	A''_g	m²	$(h_p - h_{z\,y})(a - dZ_1)$	75.95
29	Gas average flow area	$A_{g,ave}$	m²	$2A'_g A''_g/(A'_g + A''_g)$	85.99
30	Steam flow area	A_{ss}	m²	$\pi/4 \times n_1 \times Z_1 \times d_n^2$	0.113

TABLE C9 Thermal Calculation of Platen Superheater

No.	Parameter	Symbol	Unit	Formulae	Result
1	Temperature of flue gas into panel	θ'	°C	See Table C7	1115.2
2	Enthalpy of flue gas into panel	I'	kJ/kg	See Table C7	16041.96
3	Convection heat of panel zone	$Q_{c,psh}$	kJ/kg	Trial and error	1587
4	Radiation heat from flue gas to heating surface through panel	Q_{rl}	kJ/kg	Trial and error	144
5	Absorbed convection heat of furnace roof at panel zone	Q_{roof}	kJ/kg	Trial and error	116
6	Absorbed convection heat of water wall at panel zone	Q_{ww}	kJ/kg	Trial and error	224
7	Enthalpy of flue gas out of panel	I''	kJ/kg	$I' - (Q_{c,psh} + Q_{rl})/\varphi$	14304.01
8	Temperature of flue gas out of panel	θ''	°C	See Table C3	1005.02

(Continued)

TABLE C9 Thermal Calculation of Platen Superheater (*cont.*)

No.	Parameter	Symbol	Unit	Formulae	Result
9	Average flue gas tempeature	θ_{ave}	°C	$(\theta' + \theta'')/2$	1060.09
		T_{ave}	K	$T + \theta_{ave}$	1333.09
10	Absorbed convection heat of panel	Q_c	kJ/kg	$Q_{c,psh} - Q_{roof} - Q_{ww}$	1247
11	Product of p_n and s	–	m·MPa	$pr_n s$	0.0188
12	Radiant absorption coefficient of gas	k_g	l/(m·MPa)	$10.2[(0.78 + 1.6r_{H_2O})/ (10.2p_n s)^{0.5} - 0.1]$ $(1 - 0.37T_{ave}/1000)$	10.06
13	Radiant absorption coefficient of fly ash	k_{fa}	l/(m·MPa)	$48350\rho_g/ (T_{ave}^2 d_{fa}^2)^{1/3}$	76.02
14	Radiant absorption coefficient of flue gas radiation	K	l/(m·MPa)	$k_g r_n + k_{fa}\mu_{fa}$	2.87
15	Exponent of Eq. (2.73)	Kps	–	kps	0.25
16	Flue gas emissivity	a	–	$1 - e^{-kPs}$	0.220
17	Coefficient considering reradiation	β	–	Refer to Appendix D6.23	0.966
18	Relative height of window	x	–	$(h - 0.5h_p)/h$	0.865
19	Thermal load distribution coefficient	η_{psh}	–	Refer to Appendix D6.24	0.771
20	Radiation heat flow of panel zone	q_r	kW/m²	$\eta_{psh}Q_r B_{cal}/H_r$	102.49
21	Corrected radiative intensity of panel zone	$q_{r,m}$	kW/m²	βq_r	98.96
22	Radiation heat absorbed by panel directly from furnace	Q'_r	kJ/kg	$q_{r,m}H''_r/B_{cal}$	915.92
23	Configuration factor from inlet to outlet of panel	φ_{fl}	–	See Table C8	0.140
24	Furnace radiation heat leaked out of panel	Q''_r	kJ/kg	$Q'_r (1-a)\varphi_{fl}/\beta$	103.46
25	Radiation heat absorbed by panel	Q_r	kJ/kg	$Q'_r - Q''_r$	812.46
26	Total heat absorbed by panel	Q	kJ/kg	$Q_c + Q'_r - Q''_r$	2059.46
27	Radiation fuel correction coefficient	ζ_r	–	Refer to Appendix D6.25	0.5

TABLE C9 Thermal Calculation of Platen Superheater (*cont.*)

No.	Parameter	Symbol	Unit	Formulae	Result
28	Radiation heat from flue gas to heating surfaces behind panel	$Q_{rl,cal}$	kJ/kg	$5.7 \times 10^{-11} a H_r'' T^4 \zeta_r / B_{cal}$	143.94
29	Pressure of steam into panel	p'	MPa	*Determine from design	14.4
30	Pressure of steam out of panel	p''	MPa	*Determine from design	14.1
31	Temperature of steam into panel (after spray attemperator)	t'	°C	Trial and error using desuperheater spray	395
32	Enthalpy of steam into panel	i'	kJ/kg	See Standard Steam Table	2971.3
33	Primary desuperheating water flow rate	D_{ds1}	kg/h	Trial and error	8200
			t/h		8.2
34	Secondary desuperheater spray flow rate	D_{ds2}	kg/h	Trial and error	5800
			t/h		5.8
35	Enthalpy of steam out of panel	i''	kJ/kg	$h' + B_{cal}Q/(D_1 - D_{ds2})$	3166.2
36	Temperature of steam out of panel	t''	°C	See Standard Steam Table	447.56
37	Average temperature of steam in panel	t_{ave}	°C	$(t' + t'')/2$	421.3
38	Average specific volume of steam in panel	v_{ave}	m³/kg	See Standard Steam Table	0.0179
39	Average speed of steam in panel	ω_{ss}	m/s	$(D_1 - D_{ds2}) v_{ave}/(3600A_{ss})$	17.82
40	Average mass flow rate of steam in panel	$\rho\omega$	kg/(m²s)	$D_1/(3600A_{ss})$	997.19
41	Correction coefficient of tube diameter	C_d	–	Refer to Appendix D6.26	0.91
42	Thermal conductivity of steam λ	λ	W/(m·°C)	See Standard Steam Table	0.0771
43	Dynamic viscosity of steam v	v	m²/s	See Standard Steam Table	4.77E-07
44	Prandtl number of steam Pr	Pr	–	See Standard Steam Table	1.18
45	Heat transfer coefficient from tube wall to steam	α_2	w/(m²·°C)	$0.023\lambda/d_n(\omega_{ss} d_n/v)^{0.8}Pr^{0.4}C_d$	3925.5

(Continued)

TABLE C9 Thermal Calculation of Platen Superheater (*cont.*)

No.	Parameter	Symbol	Unit	Formulae	Result
46	Average flue gas velocity among panel	v_g	m/s	$V_g B_{cal} (\theta_{ave}+273)/(273A_{g,ave})$	5.59
47	Thermal conductivity of flue gas	λ	W/(m·°C)	Refer to Appendix D6.27	0.114
48	Dynamic viscosity of flue gas	ν	m²/s	Refer to Appendix D6.27	1.8E-04
49	Average Prandtl number of flue gas	Pr_{ave}	–	Refer to Appendix D6.28	0.574
50	Prandtl number of flue gas	Pr	–	$(0.94 + 0.56r_{H_2O})Pr_{ave}$	0.563
51	Correction factor of tube rows	C_z	–	Refer to Appendix D6.29	1
52	Flue gas composition and temperature correction coefficient	C_w	–	$0.92 + 0.726r_{H_2O}$	0.973
53	Correction factor for geometric arrangement	C_s	–	Refer to Appendix D6.30	0.858
54	Flue gas side convection coefficient	α_g	w/(m²·°C)	$0.2\lambda/d(v_g d/\nu)^{0.65} Pr^{0.33}C_z C_s C_w$	39.80
55	Ash deposition coefficient	ε	m²·°C/w	Refer to Appendix D6.31	9.7E-03
56	Tube wall fouling emissivity	a_w	–	Refer to Appendix D6.32	0.8
57	Fouling layer temperature of tube wall	t_w	°C	$t_{ave} + 1000(\varepsilon + 1/\alpha_2)B_{cal}(Q_c + Q_r)/H_{psh}$	840.2
58	Radiation heat transfer coefficient	α_r	w/(m²·°C)	$5.7 \times 10^{-8}(a_w + 1)2aT_{ave}^3[1-(T_w/T_{ave})^4]/[1-(T_w/T_{ave})]$	83.29
59	Utilization coefficient of panel	ξ		Refer to Appendix D6.31	0.85
60	Flue gas side heat transfer coefficient	a_l	w/(m²·°C)	$\xi [(\pi \times d \times a_g)/(2 \times s_2)+ a_r]$	117.30
61	Overall heat transfer coefficient	K	w/(m²·°C)	$a_l/[1 + (1 + Q_r/Q_c)(\varepsilon + 1/a_2) a_l]$	40.07
62	Large temperature difference	Δt_{max}	°C	$\theta' - t'$	720.2
63	Small temperature difference	Δt_{min}	°C	$\theta'' - t''$	557.5
64	Logarithmic mean temperature difference	Δt	°C	$(\Delta t_{max}-\Delta t_{min})/(Ln\Delta t_{max}/\Delta t_{min})$	635.34

TABLE C9 Thermal Calculation of Platen Superheater (*cont.*)

No.	Parameter	Symbol	Unit	Formulae	Result
65	Convection heat of panel	$Q_{c,cal}$	kJ/kg	$K\Delta t\, A_{psh}/(1000\, B_{cal})$	1245.296
66	Error	e	%	$100(Q_c-Q_{c,cal})/Q_c$	0.14
67	Water temperature of water wall at panel side	t_{sw}	°C	See Standard Steam Table (saturated temperature)	342.53
68	Average temperature difference of heat transfer	Δt	°C	$\theta_{ave}-t_{sw}$	717.56
69	Absorbed convection heat by both side water walls within panel zone	$Q_{ww,cal}$	kJ/kg	$K\Delta t\, A_{ww}/(1000 B_{cal})$	223.6
70	Error	e	%	$100(Q_{ww}-Q_{ww,cal})/Q_{ww}$	0.18
71	Relative height of furnace roof cover	x	–	Constant	1
72	Thermal load distribution coefficient	η_{roof}	–	Refer to Appendix D6.24	0.633
73	Furnace radiation heat flux absorbed by furnace roof cover	q_r	kW/m^2	$\eta_{roof}Q_r B_{cal}/H_r$	84.19
74	Furnace radiation heat absorbed by furnace roof cover	$Q_{r,roof}$	kJ/kg	$\beta q_r\, H_{roof}/B_{cal}$	560.20
75	Enthalpy increment of furnace roof cover	Δi	kJ/kg	$3600 B_{cal}Q_r'/(D-D_{ds1}-D_{ds2})$	54.12
76	Saturated steam temperature of drum outlet	t_s	°C	See Standard Steam Table (saturated temperature)	342.53
77	Dry saturated steam enthalpy of drum outlet	i_s''	kJ/kg	See Standard Steam Table ($x = 1$)	2608.86
78	Furnace roof pressure within panel zone	p_{roof}	MPa	*Determine from design	15
79	Steam enthalpy of furnace roof inlet within panel zone	i'_{roof}	kJ/kg	$i_s'' + \Delta i$	2663.0

(*Continued*)

TABLE C9 Thermal Calculation of Platen Superheater (*cont.*)

No.	Parameter	Symbol	Unit	Formulae	Result
80	Steam temperature of furnace roof inlet within panel zone	t'_{roof}	°C	See Standard Steam Table	347.1
81	Steam enthalpy increment of furnace roof within panel zone	Δi	kJ/kg	$3.6 B_{cal}\, Q_{roof}/$ $(D_1 - D_{ds1} - D_{ds2})$	11.21
82	Steam enthalpy of furnace roof outlet within panel zone	i''_{roof}	kJ/kg	$i'_{roof} + \Delta i$	2674.2
83	Steam temperature of furnace roof within panel zone	t''_{roof}	°C	See Standard Steam Table	347.9
84	Average steam temperature of furnace roof within panel zone	t_{ave}	°C	$0.5(t'_{roof} + t''_{roof})$	347.50
85	Average temperature difference of heat transfer	Δt	°C	$\theta_{ave} - t_{ave}$	712.59
86	Absorbed convection heat of furnace roof within panel zone	$Q_{roof,cal}$	kJ/kg	$K \Delta t\, A_{roof}/$ $(1000 B_{cal})$	116.1
87	Error	e	%	$100(Q_{roof} -$ $Q_{roof,cal})/Q_{roof}$	−0.06
88	Absorbed convection heat within panel zone	$Q_{c,psh,cal}$	kJ/kg	$Q_{c,cal} + Q_{ww,cal} +$ $Q_{roof,cal}$	1585.0
89	Total error	e	%	$100(Q_{c,psh} -$ $Q_{c,psh,cal})/Q_{c,psh}$	0.13

temperature between the panels is high, a portion of the heat is radiated to the furnace, then to the downstrem convection heating surface. Therefore, the heat transfer equations for these heating surfaces differ from those for general convection heating surfaces.

The platen superheater is an intermediate stage superheater, the inlet and outlet working medium parameters of which are usually unknown during thermal calculation. Item 31 of Table C9 is the assumption based on related data. Whether the assumption is correct should be verified after heating surface thermal calculation.

The transverse spacing of the platen superheater is large, flue gas velocity is low, and sweeping is nonuniform. Therefore, certain heat exchange parameters

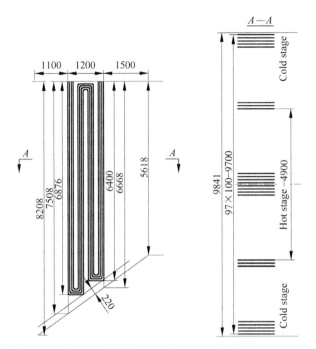

FIGURE C6   Structure diagram of high-temperature superheater.

of the platen superheater (such as the fouling factor, utilization coefficient, heat transfer coefficient, and calculation method of the heat exchange area) are different from those of general convection heating surfaces.

The structure of a high-temperature superheater is shown in Fig. C6. Its structure design and thermal calculation are shown in Tables C10 and C11.

Total pressure drop in a superheater system is usually less than 10% of the boiler rated pressure; it can be up to 15% for an ultra-high pressure unit. During thermal calculation for every stage of the heating surface, the inlet and outlet pressure of the superheater at mid-stages can be determined according to related references or estimated according to the total pressure drop of the superheater, but only at error allowances below 3%.

The characteristics of this example are in a multiple parallel-flow arrangement. The working medium flows countercurrent to the cold stage and downstream to the hot stage. The cold and hot stages can be integrated during calculation, and the average temperature difference of heat transfer must be corrected through the lookup-table or staged calculation in cases such as this. These methods do have some drawbacks. The first method fails to account for the working medium of the cold stage outlet or hot stage inlet, and the measured average temperature difference of heat transfer does not include the temperature drop by

TABLE C10 Structure/Size of High-Temperature Superheater

No.	Parameter	Symbol	Unit	Formulae and data source	Result
1	Tube diameter	d	mm	Determine from design	42
2	Tube thickness	δ	mm	Determine from design	5
3	Longitudinal tube rows	Z_2	–	Determine from design	16
4	Number of head	n	–	Determine from design	4
5	Pass number of high-temperature superheater	n_1	–	Determine from design	4
6	Sectional area of single tube	A_i	m^2	$\pi d_n^2/4$	$8.04\ 10^{-4}$
7	Steam mass flow flux of cold stage	ρw_c	$kg/(m^2 \cdot s)$	Refer to Appendix D6.10	720
8	Total steam flow area of cold stage	$A_{ss,c}$	m^2	$(D-D_{ds1})/(3600\rho w_c)$	0.155
9	Transverse rows of cold stage	Z_{1c}	–	$A_{ss,c}/(A_i\ n)$	48.19
10	Transverse rows of cold stage	Z_{1c}	–	Determine from design	48
11	Steam mass flux of hot stage	ρw_h	$kg/(m^2 \cdot s)$	Refer to Appendix D 6.10	700
12	Total steam flow area of hot stage	$A_{ss,h}$	m^2	$D_1/(3600\rho w_h)$	0.163
13	Transverse rows of hot stage	Z_{1h}	–	$A_{ss,h}/(A_i\ n)$	50.57
14	Transverse rows of hot stage	Z_{1h}	–	Determine from design	50
15	Total horizontal rows	Z_l	–	$Z_{1h} + Z_{1c}$	98
16	Transverse spacing	s_1	mm	Determine from design	100
17	Relative transverse pitch	σ_1	–	s_1/d	2.38
18	Longitudinal spacing	s_2	mm	Determine from design	60
19	Superheater depth	b	m	See Fig. C6	1.2
20	Average longtitudinal pitch	$s_{2,ave}$	mm	$b/(Z_2-1)$	80

TABLE C10 Structure/Size of High-Temperature Superheater (*cont.*)

No.	Parameter	Symbol	Unit	Formulae and data source	Result
21	Relative longitudinal pitch	σ_2	mm	$S_{2,ave}/d$	1.90
22	Minimum bending radius of superheater	R	–	Determine from design	80
23	Radiative space depth prior to superheater	l_r	m	See Fig. C6	1.1
24	Height of flue duct at high-temperature inlet	h'	m	See Fig. C6	7.508
25	Average tube calculated length	l_{ave}	m	Fig. C6, $2 \times$ (6.876 + 6.4)	26.55
26	Heating area of hot stage	A_h	m^2	$nZ_{1h} \times \pi{\times}d \times l_{ave}$	700.69
27	Heating area of cold stage	A_c	m^2	$nZ_{1c} \times \pi{\times}d \times l_{ave}$	672.66
28	Total heating area	A	m^2	$A_h + A_c$	1373.35
29	Distance from high-temperature superheater to low-temperature superheater	$l_{p\text{-}s,sh}$	m	Determine from design	1.5
30	Height of flue duct at low-temperature inlet	$h'_{p,sh}$	m	See Fig. C6	5.618
31	Length of roof tube within high-temperature zone	l_{roof}	m	$l_{p\text{-}s,sh} + b$	2.7
32	Height of flue duct at high-temperature outlet	h''	m	See Fig. C6	6.668
33	Heating area of roof tubes	A_{roof}	m^2	$al_{roof}(0.5\pi d + (s_{roof}{-}d))/s_{roof}$	34.31
34	Heating area of side water wall	A_{sw}	m^2	$2 \times (0.5 \times \pi \times d + (s_{sw}{-}d))/s_{sw} \times 0.5l_{roof}(h' + h'_{p,sh})$	50.61
35	Heating area of bottom water wall	A_{bww}	m^2	$a(0.5 \times \pi \times d + (s_{bww}{-}d))/s_{bww} \times l_{roof}/\cos\theta_{up}$	46.32
36	Heating area of water wall	A_{ww}	m^2	$A_{sw} + A_{bww}$	96.94
37	Inlet flue gas flow area	A'_g	m^2	$h'(a{-}dZ_1)$	42.99
38	Gas outlet flow area	A''_g	m^2	$h''(a{-}dZ_1)$	38.18

TABLE C10 Structure/Size of High-Temperature Superheater (*cont.*)

No.	Parameter	Symbol	Unit	Formulae and data source	Result
39	Gas average flow area	$A_{g,ave}$	m^2	$2A_g'A_g''/(A_g' + A_g'')$	40.44
40	Steam flow area of hot stage	$A_{ss,h}$	m^2	$\pi/4 \times n \times Z_{lh} \times d_i^2$	0.161
41	Steam flow area of cold stage	$A_{ss,c}$	m^2	$\pi/4 \times n \times Z_{1c} \times d_i^2$	0.154
42	Effective radiation layer thickness	S	m	$0.9d(4\sigma_1\sigma_2/\pi-1)$	0.180
43	Rear water leading tube number	N_{rw}	–	Determine from design	40
44	Rear water leading tube diameter	d_{rw}	mm	Determine from design	60
45	Rear water leading tube length	L_{rw}	m	Determine from design	7.2
46	Rear water leading tube heating area	A_{rw}	m^2	$N_{rw} \times \pi \times d_{rw} \times L_{rw}$	54.29

TABLE C11 Thermal Calculation of High-Temperature Superheater

No.	Parameter	Symbol	Unit	Formulae and data source	Result
1	Inlet flue gas temperature	θ'	°C	See Table C9	1005.0
2	Inlet flue gas enthalpy	I'	kJ/kg	See Table C9	14304.0
3	Inlet steam temperature	t'	°C	See Table C9	447.6
4	Inlet steam enthalpy	i'	kJ/kg	See Table C9	3166.2
5	Outlet steam temperature	t''	°C	Design assignment	540
6	Outlet steam enthalpy	i''	kJ/kg	See Standard Steam Table	3437.5
7	Desuperheater spray water enthalpy	i_{fw}	kJ/kg	See Table C4	1016.1
8	Radiation heat from flue gas to panel back	Q_{rl}	kJ/kg	See Table C9	143.94

TABLE C11 Thermal Calculation of High-Temperature Superheater (*cont.*)

No.	Parameter	Symbol	Unit	Formulae and data source	Result
9	Absorbed convection heat of superheated steam	Q_c	kJ/kg	$[(D_1-D_{ds2})(i''-i')+ D_{ds2}(i''-i_{fw})]/$ $(3600B_{cal})-Q_{rl}$	3089.0
10	Absorbed heat of additional heating surfaces of furnace roof	Q_{roof}	kJ/kg	Trial and error	111
11	Absorbed heat of additional heating surfaces of water wall	Q_{ww}	kJ/kg	Trial and error	319
12	Absorbed heat of rear water leading tube	Q_{rw}	kJ/kg	Trial and error	178
13	Outlet flue gas enthalpy	I''	kJ/kg	$I'-(Q_d + Q_{roof} + Q_{ww} + Q_{rw})/\varphi+\Delta\alpha I^{\circ}_{ca}$	10597.9
14	Outlet flue gas temperature	θ''	°C	See flue gas enthalpy –temperature table	750.45
15	Flue gas release heat	Q_g	kJ/kg	$\varphi(I'-I'' + \Delta\alpha I^{\circ}_{ca})$	3688.4
16	Average steam temperature	t_{ave}	°C	$(t' + t'')/2$	493.78
17	Flue gas average temperature	θ_{ave}	°C	$(\theta' + \theta'')/2$	877.73
18	Flue gas velocity	ω_g	m/s	$B_{cal}V_g(\theta_{ave} + 273)/$ $(273 \times A_{g,ave})$	10.52
19	Thermal conductivity of flue gas	λ	W/(m·°C)	Refer to Appendix D6.27	0.0980
20	Dynamic viscosity of flue gas	v	m²/s	Refer to Appendix D6.27	1.42 10^{-4}
21	Average Prandtl number of flue gas	Pr_{ave}	–	Refer to Appendix D6.28	0.582
22	Prandtl number of flue gas	Pr	–	$(0.94 + 0.56r_{H_2O})Pr_{ave}$	0.571
23	Correction factor for rows	C_z	–	Refer to Appendix D6.29	1
24	Flue gas composition and temperature correction coefficient	C_w	–	$0.92 + 0.726r_{H_2O}$	0.973
25	Correction factor for the geometric arrangement	C_s	–	Refer to Appendix D6.30	1.000
26	Flue gas side convection coefficient	α_g	w/(m²·°C)	$0.2\lambda/d(\omega_g d/v)^{0.65}$ $Pr^{0.33}C_zC_sC_w$	70.40

(*Continued*)

TABLE C11 Thermal Calculation of High-Temperature Superheater (*cont.*)

No.	Parameter	Symbol	Unit	Formulae and data source	Result
27	Cold stage flue gas mass fraction	g_c	–	Trial and error	0.525
28	Hot stage flue gas mass fraction	g_h	–	Trial and error	0.475
29	Absorbed radiation heat of hot stage	$Q_{r,h}$	kJ/kg	$g_h Q_{rl}$	68.37
30	Absorbed radiation heat of cold stage	$Q_{r,c}$	kJ/kg	$g_c Q_{rl}$	75.57
31	Inlet steam enthalpy of hot stage	i'_h	kJ/kg	$i''-3.6B_{cal}/D_1\{g_h[\varphi(I'-I'' + \Delta\alpha I^\circ_{ca})-Q_{roof}-Q_{ww}-Q_{rw}] + Q_{r,h}\}$	3294.6
32	Inlet steam pressure of hot stage	p'_h	MPa	Determine from design	13.9
33	Outlet steam pressure of hot stage	p''_h	MPa	Determine from design	13.7
34	Average steam pressure of hot stage	$p_{h,ave}$	MPa	$0.5(p'_h + p''_h)$	13.8
35	Inlet steam temperature of hot stage	t'_h	°C	See Standard Steam Table	489.41
36	Average steam temperature of hot stage	$t_{h,ave}$	°C	$0.5(t'' + t'_h)$	514.7
37	Absorbed convection heat of hot stage	$Q_{h,c}$	kJ/kg	$D_1/(3.6B_{cal}) \times (i''-i'_h)-Q_{h,r}$	1463.2
38	Outlet steam enthalpy of cold stage	i''_c	kJ/kg	$D_1/(D_1-D_{ds2})i'_h-D_{ds2}/(D_1-D_{ds2})i_{fw}$	3327.3
39	Inlet steam pressure of cold stage	p'_c	MPa	*Determine from design	14.1
40	Outlet steam pressure of cold stage	p''_c	MPa	*Determine from design	13.9
41	Average steam pressure of cold stage	$p_{c,ave}$	MPa	$0.5(p'_c + p''_c)$	14
42	Outlet steam temperature of cold stage	t''_c	MPa	See Standard Steam Table	500.29
43	Enthalpy increment of cold stage	Δi	kJ/kg	$g_c[\varphi(I'-I'' + \Delta\alpha I^\circ_{ca})-Q_{roof}-Q_{ww}-Q_{rw} + Q_{rl}] \times 3.6B_{cal}/(D_1-D_{ds2})$	160.22

TABLE C11 Thermal Calculation of High-Temperature Superheater (*cont.*)

No.	Parameter	Symbol	Unit	Formulae and data source	Result
44	Inlet steam enthalpy of cold stage	i'_c	kJ/kg	$i''_c - \Delta i$	3167.0
45	Inlet steam temperature of cold stage	t'_c	kJ/kg	See Standard Steam Table	447.81
46	Outlet steam temperature of panel	t''_{psh}	°C	See Table C9	447.56
47	Interface temperature difference	Δt	°C	$t'_c - t''_{psh}$	0.25
48	Average steam temperature of cold stage	$t_{c,ave}$	°C	$0.5(t''_c + t'_c)$	474.05
49	Absorbed convection heat of cold stage	$Q_{c,c}$	kJ/kg	$(D_1 - D_{ds2})/(3.6B_{cal})$ $(i''_c - i'_c) - Q_{r,c}$	1617.2
50	Average specific volume of steam of hot stage	$v_{h,ave}$	m³/kg	See Standard Steam Table	0.0236
51	Average speed of steam of hot stage	$\omega_{ss,h}$	m/s	$D_1 v_{h,ave}/(3.6A_{ss,h})$	16.71
52	Average specific volume of steam of cold stage	$v_{c,ave}$	m³/kg	See Standard Steam Table	0.0212
53	Average speed of steam of cold stage	$\omega_{ss,c}$	m/s	$(D_1 - D_{ds2})v_{c,ave}/$ $(3.6A_{ss,c})$	15.41
54	Correction factor for tube diameter	C_d	–	Refer to Appendix D6.26	0.91
55	Thermal conductivity of steam of cold stage λ	λ_c	W/(m·°C)	See Standard Steam Table	0.0786
56	Dynamic viscosity of steam of cold stage ν	ν_c	m²/s	See Standard Steam Table	5.99 10^{-7}
57	Prandtl number of steam of cold stage Pr	Pr_c	–	See Standard Steam Table	1.061
58	Steam side heat transfer coefficient of cold stage	$\alpha_{2,c}$	W/(m²·°C)	$0.023\lambda_c/d_n(\omega_{ss,c}$ $d_n/\nu_c)^{0.8}Pr_c^{0.4}C_d$	2850.2
59	Correction factor for tube diameter	C_d	–	Refer to Appendix D6.26	0.91
60	Thermal conductivity of steam of hot stage λ	λ_h	W/(m·°C)	See Standard Steam Table	0.0818

(*Continued*)

TABLE C11 Thermal Calculation of High-Temperature Superheater (*cont.*)

No.	Parameter	Symbol	Unit	Formulae and data source	Result
61	Dynamic viscosity of steam of hot stage v	v_h	m^2/s	See Standard Steam Table	6.92 10^{-7}
62	Prandtl number of steam of hot stage Pr	Pr_h	–	See Standard Steam Table	1.011
63	Steam side heat transfer coefficient of hot stage	$\alpha_{2,h}$	$W/(m^2 \cdot {}^\circ C)$	$0.023\lambda_h/d_n(\omega_{ss,h} d_n/v_h)^{0.8}Pr_h^{0.4}C_d$	2761.3
64	Ash deposition coefficient	ε	$m^2 \cdot {}^\circ C/W$	Refer to Appendix D6.33	0.0043
65	Fouling layer temperature of tube wall of cold stage	$t_{w,c}$	${}^\circ C$	$t_{c,ave} + 1000(\varepsilon + 1/\alpha_{2,c})B_{cal}(Q_{c,c} + Q_{r,c})/A_c$	598.43
66	Fouling layer temperature of tube wall of hot stage	$t_{w,h}$	${}^\circ C$	$t_{h,ave} + 1000(\varepsilon + 1/\alpha_{2,h})B_{cal}(Q_{c,h} + Q_{r,h})/A_h$	623.00
67	Product of p_n and s	$p_n s$	$m \cdot MPa$	$pr_n s$	0.00388
68	Radiant absorption coefficient of gas	k_g	$1/(m \cdot MPa)$	$10.2[(0.78 + 1.6r_{H_2O})/ (10.2p_n s)^{0.5} - 0.1] (1 - 0.37T_{ave}/1000)$	25.80
69	Radiant absorption coefficient of fly ash	k_{fa}	$1/(m \cdot MPa)$	$48350\rho_g/(T_{ave}^2 d_{fa}^2)^{1/3}$	83.82
70	Radiant absorption coefficient of flue gas radiation	k	$1/(m \cdot MPa)$	$k_g r_n + k_{fa}\mu_{fa}$	6.29
71	Exponent of Eq. (2.73)	kps	–	kps	0.113
72	Flue gas emissivity	a	–	$1 - e^{-kps}$	0.107
73	Radiation heat transfer coefficient of cold stage	$\alpha_{r,c}$	$w/(m^2 \cdot {}^\circ C)$	$5.7 \times 10^{-8}(a_w + 1)/2aT_{ave}^3[1 - (T_{w,c}/T_{ave})^4]/[1 - (T_{w,c}/T_{ave})]$	23.18
74	Radiation heat transfer coefficient of hot stage	$\alpha_{r,h}$	$w/(m^2 \cdot {}^\circ C)$	$5.7 \times 10^{-8}(a_w + 1)/2aT_{ave}^3[1 - (T_{w,h}/T_{ave})^4]/[1 - (T_{w,h}/T_{ave})]$	23.96
75	Fuel correction coefficient	A		Refer to Appendix D6.34	0.4
76	Radiation heat transfer coefficient of cold stage correction	$\alpha'_{r,c}$	$w/(m^2 \cdot {}^\circ C)$	$\alpha_{r,c}[1 + A(T'/1000)^{0.25} (l_r/b)^{0.07}]$	32.98

TABLE C11 Thermal Calculation of High-Temperature Superheater (*cont.*)

No.	Parameter	Symbol	Unit	Formulae and data source	Result
77	Radiation heat transfer coefficient of hot stage correction	$\alpha'_{r,h}$	w/(m²·°C)	$\alpha_{r,h}[1 + A(T'/1000)^{0.25}(I_r/b)^{0.07}]$	34.08
78	Flowing uniformity coefficient	ω	–	Determine from design	1
79	Effective coefficient	ψ	–	Refer to Appendix D6.35	0.65
80	Flue gas side heat transfer coefficient of cold stage	$\alpha_{1\,c}$	W/(m²°C)	$\omega\alpha_g + \alpha'_{r,c}$	103.38
81	Flue gas side heat transfer coefficient of hot stage	$\alpha_{1\,h}$	W/(m²°C)	$\omega\alpha_g + \alpha'_{r,h}$	104.48
82	Heat transfer coefficient of cold stage	K_c	W/(m²·°C)	$\psi\alpha_{1c}/(1 + \alpha_{1c}/\alpha_{2,c})$	64.85
83	Heat transfer coefficient of hot stage	K_h	W/(m²·°C)	$\psi\alpha_{1h}/(1 + \alpha_{1h}/\alpha_{2,h})$	65.44
84	Small temperature difference of cold stage	Δt_{min}	°C	$\theta''-t'_c$	302.6
85	Large temperature difference of cold stage	Δt_{max}	°C	$\theta'-t''_c$	504.7
86	Logarithmic mean temperature difference of cold stage	Δt	°C	$(\Delta t_{max}-\Delta t_{min})/(ln\Delta t_{max}/\Delta t_{min})$	395.11
87	Convection heat of cold stage	$Q_{c,cal}$	kJ/kg	$K A_c \Delta t/(1000\, B_{cal})$	1621.8
88	Error	e	%	$(Q_{c,c}-Q_{c,cal})/Q_{c,c} \times 100$	−0.29
89	Small temperature difference of hot stage	Δt_{min}	°C	$\theta''-t''$	210.45
90	Large temperature difference of hot stage	Δt_{max}	°C	$\theta'-t'_h$	515.61
91	Logarithmic mean temperature difference of hot stage	Δt	°C	$(\Delta t_{max}-\Delta t_{min})/(ln\Delta t_{max}/\Delta t_{min})$	340.54
92	Convection heat of hot stage	$Q_{h,cal}$	kJ/kg	$K\Delta t A_h/(1000 B_{cal})$	1469.3

(*Continued*)

TABLE C11 Thermal Calculation of High-Temperature Superheater (*cont.*)

No.	Parameter	Symbol	Unit	Formulae and data source	Result
93	Error	e	%	$(Q_{c,h}-Q_{h,cal})/Q_{c,h} \times 100$	−0.42
94	Total absorbed convection heat of superheater	$Q_{c,cal}$	kJ/kg	$Q_{c,cal} + Q_{h,cal}$	3091.1
95	Error	e	%	$(Q_c-Q'_c)/Q_c \times 100$	−0.07
96	Average heat transfer coefficient of superheater zone	K_{ave}	W/(m²·°C)	$0.5(K_c + K_h)$	65.14
97	Working medium temperature of water wall	t_{sw}	°C	See Standard Steam Table (saturated temperature)	342.53
98	Average temperature difference of heat transfer	Δt	°C	$\theta_{ave}-t_{sw}$	535.20
99	Absorbed convection heat of water wall	$Q_{ww,cal}$	kJ/kg	$K_{ave}\Delta t A_{ww}/(1000B_{cal})$	318.0
100	Error	e	%	$(Q_{ww}-Q_{ww,cal})/Q_{ww} \times 100$	0.31
101	Inlet steam enthalpy of superheater at furnace roof	i'_{roof}	kJ/kg	See Table C9	2674.2
102	Inlet steam temperature of superheater at furnace roof	t'_{roof}	°C	See Table C9	347.9
103	Steam enthalpy increment of superheater at furnace roof	Δi_{roof}	kJ/kg	$3.6Q_{roof}B_{cal}/(D_1-D_{ds1}-D_{ds2})$	10.72
104	Steam pressure of furnace roof at high-temperature superheater zone	p_{roof}	MPa	*Determine from design	14.9
105	Outlet steam enthalpy of furnace roof superheater	i''_{roof}	kJ/kg	$i'_{roof} + \Delta i$	2684.9
106	Outlet steam temperature of furnace roof superheater	t''_{roof}	°C	See Standard Steam Table	348.3

TABLE C11 Thermal Calculation of High-Temperature Superheater (*cont.*)

No.	Parameter	Symbol	Unit	Formulae and data source	Result
107	Average temperature difference	Δt	°C	$\theta_{ave}-0.5(t'_{roof} + t''_{roof})$	529.6
108	Absorbed convection heat of superheater at furnace roof	$Q_{roof,cal}$	kJ/kg	$K_{ave}\Delta tA_{roof}/(1000\ B_{cal})$	111.4
109	Error	e	%	$(Q_{roof}-Q_{roof,cal})/Q_{roof}$ $\times\ 100$	−0.34
110	Average temperature difference of heat transfer of rear water leading tube	Δt	°C	$\theta_{ave}-t_{sw}$	535.2
111	Absorbed convection heat of rear water leading tube	$Q_{rw,cal}$	kJ/kg	$K_{ave}\Delta tA_{rw}/(1000\ B_{cal})$	178.1
112	Error	e	%	$(Q_{rw}-Q_{rw,cal})/Q_{rw} \times$ 100	−0.06

desuperheater spray. The second method assumes the flue gas mass fraction, so error may result during calculation. Both methods are practiced by real-world engineers, and are applicable as long as calculation errors are identified and addressed.

There are usually determined values (θ', t', and t'') during the thermal calculation of high-temperature superheaters, thus the successive approximation method is not applicable. The convection heat is balanced usually either to rearrange the superheater and change the heat transfer area of the convection superheater, or to change the extent of desuperheating.

The structure of a low-temperature superheater is shown in Fig. C7. Its structural design and thermal calculation are shown in Tables C12 and C13.

Because the inlet and outlet steam temperature of the low-temperature superheater is unknown, further verification is necessary. For the inlet, the additional superheater should be determined after calculating the reversing chamber; for the outlet, the inlet steam temperature of the platen superheater is the temperature after desuperheating. Therefore, the quantity of desuperheater spray water and the outlet flue gas temperature should be assumed first to calculate the inlet steam temperature, then compared to the inlet weighted temperature of the low-temperature superheater obtained by the reversing chamber calculation. A temperature difference below 1°C is acceptable.

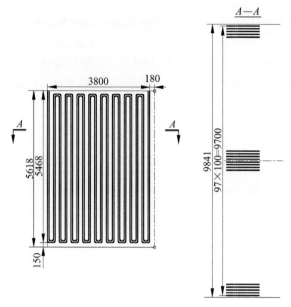

FIGURE C7 Structure diagram of low-temperature superheater.

TABLE C12 Structure of Low-Temperature Superheater

No.	Parameter	Symbol	Unit	Formulae	Value
1	Tube diameter	d	mm	Determine from design	42
2	Tube thickness	δ	mm	Determine from design	5
3	Number of head	n	–	Determine from design	2
4	Pass number of low-temperature superheater	n_1	–	Determine from design	18
5	Transverse rows	Z_1	–	Determine from design	98
6	Minimum bending radius of superheater	R	mm	Determine from design	80
7	Transverse spacing	s_1	mm	Determine from design	100
8	Relative transverse pitch	σ_1	–	s_1/d	2.38
9	Longitudinal spacing	s_2	mm	Determine from design	60

TABLE C12 Structure of Low-Temperature Superheater (*cont.*)

No.	Parameter	Symbol	Unit	Formulae	Value
10	Superheater depth	b	m	See Fig. C7	3.8
11	Radiative space depth before superheater	l_r	m	See Table C10	1.5
12	Longitudinal average spacing	$s_{2,ave}$	mm	$1000\,b/(n \times n_1-1)$	108.57
13	Relative transverse pitch	σ_2	–	$s_{2,ave}/d$	1.43
14	Flue duct height of low-temperature inlet	$h_{p,sh}$	m	See Table C10	5.618
15	Transverse flue duct height	h_g	m	See Table C10	5.618
16	Average tube calculated length	l_{ave}	m	Fig. C7, 18×5.468	98.42
17	Heating area	A	m^2	$nZ_1 \times \pi{\times}d \times l_{ave}$	2545.30
18	Distance from low-temperature superheater to reversing chamber	$l_{p,sh-rc}$	m	Determine from design	0.18
19	Roof tube length within low-temperature zone	l_{roof}	m	$l_{p,sh-rc} + b$	3.98
20	Heating area of roof tubes	A_{roof}	m^2	$al_{roof}(0.5\pi d + (s_{roof}-d))/s_{roof}$	49.55
21	Diameter of side wall tube	d	mm	Determine from design	51
22	Thickness of side wall tube	δ	mm	Determine from design	5
23	Spacing of side wall tube	$s_{s,sh}$	mm	Determine from design	100
24	Heating area of side wall	$A_{s,sh}$	m^2	$2h_g l_{roof}(0.5 \times \pi{\times}d + (s_{s,sh}-d))/s_{s,rc}$	57.73
25	Heating area of bottom water wall	A_{bww}	m^2	$al_{roof}(0.5 \times \pi{\times}d + (s_{bww}-d))/s_{bww}$	55.94
26	Gas average flow area	A_g	m^2	$h_g(a-dZ_1)$	32.16
27	Steam flow area	A_{ss}	m^2	$\pi/4 \times n \times Z_1 \times d_n^2$	0.158
28	Effective radiation layer thickness	s	m	$0.9d(4\sigma_1\sigma_2/\pi-1)$	0.126
29	Water leading tube diameter	d	mm	Determine from design	108

(Continued)

TABLE C12 Structure of Low-Temperature Superheater (*cont.*)

No.	Parameter	Symbol	Unit	Formulae	Value
30	Water leading tube number	n_{rw}	–	Determine from design	12
31	Water leading tube heating area	A_{rw}	m^2	$n_{rw} \times \pi \times d \times h_g$	22.87
32	Roof tube number	n_{roof}	–	See Table D5	98
33	Roof steam flow area	A_{roof}	m^2	$\pi/4 d_{n,roof}^2 n_{roof}$	0.129
34	Front wall connecting pipe diameter	d	mm	Determine from design	133
35	Front wall connecting pipe thickness	δ	mm	Determine from design	10
36	Front wall connecting pipe number	$Z_{f,sh}$	–	Determine from design	12
37	Front wall connecting pipe steam flow area	$A_{f,sh}$	m^2	Determine from design	0.120
38	Total steam flow rate	D	t/h	$(D_1 - D_{ds2} - D_{ds1})$	396
39	Roof steam flow rate	D_{roof}	t/h	$(D_1 - D_{ds2} - D_{ds1}) A_{roof}/(A_{roof} + A_{f,sh})$	205.17
40	Side wall steam flow rate	$D_{s,sh}$	t/h	$(D_1 - D_{ds2} - D_{ds1}) A_{f,sh}/(A_{roof} + A_{f,sh})$	190.83

TABLE C13 Thermal Calculation of Low-Temperature Superheater

No.	Parameter	Symbol	Unit	Formulae and data source	Result
1	Inlet flue gas temperature	θ'	°C	See Table C11	750.45
2	Inlet flue gas enthalpy	I'	kJ/kg	See Table C11	10597.9
3	Inlet flue gas temperature	θ''	°C	Trial and error	522
4	Outlet flue gas enthalpy	I''	kJ/kg	See Table C3	7318.4
5	Desuperheating water (main feeding water) enthalpy	i_{fw}	kJ/kg	See Table C4	1016.1

TABLE C13 Thermal Calculation of Low-Temperature Superheater (*cont.*)

No.	Parameter	Symbol	Unit	Formulae and data source	Result
6	Outlet steam enthalpy (without superheating)	i''	kJ/kg	$(D_1-D_{ds2})/(D_1-D_{ds1}-D_{ds2})i'_{psh}-D_{ds1}/(D_1-D_{ds1}-D_{ds2})i_{fw}$	3011.8
7	Outlet steam pressure	p''	°C	See Table C9	14.4
8	Outlet steam temperature	t''	°C	See Standard Steam Table	407.5
9	Absorbed convection heat of furnace roof	Q_{roof}	kJ/kg	Trial and error	71
10	Absorbed convection heat of side wall	$Q_{s,sh}$	kJ/kg	Trial and error	82
11	Absorbed convection heat of water leading tube	Q_{rw}	kJ/kg	Trial and error	33
12	Absorbed convection heat of bottom water wall	Q_{bww}	kJ/kg	Trial and error	81
13	Absorbed convection heat of superheater	Q_c	kJ/kg	$\varphi(I''-I' + \Delta\alpha I°_{ca})-Q_{roof}-Q_{s,sh}-Q_{bw}-Q_{bww}$	3105.1
14	Inlet steam enthalpy	i'	kJ/kg	$I''-3.6B_{cal}Q_c/(D_1-D_{ds2}-D_{ds1})$	2711.8
15	Inlet steam pressure	p'	°C	Determine from design	14.7
16	Inlet steam temperature	t'	°C	See Standard Steam Table	350.0
17	Weighted temperature of inlet steam flow rate	$t_{w,ave}$	°C	$(D_{roof}/D)t''_{rc} + (D_{s,sh}/D)t''_{s,sh}$	349.7
18	Temperature difference of low-temperature inlet interface	Δt_{li}	°C	$t'-t_{w,ave}$	0.2
19	Average steam temperature	t_{ave}	°C	$(t' + t'')/2$	378.7
20	Flue gas average temperature	θ_{ave}	°C	$(\theta' + \theta'')/2$	636.2

(Continued)

TABLE C13 Thermal Calculation of Low-Temperature Superheater (*cont.*)

No.	Parameter	Symbol	Unit	Formulae and data source	Result
21	Flue gas velocity	ω_g	m/s	$B_{cal}V_g(\theta_{ave} + 273)/(273A_g)$	10.69
22	Thermal conductivity of flue gas	λ	W/(m·°C)	Refer to Appendix D6.27	0.0771
23	Dynamic viscosity of flue gas	v	m²/s	Refer to Appendix D6.27	9.60 10⁻⁵
24	Average Prandtl number of flue gas	Pr_{ave}	–	Refer to Appendix D6.28	0.606
25	Prandtl number of flue gas	Pr	–	$(0.94 + 0.56r_{H_2O})Pr_{ave}$	0.594
26	Correction factor for rows	C_z	–	Refer to Appendix D 6.29	1
27	Flue gas composition and temperature correction coefficient	C_w	–	$0.92 + 0.726r_{H_2O}$	0.972
28	Correction factor for the geometric arrangement	C_s	–	Refer to Appendix D6.30	0.923
29	Flue gas side convection coefficient	α_g	w/(m²·°C)	$0.2\lambda/d(\omega_g/v)^{0.65}Pr^{0.33}C_zC_sC_w$	67.33
30	Average steam pressure	p_{ave}	MPa	$0.5(p' + p'')$	14.55
31	Average specific volume of steam	v_{ave}	m³/kg	See Standard Steam Table	0.015
32	Average velocity of steam	v_{ss}	m/s	$(D_1-D_{ds2}-D_{ds1})v_{ave}/(3600A_{ss})$	10.47
33	Correction factor for tube diameter	C_d	–	Refer to Appendix D6.26	0.91
34	Thermal conductivity of steam	λ	W/(m·°C)	See Standard Steam Table	0.081
35	Dynamic viscosity of steam	v	m²/s	See Standard Steam Table	3.81E-07
36	Prandtl number of steam	Pr	–	See Standard Steam Table	1.41
37	Steam side heat transfer coefficient	α_2	W/(m²·°C)	$0.023\lambda/d_n(\omega_{ss}d_n/v)^{0.8}Pr^{0.4}C_d$	3459.4
38	Ash deposition coefficient	ε	m²·°C/W	Refer to Appendix D 6.33	0.0043

TABLE C13 Thermal Calculation of Low-Temperature Superheater (*cont.*)

No.	Parameter	Symbol	Unit	Formulae and data source	Result
39	Fouling layer temperature of tube wall	t_w	°C	$t_{ave} + 1000(\varepsilon+1/\alpha_2)B_{cal}Q_c/A$	438.2
40	Product of p_n and s	p_ns	m·MPa	pr_ns	0.00265
41	Radiant absorption coefficient of gas	k_g	1/ (m·MPa)	$10.2[(0.78 + 1.6r_{H_2O})/ (10.2p_ns)^{0.5}-0.1] (1-0.37T_{ave}/1000)$	36.14
42	Radiant absorption coefficient of fly ash	k_{fa}	1/ (m·MPa)	$48350\rho_g/(T_{ave}^2 d_{fa}^2)^{1/3}$	97.99
43	Radiant absorption coefficient of flue gas radiation	k	1/ (m·MPa)	$k_g r_n + k_{fa}\mu_{fa}$	8.45
44	Exponent of Eq. (2.73)	kps	–	Kps	0.11
45	Flue gas emissivity	a	–	$1-e^{-kps}$	0.10
46	Radiation heat transfer coefficient	α_r	w/ (m²·°C)	$5.7 \times 10^{-8}(a_w + 1)/2aT_{ave}^3[1-(T_w/T_{ave})^4]/ [1-(T_w/T_{ave})]$	11.18
47	Fuel correction coefficient	A	–	Refer to Appendix D6.34	0.4
48	Radiation heat transfer coefficient correction	α'_r	w/ (m²·°C)	$\alpha_r[1 + A(T'/1000)^{0.25} (l_r/b)^{0.07}]$	15.39
49	Flowing uniformity coefficient	ω	–	Determine from design (constant 1)	1
50	Flue gas side heat transfer coefficient	α_1	W/ (m²·°C)	$\omega\alpha_g + \alpha'_r$	82.72
51	Effective coefficient	ψ	–	Refer to Appendix D6.35	0.65
52	Overall heat transfer coefficient	K	W/ (m²·°C)	$\psi\alpha_1\alpha_2/(\alpha_2 + \alpha_1)$	52.51
53	Counterflow small temperature difference	Δt_{min}	°C	$\theta''-t'$	172.1
54	Counterflow large temperature difference	Δt_{max}	°C	$\theta'-t''$	343.0
55	Logarithmic mean temperature difference	Δt	°C	$(\Delta t_{max}-\Delta t_{min})/(ln\Delta t_{max}/\Delta t_{min})$	247.8
56	Convection heat	$Q_{c,cal}$	kJ/kg	$K\Delta tA/(1000B_{cal})$	3116.2

(Continued)

TABLE C13 Thermal Calculation of Low-Temperature Superheater (*cont.*)

No.	Parameter	Symbol	Unit	Formulae and data source	Result
57	Error	E	%	$(Q_c - Q_{c,cal})/Q_c \times 100$	−0.36
58	Working medium temperature of bottom water wall	t_{sw}	°C	See Standard Steam Table (saturated temperature)	342.53
59	Average temperature difference of heat transfer	Δt	°C	$\theta_{ave} - t_{sw}$	293.7
60	Absorbed convection heat of bottom water wall	$Q_{bww,cal}$	kJ/kg	$K\Delta t A_{bww}/(1000 B_{cal})$	81.2
61	Error	e	%	$(Q_{bww} - Q_{bww,cal})/Q_{bww} \times 100$	−0.22
62	Inlet steam enthalpy of furnace roof superheater	i'_{roof}	kJ/kg	See Table C11	2684.9
63	Inlet steam temperature of furnace roof superheater	t'_{roof}	°C	See Table C11	348.3
64	Steam enthalpy increment of furnace roof superheater	Δi_{roof}	kJ/kg	$3.6 Q_{roof} B_{cal}/(D_1 - D_{ds1} - D_{ds2})$	6.86
65	Outlet steam enthalpy of furnace roof superheater	i''_{roof}	kJ/kg	$i'_{roof} + \Delta i_{roof}$	2691.8
66	Pressure of furnace roof superheater	p_{roof}	Mpa	Determine from design	14.8
67	Outlet steam temperature of furnace roof superheater	t''_{roof}	°C	See Standard Steam Table	348.5
68	Average temperature difference	Δt	°C	$\theta_{ave} - 0.5(t'_{roof} + t''_{roof})$	287.8
69	Absorbed convection heat of furnace roof superheater	$Q_{roof,cal}$	kJ/kg	$K\Delta t A_{roof}/(1000 B_{cal})$	70.5
70	Error	e	%	$(Q_{roof} - Q_{roof,cal})/Q_{roof} \times 100$	0.73
71	Average temperature difference of heat transfer of rear water leading tube	Δt_{rw}	°C	$\theta_{ave} - t_{sw}$	293.7

TABLE C13 Thermal Calculation of Low-Temperature Superheater (*cont.*)

No.	Parameter	Symbol	Unit	Formulae and data source	Result
72	Rear water leading tube convection heat	$Q_{rw,cal}$	kJ/kg	$K\Delta tA_{rw}/(1000B_{cal})$	33.2
73	Error	e	%	$(Q_{rw}-Q_{rw,cal})/Q_{rw} \times 100$	−0.59
74	Side wall temperature pressure	$p_{s,sh}$	Mpa	Determine from design	14.7
75	Side wall steam flow rate	$D_{s,sh}$	t/h	See Table C12 (paralled two paths)	190.83
76	Side wall steam enthalpy increment	$\Delta i_{s,sh}$	kJ/kg	$3.6Q_{s,sh}B_{cal}/D_{s,sh}$	16.44
77	Side wall inlet steam enthalpy	$i'_{s,sh}$	kJ/kg	$i'_{s,sh} = i''_{roof}$	2691.8
78	Side wall inlet steam temperature	$t'_{s,sh}$	°C	$t'_{s,sh} = t''_{roof}$	348.5
79	Side wall outlet steam enthalpy	$i''_{s,sh}$	kJ/kg	$i'_{s,sh} + \Delta i_{s,sh}$	2708.2
80	Side wall outlet steam temperature	$t''_{s,sh}$	°C	See Standard Steam Table	349.57
81	Side wall average steam temperature	$t_{s,sh,ave}$	°C	$0.5(t'_{s,sh} + t''_{s,sh})$	349.04
82	Average temperature difference	Δt	°C	$\theta_{ave}-t_{s,sh,ave}$	287.19
83	Side wall absorbed convection heat	$Q_{s,sh,cal}$	kJ/kg	$K\Delta tA_{s,sh}/(1000 B_{cal})$	81.9
84	Error	e	%	$(Q_{s,sh}-Q_{s,sh,cal})/Q_{s,sh} \times 100$	0.08

The structure design and thermal calculation of the reversing chamber are shown in Tables C14 and C15.

Recall that the connecting space between the horizontal flue ductwork and back-end shaft is the reversing chamber. The flue gas flow field is uneven, and the flue gas sweeping to heating surfaces is incomplete, thus heating surfaces are generally not arranged in the reversing chamber. For modern large- and mid-size boilers, which typically adopt a pendant structure, the enclosure wall superheater, pendant tube, and steam and water leading tubes are usually arranged in the reversing chamber. Due to the low flue gas velocity in the reversing chamber, the convection heat is low enough to be ignored; only radiation heat must be accounted for.

TABLE C14 Structure Calculation of Reversing Chamber

No.	Parameter	Symbol	Unit	Formulae or data source	Value
1	Tube diameter	d	mm	Determine from design	51
2	Tube thickness	δ	mm	Determine from design	5
3	Reversing chamber height	h_{rc}	m	Determine from design	5.618
4	Reversing chamber depth	b_{rc}	m	Determine from design	5.6
5	Reversing chamber width	w_{rc}	m	Determine from design	9.841
6	Total side wall area	$H_{s,sh}$	m^2	$2h_{rc}b_{rc}$	62.92
7	Rear wall area	$H_{r,sh}$	m^2	$w_{rc}h_{rc}$	55.28
8	Roof area	H_{roof}	m^2	$w_{rc}L_{rc}$	55.11
9	Total heating area	H_{rc}	m^2	$H_{s,sh} + H_{r,sh} + H_{roof}$	173.31
10	Reversing chamber configuration factor	x_{rc}	m^2	Determine from design	1
11	Radiation heating area	H	m^2	$x_{rc}H_{rc}$	173.31
12	Reversing chamber peripheral area	H_{sur}	m^2	$H_{s,sh} + 2H_{r,sh} + 2H_{roof}$	283.71
13	Reversing chamber volume	V	m^3	$0.5H_{s,sh}w_{rc}$	309.59
14	Effective radiation layer thickness	s	m	$3.6V/H_{sur}$	3.93

TABLE C15 Thermal Calculation of Reversing Chamber

No.	Parameter	Symbol	Unit	Formulae or data source	Value
1	Inlet flue gas temperature	θ'	°C	See Table C13	522
2	Inlet flue gas enthalpy	I'	kJ/kg	See Table C13	7318.4
3	Side wall absorbed heat	$Q_{s,sh}$	kJ/kg	Trial and error	38

TABLE C15 Thermal Calculation of Reversing Chamber (*cont.*)

No.	Parameter	Symbol	Unit	Formulae or data source	Value
4	Rear wall absorbed heat	$Q_{r,sh}$	kJ/kg	Trial and error	33.5
5	Roof absorbed heat	Q_{roof}	kJ/kg	Trial and error	33.3
6	Total absorbed heat	Q_r	kJ/kg	$Q_{s,sh} + Q_{r,sh} + Q_{roof}$	104.8
7	Outlet flue gas enthalpy	I''	kJ/kg	$I' - Q_r/\varphi$	7213.2
8	Outlet flue gas temperature	θ''	°C	See Table C3	515.0
9	Average flue gas temperature	θ_{ave}	°C	$0.5(\theta' + \theta'')$	518.5
10	Inlet steam temperature	t'	°C	See Table C13, t''_{roof}	348.5
11	Inlet steam enthalpy	i'	kJ/kg	See Table C13	2691.8
12	Steam enthalpy increment	Δi	kJ/kg	$3.6Q_r B_{cal}/D_{tn}$	19.54
13	Reversing chamber outlet pressure	p''	Mpa	Determine from design	14.7
14	Reversing chamber outlet steam enthalpy	i''	kJ/kg	$i' + \Delta i$	2711.3
15	Reversing chamber outlet steam temperature	t''	°C	See Standard Steam Table	349.86
16	Average steam temperature	t_{ave}	°C	$0.5(t' + t'')$	349.18
17	Ash deposition coefficient	ε	m²·°C/W	Refer to Appendix D6.33	0.0043
18	Fouling layer temperature of tube wall	t_w	°C	$t_{ave} + 1000\varepsilon B_{cal}Q_r/H$	376.8
19	Product of p_n and s	$p_n s$	m·MPa	$p r_n s$	0.0817
20	Radiant absorption coefficient of gas	k_g	l/(m·MPa)	$10.2[(0.78 + 1.6r_{H2O})/ (10.2p_n s)^{0.5} - 0.1] (1 - 0.37T_{ave}/1000)$	6.33

(*Continued*)

TABLE C15 Thermal Calculation of Reversing Chamber (*cont.*)

No.	Parameter	Symbol	Unit	Formulae or data source	Value
21	Radiant absorption coefficient of fly ash	k_{fa}	l/(m·MPa)	$48350\rho_g/(T_{ave}^2 d_{fa}^2)^{1/3}$	107.43
22	Radiant absorption coefficient of flue gas radiation	k	l/(m·MPa)	$k_g r_n + k_{fa}\mu_{fa}$	2.24
23	Exponent of Eq. (2.73)	kps	–	Kps	0.879
24	Flue gas emissivity	a	–	$1-e^{-kps}$	0.585
25	Radiation heat transfer coefficient	α_r	w/(m²·°C)	$5.7 \times 10^{-8}(a_w + 1)/2aT_{ave}^3[1-(T_w/T_{ave})^4]/[1-(T_w/T_{ave})]$	45.35
26	Average temperature difference	Δt	°C	$\theta_{ave}-t_w$	141.7
27	Absorbed heat of roof superheater	$Q_{roof,cal}$	kJ/kg	$\alpha_r\Delta t H_{roof}/(1000B_{cal})$	33.3
28	Error	e	%	$(Q_{roof}-Q_{roof,cal})/Q_{roof} \times 100$	–0.07
29	Absorbed heat of rear wall superheater	$Q_{r,sh,cal}$	kJ/kg	$\alpha_r\Delta t H_{r,sh}/(1000B_{cal})$	33.4
30	Error	e	%	$(Q_{r,sh}-Q_{r,sh,cal})/Q_{r,sh} \times 100$	0.21
31	Absorbed heat of side wall superheater	$Q_{s,sh}$	kJ/kg	$\alpha_r\Delta t H_{r,sh}/(1000B_{cal})$	38.0
32	Error	e	%	$(Q_{s,sh}-Q_{s,sh,cal})/Q_{s,sh} \times 100$	–0.12

C7   HEAT DISTRIBUTION

The above calculation is only the first step. Next, it is necessary to validate the accuracy of the boiler heat distribution and exhaust gas temperature. If the error is acceptable by specified requirements, the other heating surfaces can be calculated—if not, the structure design and thermal calculation of the heating surfaces must be recalculated. Table C16 shows necessary information related to heat distribution.

TABLE C16 Heat Distribution

No.	Parameter	Symbol	Unit	Formulae and data source	Result
1	Heat utilized by boiler	Q_{ut}	kJ/kg	Q_b/B_{cal}	26014.1
2	Radiation heat of furnace	Q_r	kJ/kg	See Table C7	15205.4
3	Absorbed convection heat of platen superheater	$Q_{psh,c,cal}$	kJ/kg	See Table C9	1245.3
4	Radiation heat from flue gas to heating surface through panel	$Q_{rl,cal}$	kJ/kg	See Table C9	143.9
5	Absorbed convection heat of high-temperature superheater	$Q_{ssh,c,cal}$	kJ/kg	See Table C11	3091.1
6	Absorbed heat of low-temperature superheater	$Q_{psh,c,cal}$	kJ/kg	See Table C13	3116.2
7	Absorbed heat of additional heating surfaces	ΣQ_{add}	kJ/kg	$\Sigma Q_{roof} + \Sigma Q_{ww} + \Sigma Q_{rw} + \Sigma Q_{s,sh} + \Sigma Q_{r,sh} = 223.6 + 116.1 + 318 + 111.4 + 178.1 + 81.2 + 70.5 + 33.2 + 81.9 + 33 + 33.4 + 37.7$	1318.7
8	Total absorbed heat after economizer	ΣQ	kJ/kg	$Q_r + Q_{psh,c,cal} + Q_{rl,cal} + Q_{ssh,c,cal} + Q_{psh,c,cal} + \Sigma Q_{add}$	24120.7
9	Adiabatic released heat before economizer	ΣQ_g	kJ/kg	$\varphi(Q_l{-}I_{rc}'' + \Delta\alpha I{^\circ}_{ca}) = 0.996 \times [31308.5 - 7213.2 + (1.26{-}1.2) \times 193.3]$	24010.5
10	Error	e	%	$1000(\Sigma Q_g{-}\Sigma Q)/\Sigma Q_g$	−0.46
11	Outlet steam enthalpy of boiler	i_{ss}''	kJ/kg	See Table C4	3437.47
12	Boiler blowdown percentage	δ_{bd}	%	See Table C4	1
13	Economizer water flow rate	D_{eco}	t/h	$D_1 + D_{bd}{-}D_{ds}$	400.1
14	Economizer outlet water enthalpy	i_{eco}''	kJ/kg	$D_1/D_{eco}i_{ss}'' - B_{cal}\Sigma Q/D_{eco} - D_{bd}/D_{eco}i_s' - D_{ds}/D_{eco}i_{fw}$	1207.4

(Continued)

TABLE C16 Heat Distribution (*cont.*)

No.	Parameter	Symbol	Unit	Formulae and data source	Result
15	Economizer absorbed heat (by balance)	Q_{eco}	kJ/kg	$[D_{eco}(i''_{eco}-i'_{eco})]/B_{cal}$	2000.1
16	Air preheater absorbed heat	Q_{aph}	kJ/kg	$(\beta''_{aph} + 0.5\Delta\alpha)(I^0_{ha}-I^0_{ca})$	3368.1
17	Exhaust gas enthalpy verification	$I_{ex,cal}$	kJ/kg	$I''_{rc}-(Q_{eco} + Q_{aph})/\varphi+0.5\Delta\alpha_{aph}(I^0_{ha} + I^0_{ca}) + \Delta\alpha_{eco}I^0_{ha}$	1952.9
18	Exhaust gas temperature verification	$\theta_{ex,cal}$	°C	See Table C3	136.3

Note: The total error of verification of heat transfer of the heating surface posteconomizer (in terms of steam and water flow direction) should be less than ±0.5%; the outlet working medium temperature of the upper stage economizer can be calculated according to the heat balance to verify the exhaust gas temperature.

C8 ECONOMIZER STRUCTURE DESIGN AND THERMAL CALCULATION

The structure of the upper stage economizer is shown in Fig. C8 and Table C17. The thermal calculation of the upper stage economizer is detailed in Table C18. The structure of the lower stage economizer is shown in Fig. C9 and Table C19, and its thermal calculation process is detailed in Table C20.

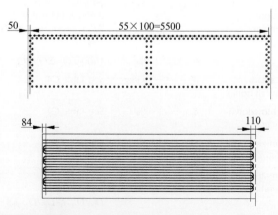

FIGURE C8   Structure diagram of upper stage economizer.

TABLE C17 Structure of Upper Economizer

No.	Parameter	Symbol	Unit	Formulae or data source	Value
1	Tube diameter	d	mm	Determine from design	32
2	Tube thickness	δ	mm	Determine from design	4
3	Number of transverse even rows	$Z_{1,even}$	–	Determine from design	56
4	Number of transverse odd rows	$Z_{1,odd}$	–	Determine from design	55
5	Average transverse rows	$Z_{1,ave}$	–	$0.5(Z_{1,even} + Z_{1,odd})$	55.5
6	Longitudinal rows (longitudinal direction)	Z_2	–	Determine from design	20
7	Number of total transverse rows	n	–	$Z_{1,odd} + Z_{1,even}$	111
8	Bending radius	R	mm	Determine from design	60
9	Transverse spacing	s_1	mm	Determine from design	100
10	Longitudinal spacing	s_2	mm	Determine from design	60
11	Relative transverse pitch	σ_1	–	s_1/d	3.13
12	Relative vertical pitch	σ_2	–	s_2/d	1.88
13	Distance from economizer to outlet header	l_{tank}	m	Determine from design	1.2
14	Radiative space depth before economizer	l_r	m	$h_{tn} + l_{tank}$	6.818
15	Economizer height	h	m	$(Z_2-1)s_2$	1.140
16	Distance from economizer to side wall	δ_1	mm	Determine from design	24
17	Distance from economizer tubes to central line	δ_2	mm	Determine from design	50
18	Distance from economizer to front and rear walls	δ_3	mm	Determine from design	50
19	Flue duct width	a	m	Determine from design	9.841
20	Flue duct depth	b_g	m	Determine from design	5.6
21	Number of sets of economizer tubes	–	–	Determine from design	2

(Continued)

TABLE C17 Structure of Upper Economizer (*cont.*)

No.	Parameter	Symbol	Unit	Formulae or data source	Value
22	Tube length of each row	l_i	m	$a-2(\delta_1+\delta_2)$	9.693
23	Total tube length of heating surface	l'	m	$Z_{1,ave}\,Z_2 l_i + n\pi R(Z_2-2)/2$	10947.8
24	Tube length of the top two rows	l_1	m	nl_i	1075.9
25	Tube length near wall and central line	l_2	m	$2(Z_2-4)l_i$	310.2
26	Tube length in the clearance (gas corridor) adjacent to side wall	l_3	m	$2n\delta_1$	5.33
27	Tube length of elbow and middle section	l_4	m	$2n\pi R(Z_2/2-1)$	376.6
28	Tube length of effective heating surface	l	m	$l'-(l_1+l_2+l_4)/2+l_3/2$	10069.1
29	Heating area	A	m²	$\pi d l$	1012.3
30	Flue gas flow area	A_g	m²	$ab-Z_1 d(l_i+4R)$	37.5
31	Water flow area	A_w	m²	$2n(\pi d_n^2/4)$	0.10
32	Effective radiation layer thickness	s	m	$0.9d(4\sigma_1\sigma_2/\pi-1)$	0.186

TABLE C18 Thermal Calculation of Upper Stage Economizer

No.	Parameter	Symbol	Unit	Formulae or data source	Value
1	Inlet flue gas enthalpy	I'	kJ/kg	See Table C15	7213.2
2	Inlet flue gas temperature	θ'	°C	See Table C15	515.0
3	Outlet flue gas temperature	θ''	°C	Trial and error	421
4	Outlet flue gas enthalpy	I''	kJ/kg	See Table C3	5938.7
5	Absorbed convection heat of economizer	Q_c	kJ/kg	$\varphi(I''-I'+\Delta\alpha I^o_{ca})$	1273.2
6	Boiler blowdown percentage	δ_{bd}	%	See Table C16	1

TABLE C18 Thermal Calculation of Upper Stage Economizer (*cont.*)

No.	Parameter	Symbol	Unit	Formulae or data source	Value
7	Economizer water flow rate	D_{eco}	t/h	See Table C16	400.1
8	Economizer outlet water enthalpy	i''	kJ/kg	See Table C16	1207.4
9	Outlet water pressure	p''	MPa	Determine from design	15.07
10	Outlet water temperature	t''	°C	See Standard Steam Table	275.0
11	Inlet water enthalpy	i'	kJ/kg	$i''-3.6B_{cal}Q_c/D_{eco}$	1085.6
12	Enthalpy increment	Δi	kJ/kg	$i''-i'$	121.7
13	Inlet water pressure	p'	MPa	Determine from design	15.3
14	Inlet water temperature	t''	°C	See Standard Steam Table	249.9
15	Counterflow large temperature difference	Δt_{max}	°C	$\theta'-t''$	240.0
16	Counterflow small temperature difference	Δt_{min}	°C	$\theta''-t'$	171.1
17	Counterflow logarithmic mean temperature difference	Δt	°C	$(\Delta t_{max}-\Delta t_{min})/(ln\Delta t_{max}/\Delta t_{min})$	203.6
18	Average flue gas temperature	θ_{ave}	°C	$(\theta' + \theta'')/2$	468.0
19	Average water temperature	t_{ave}	°C	$(t' + t'')/2$	262.4
20	Working medium mass flux	$\rho\omega$	kg/(m²s)	$D_{eco}/(3.6A_w)$	1106.6
21	Flue gas velocity	ω_g	m/s	$B_{cal}V_g(\theta_{ave} + 273)/(273A_g)$	7.54
22	Thermal conductivity of flue gas λ	λ	W/(m·°C)	Refer to Appendix D6.27	0.0626
23	Dynamic viscosity of flue gas ν	ν	m²/s	Refer to Appendix D6.27	6.79 10^{-5}
24	Average Prandtl number of flue gas	Pr_{ave}	–	Refer to Appendix D6.28	0.623
25	Prandtl number of flue gas	Pr	–	$(0.94 + 0.56r_{H_2O})$ Pr_{ave}	0.610
26	Relative diagonal pitch	s_2'/d	–	$[1/4(s_1/d)^2 + (s_2/d)^2]^{1/2}$	2.44

(Continued)

TABLE C18 Thermal Calculation of Upper Stage Economizer (*cont.*)

No.	Parameter	Symbol	Unit	Formulae or data source	Value
27	Parameter	φ_σ	–	$(s_1/d-1)/(s_2'/d-1)$	1.47
28	Correction factor for rows	C_z	–	Refer to Appendix D6.36	1
29	Flue gas composition and temperature correction coefficient	C_w	–	$0.92 + 0.726 r_{H_2O}$	0.971
30	Correction factor for the geometric arrangement	C_s	–	$0.95\varphi_\sigma^{0.1}$	0.988
31	Flue gas side convection coefficient	α_c	w/(m²·°C)	$0.358\lambda/d(\omega_g/\nu)^{0.6}$ $Pr^{0.33}C_zC_sC_w$	77.02
32	Fouling layer temperature of tube wall	t_w	°C	$t_{ave} + 60$	322.4
33	Product of p_n and s	$p_n s$	m·MPa	$pr_n s$	3.84 10^{-3}
34	Radiant absorption coefficient of gas	k_g	l/(m·MPa)	$10.2[(0.78 + 1.6r_{H_2O})/ (10.2p_n s)^{0.5}-0.1]$ $(1-0.37T_{ave}/1000)$	32.62
35	Radiant absorption coefficient of fly ash	k_{fa}	l/(m·MPa)	$48350\rho_g/ (T_{ave}^2 d_{fa}^2)^{1/3}$	112.2
36	Radiant absorption coefficient of flue gas radiation	k	l/(m·MPa)	$k_g r_n + k_{fa}\mu_{fa}$	7.69
37	Exponent of Eq. (2.73)	kps	–	Kps	0.143
38	Flue gas emissivity	a	–	$1-e^{-kps}$	0.133
39	Radiation heat transfer coefficient	α_r	w/(m²·°C)	$5.7 \times 10^{-8}(a_w + 1)/2aT_{ave}^3[1- (T_w/T_{ave})^4]/[1- (T_w/T_{ave})]$	8.26
40	Fuel correction coefficient	A	–	Refer to Appendix D6.34	0.4
41	Radiation heat transfer coefficient correction	α'_r	w/(m²·°C)	$\alpha_r[1 + A(T'/1000)^{0.2} {}^5(l_r/b)^{0.07}]$	13.99
42	Basic ash deposition coefficient	ε	m²·°C/w	Refer to Appendix D6.37	0.0032
43	Additional correction factor for ash deposition	$\Delta\varepsilon$	m²·°C/w	Refer to Appendix D6.38	0.0017
44	Correction factor for tube diameter	C_d	–	Refer to Appendix D6.39	0.75

TABLE C18 Thermal Calculation of Upper Stage Economizer (*cont.*)

No.	Parameter	Symbol	Unit	Formulae or data source	Value
45	Ash deposition coefficient	ε	$m^2 \cdot {}^\circ C/w$	$C_d \varepsilon + \Delta \varepsilon$	0.0041
46	Flue gas side heat transfer coefficient	α_1	$W/(m^2 \cdot {}^\circ C)$	$\alpha_d + \alpha'_f$	91.00
47	Overall heat transfer coefficient	K	$W/(m^2 \cdot {}^\circ C)$	$\alpha_1/(\varepsilon \alpha_1 + 1)$	66.27
48	Convection heat	$Q_{c,cal}$	kJ/kg	$K \Delta t A/(1000 B_{cal})$	1285.4
49	Error	e	%	$(Q_c - Q_{c,cal})/Q_c \times 100$	−0.96

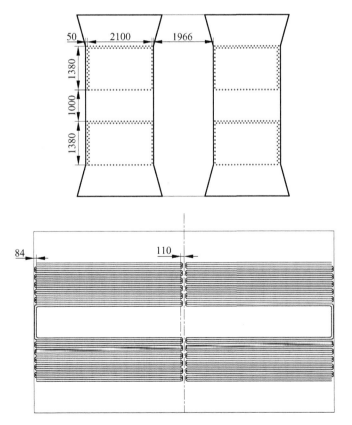

FIGURE C9 Structure diagram of lower stage economizer.

TABLE C19 Structure of Lower Stage Economizer

No.	Parameter	Symbol	Unit	Formulae or data source	Value
1	Tube diameter	d	mm	Determine from design	32
2	Tube thickness	δ	mm	Determine from design	4
3	Number of transverse even rows	$Z_{1,even}$	–	Determine from design	44
4	Number of transverse odd rows	$Z_{1,odd}$	–	Determine from design	43
5	Average transverse rows	$Z_{1,ave}$	–	$0.5(Z_{1,even} + Z_{1,odd})$	43.5
6	Longitudinal rows (vertical direction)	Z_2	–	Determine from design	48
7	Number of total transverse rows	n	–	$(Z_{1,even} + Z_{1,odd})$	87
8	Bending radius	R	mm	Determine from design	60
9	Transverse spacing	s_1	mm	Determine from design	100
10	Longitudinal spacing	s_2	mm	Determine from design	60
11	Radiative space depth before economizer	l_r	m	Determine from design	1
12	Economizer partition depth	l_{spa}	m	Determine from design	1
13	Economizer height	h	m	Determine from design	3.76
14	Relative transverse pitch	σ_1	–	s_1/d	3.125
15	Relative longitudinal pitch	σ_2	–	s_2/d	1.875
16	Distance from economizer tubes to side wall	δ_1	mm	Determine from design	24
17	Distance from economizer tubes to central line	δ_2	mm	Determine from design	50
18	Distance from economizer to front or rear wall	δ_3	mm	Determine from design	50
19	Flue duct width	a	m	Determine from design	9.841
20	Flue duct depth	b	m	Determine from design	4.4
21	Number of sets of economizer tubes	–	–	Determine from design	2

TABLE C19 Structure of Lower Stage Economizer (*cont.*)

No.	Parameter	Symbol	Unit	Formulae or data source	Value
22	Tube length of each row	l_i	m	$a-2(\delta_1 + \delta_2)$	9.693
23	Tube length of heating surfaces	l'	m	$Z_{1,ave}Z_2l_i + n\pi R(Z_2-2)/2$	20616.7
24	Tube length of the top two rows	l_1	m	$2nl_i$	1686.6
25	Tube length near wall and central line	l_2	m	$4(Z_2-4)l_i$	1706.0
26	Tube length in the clearance (gas corridor) adjacent to side wall	l_3	m	$2n\delta_1$	4.18
27	Bend and middle section	l_4	m	$2n\pi R(Z_2/2-1)$	754.4
28	Tube length of effective heating surface	l	m	$l'-(l_1 + l_2 + l_4)/2 + l_3/2$	18545.3
29	Heating area	A	m^2	$\pi d l$	1864.4
30	Flue gas flow area	A_g	m^2	$ab-Z_1d(l_i + 4R)$	29.47
31	Water flow area	A_w	m^2	$2n(\pi d_n^2/4)$	0.0787
32	Effective radiation layer thickness	s	m	$0.9d(4\sigma_1\sigma_2/\pi-1)$	0.186

TABLE C20 Thermal Calculation of Lower Stage Economizer

No.	Parameter	Symbol	Unit	Formulae or data source	Value
1	Inlet flue gas enthalpy	I'	kJ/kg	See Table C22	4625.3
2	Inlet flue gas temperature	θ'	°C	See Table C22	326.8
3	Inlet water pressure	p'	MPa	Determine from design	15.6
4	Inlet water temperature	t''	°C	Determine from design	235
5	Inlet water enthalpy	i'	kJ/kg	See Standard Steam Table	1016.1

(Continued)

TABLE C20 Thermal Calculation of Lower Stage Economizer (*cont.*)

No.	Parameter	Symbol	Unit	Formulae or data source	Value
6	Economizer water flow rate	D_{eco}	t/h	See Table C16	400.1
7	Outlet flue gas temperature	θ''	°C	Trial and error	275
8	Outlet flue gas enthalpy	I''	kJ/kg	See Table C3	3919.7
9	Absorbed convection heat of economizer	Q_d	kJ/kg	$\varphi(I'-I'' + \Delta\alpha I^{\circ}_{ca})$	706.6
10	Economizer outlet water enthalpy	i''	kJ/kg	$i' + 3.6B_{cal}Q_c/D_{eco}$	1083.7
11	Outlet water pressure	p''	MPa	Determine from design	15.3
12	Outlet water temperature	t'	°C	See Standard Steam Table	249.5
13	Temperature difference	–	°C	–	–0.4
14	Counterflow large temperature difference	Δt_{max}	°C	$\theta'-t''$	77.3
15	Counterflow small temperature difference	Δt_{min}	°C	$\theta''-t'$	40.0
16	Counterflow logarithmic mean temperature difference	Δt	°C	$(\Delta t_{max}-\Delta t_{min})/(ln\Delta t_{max}/\Delta t_{min})$	56.6
17	Average flue gas temperature	θ_{ave}	°C	$(\theta' + \theta'')/2$	300.9
18	Average water temperature	t_{ave}	°C	$(t' + t'')/2$	242.3
19	Flue gas velocity	ω_g	m/s	$B_{cal}V_g(\theta_{ave} + 273)/(273A_g)$	7.70
20	Working medium mass flux	$\rho\omega$	kg/(m²s)	$D_{eco}/(3.6A_w)$	1411.9
21	Thermal conductivity of flue gas λ	λ	W/(m·°C)	Refer to Appendix D6.27	0.0483
22	Dynamic viscosity of flue gas v	v	m²/s	Refer to Appendix D6.27	4.37 10^{-5}
23	Average Prandtl number of flue gas	Pr_{ave}	–	Refer to Appendix D6.27	0.650
24	Prandtl number of flue gas	Pr	–	$(0.94 + 0.56r_{H_2O})Pr_{ave}$	0.636

TABLE C20 Thermal Calculation of Lower Stage Economizer (*cont.*)

No.	Parameter	Symbol	Unit	Formulae or data source	Value
25	Relative diagonal pitch	s_2'/d	–	$[1/4(s_1/d)^2 + (s_2/d)^2]^{1/2}$	2.44
26	Parameter	φ_σ	–	$(s_1/d{-}1)/(s_2'/d{-}1)$	1.47
27	Correction factor for rows	C_z	–	Refer to Appendix D6.36	1
28	Flue gas composition and temperature correction coefficient	C_w	–	$0.92 + 0.726 r_{H_2O}$	0.97
29	Correction factor for the geometric arrangement	C_s	–	$0.768\varphi_\sigma^{0.5}$	0.93
30	Flue gas side convection coefficient	α_c	w/(m$^2\cdot$°C)	$0.358\lambda/d(\omega_g/\nu)^{0.6}$ $Pr^{0.33}C_zC_sC_w$	74.94
31	Fouling layer temperature of tube wall	t_w	°C	$t_{ave} + 25$	267.3
32	Product of p_n and s	$p_n s$	m·MPa	$p r_n s$	0.0037
33	Radiant absorption coefficient of gas	k_g	l/(m·MPa)	$10.2[(0.78 + 1.6 r_{H_2O})/$ $(10.2 p_n s)^{0.5}{-}0.1]$ $(1{-}0.37 T_{ave}/1000)$	35.89
34	Radiant absorption coefficient of fly ash	k_{fa}	l/(m·MPa)	$48350\rho_g/(T_{ave}^2 d_{fa}^2)^{1/3}$	132.91
35	Radiant absorption coefficient of flue gas radiation	k	l/(m·MPa)	$k_g r_n + k_{fa}\mu_{fa}$	8.25
36	Exponent of Eq. (2.73)	kps	–	kps	0.154
37	Flue gas emissivity	a	–	$1{-}e^{-kps}$	0.142
38	Radiation heat transfer coefficient	α_r	w/(m$^2\cdot$°C)	$5.7 \times 10^{-8}(a_w +$ $1)/2a T_{ave}^3[1{-}(T_w/T_{ave})^4]/$ $[1{-}(T_w/T_{ave})]$	5.06
39	Fuel correction coefficient	A	–	Refer to Appendix D6.34	0.4
40	Radiation heat transfer coefficient	α_r'	w/(m$^2\cdot$°C)	$\alpha_r[1 + A(T'/1000)^{0.25}(l$ $_t/b)^{0.07}]$	7.69
41	Basic ash deposition coefficient	ε	m$^2\cdot$°C/w	Refer to Appendix D6.37	0.0025

(Continued)

TABLE C20 Thermal Calculation of Lower Stage Economizer (*cont.*)

No.	Parameter	Symbol	Unit	Formulae or data source	Value
42	Additional correction factor for ash deposition	$\Delta\varepsilon$	$m^2 \cdot °C/w$	Refer to Appendix D6.38	0
43	Correction factor for tube diameter	C_d	–	Refer to Appendix D6.39	0.75
44	Ash deposition factor	ε	$m^2 \cdot °C/w$	$C_d\varepsilon + \Delta\varepsilon$	1.88 10^{-3}
45	Flue gas side heat transfer coefficient	α_1	$W/(m^2 \cdot °C)$	$\alpha_d + \alpha'_f$	82.62
46	Overall heat transfer coefficient	K	$W/(m^2 \cdot °C)$	$\alpha_1/(\varepsilon\alpha_1 + 1)$	71.54
47	Convection heat	$Q_{c,cal}$	kJ/kg	$K\Delta t A/(1000 B_{cal})$	710.8
48	Error	e	%	$(Q_c - Q_{c,cal})/Q_c \times 100$	−0.59

The economizer is arranged in two stages and calculated by stage. Considering the ash content of coal, guard tiles should be arranged for the first and second rows of tubes at the flue gas inlet of the tube bundles, tube bend, and tubes near the wall.

The upper stage economizer should be smaller than the lower stage economizer to leave the upper air preheater sufficient temperature difference.

Flue gas velocity of convection heating surfaces is related to both heat transfer intensity of heating surfaces and flue gas side flow resistance and abrasion. Enhancing flue gas velocity improves the heat transfer and reduces the consumption of steel resources, but increases flow resistance and abrasion of heating surfaces. This example refers to practical experience provided by Beijing Boiler Company for heating surfaces arrangements designed to ensure reasonable flue gas velocity.

C9   AIR PREHEATER STRUCTURE DESIGN AND THERMAL CALCULATION

The structure of the upper air preheater is shown in Fig. C10 and Table C21, and its thermal calculation process is detailed in Table C22. The structure of the lower air preheater is shown in Fig. C11 and Table C23, and its thermal calculation process is detailed in Table C24.

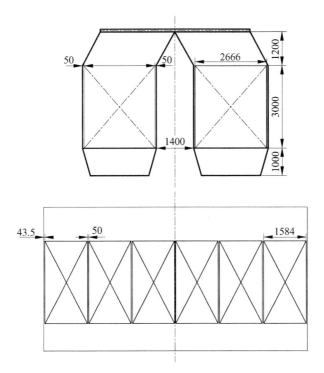

FIGURE C10   Structure design of upper stage air preheater.

TABLE C21 Structure of Upper Stage Air Preheater

No.	Parameter	Symbol	Unit	Formulae or data source	Value
1	Tube diameter	d	mm	Determine from design	40
2	Tube thickness	δ	mm	Determine from design	1.5
3	Number of pass	n	–	Determine from design	1
4	Number of tube box	B	–	Determine from design	12
5	Number of transverse tube box	B_1	–	Determine from design	6
6	Rows of tube box of air flow direction	B_2	–	Determine from design	2
7	Tube box height	H	m	Determine from design	3
8	Transverse spacing	s_1	–	Determine from design	72
9	Longitudinal spacing	s_2	–	Determine from design	43

(Continued)

TABLE C21 Structure of Upper Stage Air Preheater (*cont.*)

No.	Parameter	Symbol	Unit	Formulae or data source	Value
10	Relative transverse pitch	σ_1	–	s_1/d	1.8
11	Relative vertical pitch	σ_2	–	s_2/d	1.075
12	Transverse rows of single casing	Z_{i1}	–	Determine from design	23
13	Rows of single casing in air flow direction	Z_{i2}	–	Determine from design	63
14	Transverse rows	Z_1	–	$B_1 Z_{i1}$	138
15	Rows of air flow direction	Z_2	–	Determine from design	126
16	Number of tubes of a casing	n_i	–	Determine from design	1418
17	Total number of tubes	Σn	–	$n_i B$	17016
18	Casing spacing	δ_1	mm	Determine from design	50
19	Distance from casing to side wall	δ_2	mm	Determine from design	43.5
20	Distance from casing to front and rear walls	δ_3	mm	Determine from design	50
21	Flue duct width	a	m	Determine from design	9.841
22	Heating area	A	m^2	$\Sigma nh\pi(d-\delta)/1000$	6174.3
23	Flue gas flow area	A_g	m^2	$\Sigma n\pi/4d_n^2$	18.30
24	Air flow area	A_a	m^2	$2h(a-Z_1 d-5\delta_1)$	24.43
25	Effective radiation layer thickness	s	m	$0.9d_n$	0.0333

TABLE C22 Thermal Calculation of Upper Stage Air Preheater

No.	Parameter	Symbol	Unit	Formulae or data source	Value
1	Inlet flue gas temperature	θ'	°C	See Table C18	421
2	Inlet flue gas enthalpy	I'	kJ/kg	See Table C18	5938.696
3	Excess air coefficient at air preheater outlet	β''_{pah}	–	See Table C7	1.11

TABLE C22 Thermal Calculation of Upper Stage Air Preheater (*cont.*)

No.	Parameter	Symbol	Unit	Formulae or data source	Value
4	Excess air coefficient at air preheater inlet	β'_{pah}	–	$\beta''_{ah} + \Delta\alpha$	1.14
5	Outlet air temperature	t''	°C	Determine from design	320
6	Outlet air enthalpy	I''_a	kJ/kg	See Table C3	3147.8
7	Inlet air temperature	t'	°C	Trial and error	197
8	Inlet air enthalpy	I'_a	kJ/kg	See Table C3	1917.7
9	Air preheater absorbed convection heat	Q_c	kJ/kg	$(I''_a - I'_a)(\beta''_{aph} + 0.5\Delta\alpha)$	1383.8
10	Average air temperature	t_{ave}	°C	$(t' + t'')/2$	258.5
11	Average air enthalpy	i_{ave}	°C	$(I''_a + I'_a)/2$	2532.8
12	Outlet flue gas enthalpy	I''	kJ/kg	$I' - Q_c/\varphi + \Delta\alpha i_{ave}$	4625.3
13	Outlet flue gas temperature	θ''	°C	See Table C3	326.8
14	Flue gas average temperature	θ_{ave}	°C	$(\theta' + \theta'')/2$	373.9
15	Counterflow small temperature difference	Δt_{min}	°C	$\theta' - t''$	101
16	Counterflow large temperature difference	Δt_{max}	°C	$\theta'' - t'$	129.8
17	Counterflow logarithmic mean temperature difference	Δt	°C	$(\Delta t_{max} - \Delta t_{min})/$ $(ln\Delta t_{max}/\Delta t_{min})$	114.8
18	Large temperature drop	t_{max}	°C	$t'' - t'$	123
19	Small temperature drop	t_{min}	°C	$\theta' - \theta''$	94.17
20	Parameter	P	–	$t_{min}/(\theta' - t')$	0.420
21	Parameter	R	–	t_{max}/t_{min}	1.31
22	Temperature difference correction coefficient	ψ	–	Refer to Appendix D6.40	0.93

(Continued)

TABLE C22 Thermal Calculation of Upper Stage Air Preheater (*cont.*)

No.	Parameter	Symbol	Unit	Formulae or data source	Value
23	Average temperature difference	Δt_{ave}	°C	$\psi \Delta t$	106.8
24	Average air velocity	ω_a	m/s	$B_{cal}V^o(\beta''_{aph}+0.5\Delta\alpha)$ $(t_{ave}+273)/$ $(273 \times A_a)$	6.98
25	Average flue gas velocity	ω_g	m/s	$B_{cal}V_g(\theta_{ave}+273)/$ $(273A_g)$	13.73
26	Tube wall temperature	t_w	°C	$0.5(t_{ave}+\theta_{ave})$	316.2
27	Product of p_n and s	$p_n s$	m·MPa	$pr_n s$	$6.76\ 10^{-4}$
28	Radiant absorption coefficient of gas	k_g	$1/$ (m·MPa)	$10.2[(0.78 +$ $1.6r_{H_2O})/$ $(10.2p_n s)^{0.5}-0.1]$ $(1-0.37T_{ave}/1000)$	82.42
29	Radiant absorption coefficient of fly ash	k_{fa}	$1/$ (m·MPa)	$48350\rho_g/(T_{ave}^2 d_{fa}^2)^{1/3}$	122.79
30	Radiant absorption coefficient of flue gas radiation	k	$1/$ (m·MPa)	$k_g r_n + k_{fa}\mu_{fa}$	17.76
31	Exponent of Eq. (2.73)	kps	–	Kps	0.0591
32	Flue gas emissivity	a	–	$1-e^{-kps}$	0.0574
33	Radiation heat transfer coefficient	α_r	$w/$ (m^2·°C)	$5.7 \times 10^{-8}(a_w + 1)/$ $2aT_{ave}^3[1-(T_w/T_{ave})^4]/$ $[1-(T_w/T_{ave})]$	2.79
34	Correction factor for tube length	C_l	–	Refer to Appendix D6.41	1
35	Flue gas temperature and composition correction coefficient	C_w	–	Refer to Appendix D6.42	0.98
36	Flue gas temperature and wall temperature correction coefficient	C_t	–	Refer to Appendix D6.43	1
37	Thermal conductivity of flue gas	λ	–	Refer to Appendix D6.27	0.055
38	Dynamic viscosity of flue gas	ν	–	Refer to Appendix D6.27	5.38×10^{-5}

TABLE C22 Thermal Calculation of Upper Stage Air Preheater (*cont.*)

No.	Parameter	Symbol	Unit	Formulae or data source	Value
39	Average Prandtl number of flue gas	Pr_{ave}	–	Refer to Appendix D6.28	0.635
40	Prandtl number of flue gas	Pr	–	$(0.94 + 0.56r_{H_2O})Pr_{ave}$	0.622
41	Flue gas side convection heat transfer coefficient	α_c	W/ (m^2·°C)	$0.023\lambda/d(\omega_g d/\nu)^{0.8}Pr^{0.4}C_tC_lC_w$	41.60
42	Flue gas side heat transfer coefficient	α_1	W/ (m^2·°C)	$\alpha_c + \alpha_r$	44.39
43	Air thermal conductivity	λ	W/ (m·°C)	Refer to Appendix D6.27	0.0422
44	Air dynamic viscosity	ν	m^2/s	Refer to Appendix D6.27	4.28 10^{-5}
45	Air Prandtl number	Pr	–	Refer to Appendix D6.27	0.69
46	Relative diagonal pitch	s_2'/d	–	$[1/4(s_1/d)^2 + (s_2/d)^2]^{1/2}$	1.40
47	Parameter	φ_σ	–	$(s_1/d-1)/(s_2'/d-1)$	1.99
48	Correction factor for rows	C_z	–	Refer to Appendix D6.36	1
49	Flue gas composition and temperature correction coefficient	C_w	–	Refer to Appendix D6.44	0.92
50	Correction factor for the geometric arrangement	C_s	–	$0.768\varphi_\sigma^{0.5}$	1.08
51	Air side convection heat transfer coefficient	α_2	W/ (m^2•°C)	$0.358\lambda/d(\omega_a d/\nu)^{0.6}Pr^{0.33}C_zC_sC_w$	64.73
52	Utilization coefficient	ξ	–	Refer to Appendix D6.45	0.85
53	Overall heat transfer coefficient	K	W/ (m^2•°C)	$\xi\alpha_1\alpha_2/(\alpha_1 + \alpha_2)$	22.38
54	Convection heat	$Q_{c,cal}$	kJ/kg	$K\Delta tA/(1000B_{cal})$	1388.535
55	Error	e	%	$(Q_c-Q_{c,cal})/Q_c \times 100$	−0.34

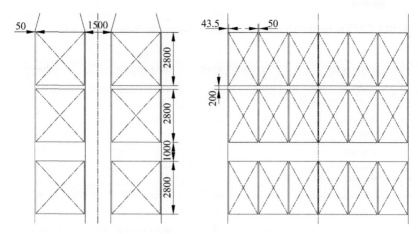

FIGURE C11   Structure diagram of lower stage air preheater.

TABLE C23 Structure of Lower Stage Air Preheater

No.	Parameter	Symbol	Unit	Formulae or data source	Value
1	Tube diameter	d	mm	Determine from design	40
2	Tube thickness	δ	mm	Determine from design	1.5
3	Number of pass	n	–	Determine from design	3
4	Number of tube box	B	–	Determine from design	12
5	Number of transverse tube box	B_1	–	Determine from design	6
6	Rows of tube box of air flow direction	B_2	–	Determine from design	2
7	Tube box height	h	m	Determine from design	2.8
8	Transverse spacing	s_1	mm	Determine from design	72
9	Longitudinal spacing	s_2	mm	Determine from design	43
10	Relative transverse pitch	σ_1	–	s_1/d	1.8
11	Relative longitudinal pitch	σ_2	–	s_2/d	1.075
12	Transverse rows of single casing	Z_{i1}	–	Determine from design	23
13	Rows of single casing in air flow direction	Z_{i2}	–	Determine from design	63
14	Transverse rows	Z_1	–	$B_1 Z_{i1}$	138

TABLE C23 Structure of Lower Stage Air Preheater (*cont.*)

No.	Parameter	Symbol	Unit	Formulae or data source	Value
15	Rows of air flow direction	Z_2	–	Determine from design	126
16	Number of tubes of a casing	n_i	–	Determine from design	1418
17	Total number of tubes	Σn	–	$n \cdot B$	17016
18	Casing spacing	δ_1	mm	Determine from design	50
19	Distance from casing to wall	δ_2	mm	Determine from design	43.5
20	Flue duct width	a	m	Determine from design	9.841
21	Heating area	A	m^2	$n \cdot \Sigma n \cdot h \pi (d-\delta)/1000$	17288.1
22	Flue gas flow area	A_g	m^2	$\Sigma n \pi/4 d_n^2$	18.30
23	Air flow area	A_a	m^2	$2h(a-Z_1 d-5\delta_1)$	22.80
24	Effective radiation layer thickness	s	m	$0.9 d_n$	0.0333

TABLE C24 Thermal Calculation of Lower Stage Air Preheater

No.	Parameter	Symbol	Unit	Formulae or data source	Value
1	Inlet flue gas temperature	θ'	°C	See Table C20	275
2	Inlet flue gas enthalpy	I'	kJ/kg	See Table C20	3919.7
3	Inlet air temperature	t'	°C	Determine from design	20
4	Inlet air enthalpy	I'_a	kJ/kg	See Table C3	193.3
5	Outlet air temperature	t''	°C	Trial and error	198
6	Outlet air enthalpy	I''_a	kJ/kg	See Table C3	1927.5
7	Average air enthalpy	$I_{a,ave}$	kJ/kg	$(I'_a + I''_a)/2$	1060.4
8	Excess air coefficient at air preheater outlet	β''_{ah}	–	See Table C20	1.14
9	Air preheater absorbed convection heat	Q_c	kJ/kg	$(\beta''_{aph} + 0.5\Delta\alpha)(I''_a - I'_a)$	2003.0

(Continued)

TABLE C24 Thermal Calculation of Lower Stage Air Preheater (*cont.*)

No.	Parameter	Symbol	Unit	Formulae or data source	Value
10	Outlet flue gas enthalpy	I''	kJ/kg	$I' + \Delta\alpha I_{a,ave} - Q_c/\varphi$	1940.4
11	Outlet flue gas temperature	θ''	°C	See Table C3	135.5
12	Average air temperature	t_{ave}	°C	$(t' + t'')/2$	109
13	Flue gas average temperature	θ_{ave}	°C	$(\theta' + \theta'')/2$	205.2
14	Counterflow small temperature difference	Δt_{min}	°C	$\theta' - t''$	77
15	Counterflow large temperature difference	Δt_{max}	°C	$\theta'' - t'$	115.5
16	Counterflow logarithmic mean temperature difference	Δt	°C	$(\Delta t_{max} - \Delta t_{min})/$ $(ln\Delta t_{max}/\Delta t_{min})$	94.9
17	Large temperature drop	t_{max}	°C	$t'' - t'$	178
18	Small temperature drop	t_{min}	°C	$\theta' - \theta''$	139.5
19	Parameter	P	–	$t_{min}/(\theta' - t')$	0.547
20	Parameter	R	–	t_{max}/t_{min}	1.28
21	Temperature difference correction coefficient	ψ	–	Refer to Appendix D6.40	0.96
22	Average temperature difference	Δt_{ave}	°C	$\psi\Delta t$	91.14
23	Average air velocity	ω_a	m/s	$B_{cal}V^0(\beta''_{aph} +$ $0.5\Delta\alpha)(t_{ave} + 273)/$ $(273 \times A_a)$	5.52
24	Average flue gas velocity	ω_g	m/s	$B_{cal}V_g(\theta_{ave} + 273)/$ $(273A_g)$	10.53
25	Tube wall temperature	t_w	°C	$0.5(t_{ave} + \theta_{ave})$	157.1
26	Product of p_n and s	$p_n s$	m·MPa	$pr_n s$	0.000654

TABLE C24 Thermal Calculation of Lower Stage Air Preheater (*cont.*)

No.	Parameter	Symbol	Unit	Formulae or data source	Value
27	Radiant absorption coefficient of gas	k_g	1/ (m·MPa)	$10.2[(0.78 + 1.6r_{H_2O})/ (10.2p_n s)^{0.5}-0.1] (1-0.37T_{ave}/1000)$	90.40
28	Radiant absorption coefficient of fly ash	k_{fa}	1/ (m·MPa)	$48350\rho_g/(T_{ave}^2 d_{fa}^2)^{1/3}$	150.00
29	Radiant absorption coefficient of flue gas radiation	k	1/ (m·MPa)	$k_g r_n + k_{fa}\mu_{fa}$	18.96
30	Exponent of Eq. (2.73)	kps	–	kps	0.0631
31	Flue gas emissivity	a	–	$1-e^{-kps}$	0.0612
32	Radiation heat transfer coefficient	α_r	w/ (m².°C)	$5.7 \times 10^{-8}(a_w + 1)/2aT_{ave}^3[1-(T_w/ T_{ave})^4]/[1-(T_w/ T_{ave})]$	1.18
33	Correction factor for tube length	C_l	–	Refer to Appendix D6.41	1
34	Flue gas temperature and composition correction coefficient	C_w	–	Refer to Appendix D6.42	0.98
35	Flue gas temperature and wall temperature correction coefficient	C_t	–	Refer to Appendix D6.43	1
36	Thermal conductivity of flue gas	λ	–	Refer to Appendix D6.27	0.040
37	Dynamic viscosity of flue gas	ν	–	Refer to Appendix D6.27	3.19×10^{-5}
38	Average Prandtl number of flue gas	Pr_{pj}	–	Refer to Appendix D6.28	0.669
39	Prandtl number of flue gas	Pr	–	$(0.94 + 0.56r_{H_2O}) Pr_{ave}$	0.654
40	Flue gas side convection heat transfer coefficient	α_c	W/ (m².°C)	$0.023\lambda/d(\omega_g d/ \nu)^{0.8}Pr^{0.4}C_t C_l C_w$	38.43
41	Flue gas side heat transfer coefficient	α_1	W/ (m².°C)	$\alpha_c + \alpha_r$	39.61

(*Continued*)

TABLE C24 Thermal Calculation of Lower Stage Air Preheater (*cont.*)

No.	Parameter	Symbol	Unit	Formulae or data source	Value
42	Air thermal conductivity	λ	W/(m·°C)	Refer to Appendix D6.27	0.12
43	Air dynamic viscosity	ν	m^2/s	Refer to Appendix D6.27	4.49×10^{-4}
44	Air Prandtl number	Pr	–	Refer to Appendix D6.27	0.7
45	Relative diagonal pitch	s_2'/d	–	$[1/4(s_1/d)^2 + (s_2/d)^2]^{1/2}$	1.40
46	Parameter	φ_σ	–	$(s_1/d-1)/(s_2'/d-1)$	1.99
47	Correction factor for rows	C_z	–	Refer to Appendix D6.36	1
48	Flue gas composition and temperature correction coefficient	C_w	–	Refer to Appendix D6.44	0.92
49	Correction factor for the geometric arrangement	C_s	–	$0.768\varphi_\sigma^{0.5}$	1.08
50	Air side convection heat transfer coefficient	α_2	W/(m²·°C)	$0.358\lambda/d(\omega_a d/\nu)^{0.6}Pr^{0.33}C_zC_sC_w$	38.58
51	Utilization coefficient	ξ	–	Refer to Appendix D6.45	0.7
52	Overall heat transfer coefficient	K	W/(m²·°C)	$\xi\alpha_1\alpha_2/(\alpha_1 + \alpha_2)$	13.68
53	Convection heat	$Q_{c,cal}$	kJ/kg	$K\Delta tA/(1000B_{cal})$	2028.6
54	Error	e	%	$(Q_c-Q_{c,cal})/Q_c \times 100$	−1.28

The above example adopts two flue ducts to arrange back-end heating surfaces, as recommended by the Beijing Boiler Works. The aim is to arrange back-end heating surfaces reasonably and sustain reasonable flue gas velocity.

Air inflation from the air preheater is different from that of other heating surfaces. Hot air is leaked into the flue gas side, so average hot air temperature and enthalpy should be adopted during the course of calculation.

The heat transfer condition of air preheaters is different from that of other heating surfaces. The heat transfer coefficient on both sides of the heating surface is a very important part of overall heat transfer. To ensure the structure is

TABLE C25 Summary of the Main Parameters of Thermal Calculation

Parameter	Symbol	Unit	Platen superheater	High-temperature superheater hot stage	High-temperature superheater cold stage	Low-temperature superheater	Reversing chamber	Upper economizer	Upper air preheater	Lower economizer	Lower air preheater
Outlet flue gas temperature	θ'	°C	1005.0	750.4	750.4	522.0	515.0	421.0	326.8	275.0	135.5
Working medium inlet temperature	t'	°C	395.0	489.4	447.8	350.0	348.5	249.9	197.0	235.0	20.0
Working medium outlet temperature	t''	°C	447.6	540.0	500.3	407.5	349.9	275.0	320.0	249.5	198.0
Average flue gas velocity	$\omega_{g,ave}$	m/s	5.59	10.52	10.52	10.69	–	7.54	13.73	7.70	10.53
Average working medium velocity	ω	m/s	17.82	16.71	15.41	10.47	–	1106.6	6.98	1411.9	5.52
Heating area	A/H	m²	519.8	700.7	672.7	2545.3	173.3	1012.3	6174.3	1864.4	17288.1
Average temperature difference	Δt	°C	635.3	340.5	395.1	247.8	141.7	203.6	106.8	56.6	91.1
Heat transfer coefficient	K	W/m²°C	40.07	65.44	64.85	52.51	45.35	66.27	22.38	71.54	13.68
Absorbed convection heat	$Q_{c,cal}$	kJ/kg	1245.3	1469.3	1621.8	3116.2	104.8	1285.4	1388.5	710.8	2028.6

Note: the average working medium velocity of economizers is mass flux [kg/(m² · s)].

economical and reasonable, a reasonable ratio of air velocity and flue gas velocity and reasonable flue gas velocity must be managed appropriately. Again, this example was established according to practical information provided by the Beijing Boiler Works. In their case, the flue gas velocity of the high-temperature air preheater is 12–14 m/s, the flue gas velocity of the low-temperature air preheater is 10–11 m/s, and the ratio of air velocity and flue gas velocity is approximately 0.5.

C10   MAIN THERMAL CALCULATION PARAMETERS IN SUMMARY

The main parameters of boiler thermal calculation are summarized in Table C25.

Appendix D

Supplementary Materials [11,17]

This appendix provides supplementary materials to the principles and calculation of heat transfer in furnaces. An introduction to fuel and fuel types is provided first, followed by fuel composition basics, including conversion of composition. Section D2 introduces quantities of air related to fuel combustion. Important terms, such as the theoretical air requirement for fuel combustion and actual air supply for fuel combustion, are defined. The theoretical volumes of combustion products and actual volume of combustion products are introduced step-by-step, as well. Section D5 provides an enthalpy–temperature table for reference during thermal calculation. The section concludes with some information that complements Appendix C.

Boilers are used to produce steam or hot water. Fuel is the energy source for the boiler. During boiler operation, fuel is continually supplied to the furnace to ensure continuous combustion and smooth boiler function. It is necessary to know the type and properties of the main products of combustion, and to know how to apply this data most effectively to boiler design.

D1 FUEL

Boiler fuel types include solid fuel (mainly referring to coal), liquid fuel (mainly heavy oil), and gas fuel (coal gas). In China, coal is the most popular boiler fuel.

The common classification for boiler coal is based on the volatile matter content on a dry ash-free basis (V_{daf}) and heating value (HV). Coal is classified into anthracite, lean coal, low-volatility bituminous coal, high-volatility bituminous coal, and lignite (Table D1). Based on the classification of V_{daf}, anthracite and bituminous coal fall into Subclass I, II, and III according to heating value. Subclass I is inferior coal with poor combustion conditions, Subclass II is medium-quality coal with better combustion conditions than Subclass I, and Subclass III is superior coal. Stone coal and gangue fall into an additional set of subclasses, also classified as I, II, and III according to HV.

Solid, liquid, and gas fuels are organic fuels comprised primarily of complex, high-molecular hydrocarbons. The main elements of the combustible portion of these fuels include carbon (C), hydrogen (H), oxygen (O), nitrogen (N), and sulfur (S). The incombustible portion includes ash (A) and moisture (M). C is the main combustible element of fuel, accounting for 15–90% of the

Theory and Calculation of Heat Transfer in Furnaces. http://dx.doi.org/10.1016/B978-0-12-800966-6.00017-X

TABLE D1 Coal Classification in Industrial Boilers

Classification		Volatile matter content (dry ash-free basis) V_{daf}/%	Lower heating value (LHV) (as-received basis) $Q_{ar,net,p}$/ (MJ/kg)
Stone coal and gangue	I		≤5.4
	II		>5.4~8.4
	III		>8.4~11.5
Lignite		>37.0	≥11.5
Anthracite	I	6.5~10.0	<21.0
	II	<6.5	≥21.0
	III	6.5~10.0	≥21.0
Lean coal		>10.0~20.0	≥17.7
Bituminous coal	I	>20.0	>14.4~17.7
	II	>20.0	>17.7~21.0
	III	>20.0	>21.0

Note: This table refers to "Thermal Calculation Method of Grate-firing and Fluidized Bed Combustion Industrial Boiler, General Technical Conditions of Industrial Boiler" (JB/T10094–2002), and "Technical Conditions of Power Plant Boiler" (JB/T 6696–93).

combustibles and 50–90% of the coal. H is a combustible element whose heating value is 120,370 kJ/kg, about four times that of C. H content in dry and ash-free coal is around 2–10%. O and N belong to incombustible parts of the fuel combustible portion that reduces the HV relative to the theoretical calculation of C and H. The S is present in two forms: sulphates such as $CaSO_4$ and $MgSO_4$ called "inorganic sulfur," which form incombustible ash, and organic sulfur and pyrite sulfur, which are combustible and have HV of 9100 kJ/kg.

"Ash" is characterized by incombustible minerals, and ash content differs by fuel type. Gas and liquid fuels, for example, contain relatively little ash—the ash content of solid coal, conversely, is 10–30%. (For certain types of inferior coal, the ash content can be up to 50%.) Moisture is another incombustible element in fuel that varies according to fuel type—the moisture content of liquid fuel is 1–3%, the moisture content of lignite is 40–50%, and the moisture content of anthracite is as low as 5%.

D2   THE BASIS OF FUEL COMPOSITION

Fuel composition can be analyzed based on four diferent bases, they are: as-received basis, dry basis, air-dry basis, or dry and ash-free basis.

D2.1   The Basis of Fuel Composition Analysis

The composition of gas fuel is denoted by volume fraction, while the composition of solid and liquid fuels is usually expressed by mass fraction. See the following:

$$C + H + O + N + S + A + M = 100\%$$ (D1)

where C, H, O, N, and S are the mass percentages of carbon, hydrogen, oxygen, nitrogen, and sulfur, and A and M are the mass percentages of ash and moisture.

Based on different bases, Eq. (D1) can be rewritten as follows.

1. On as-received basis:

$$C_{ar} + H_{ar} + O_{ar} + N_{ar} + S_{ar} + A_{ar} + M_{ar} = 100$$ (D2a)

2. On air-dry basis:

$$C_{ad} + H_{ad} + O_{ad} + N_{ad} + S_{ad} + A_{ad} + M_{ad} = 100$$ (D2b)

3. On dry basis:

$$C_{d} + H_{d} + O_{d} + N_{d} + S_{d} + A_{d} = 100$$ (D2c)

4. On dry and ash-free basis:

$$C_{daf} + H_{daf} + O_{daf} + N_{daf} + S_{daf} = 100$$ (D2d)

D2.2   Conversion of Composition Basis

The air-dry basis is commonly used for lab analysis, the as-received basis is typically used for the cases of fuel samples as received from an industrial site, and the dry and ash-free basis is used to judge how to selecte and operate burners during boiler design and operation. The difference between the different bases is the influence of moisture and ash. The basis of fuel can be mutually converted as shown in Table D2 and Fig. D1, which illustrates the relationship between the bases. S_S denotes incombustible sulfur, S_R denotes combustible sulfur, M_{nz} denotes inherent moisture, M_{wz} denotes surface moisture, and M_R denotes crystallized moisture.

D3   AIR AMOUNT FOR FUEL COMBUSTION

To design boilers based on known fuel, first it is necessary to calculate the theoretical air amount and actual air amount for combustion from the C, H, O, N, S, M, and A of the fuel. The theoretical flue gas amount and actual flue gas amount can then be calculated to compile the gas enthalpy–temperature table, which serves as a reference for calculation of mass balance, energy balance, and heat transfer in the boiler.

TABLE D2 Conversion Factor of Different Basis of Fuel

Basis	As-received basis	Air dry basis	Dry basis	Dry and ash-free basis
As-received basis	1	$\dfrac{100-M_{ad}}{100-M_{ar}}$	$\dfrac{100}{100-M_{ar}}$	$\dfrac{100}{100-M_{ar}-A_{ar}}$
Air dry basis	$\dfrac{100-M_{ar}}{100-M_{ad}}$	1	$\dfrac{100}{100-M_{ad}}$	$\dfrac{100}{100-M_{ad}-A_{ad}}$
Dry basis	$\dfrac{100-M_{ar}}{100}$	$\dfrac{100-M_{ad}}{100}$	1	$\dfrac{100}{100-A_d}$
Dry and ash-free basis	$\dfrac{100-M_{ar}-A_{ar}}{100}$	$\dfrac{100-M_{ad}-A_{ad}}{100}$	$\dfrac{100-A_d}{100}$	1

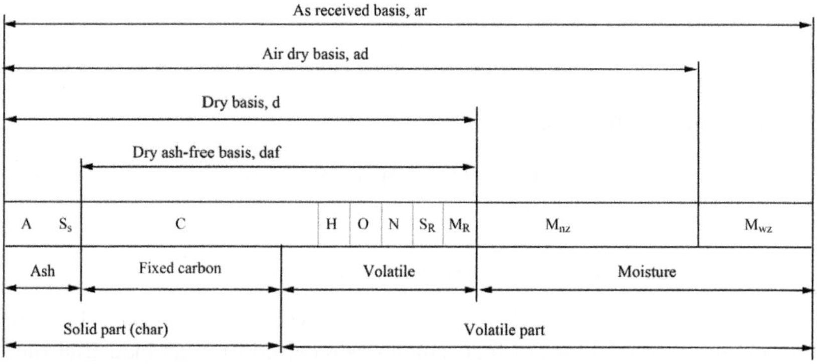

FIGURE D1   Relationship of different bases of solid fuel.

D3.1   Theoretical Air Amount for Fuel Combustion

For boiler design in China, the standard approximate composition of air is $O_2 =$ 21 vol. %, $N_2 = 79$ vol.% and the density of dry air under standard conditions is 1.293 kg/m^3. It is generally assumed that there is 0.0161 Nm^3 steam per Nm^3 of dry air.

The theoretical air amount is the calculation basis for flue gas quantity, and is defined as the minimum air amount for the complete combustion of 1 kg of fuel (solid or liquid). The theoretical air amount can be expressed as follows:

$$V_0 = 0.0889(C_{ar} + 0.375S_{ar}) + 0.265H_{ar} - 0.0333O_{ar} \qquad (D3)$$

where V_0 is the theoretical dry air volume for the complete combustion of 1 kg of fuel, in Nm^3/kg fuel.

D3.2   Actual Air Amount for Fuel Combustion

To achieve complete combustion of fuel, excess air is needed to compensate for imperfect mixing of the air and fuel. The ratio of actual air supply amount V_a and theoretical air requirement amount V_0 is α (β for air side), the excess air coefficient.

$$\frac{V_a}{V_0} = \alpha(or\beta) \tag{D4}$$

The excess air coefficient α is generally larger than 1. The α in a furnace is related to the furnace's combustion type and fuel characteristics. For a boiler with coal gas and heavy oil, $\alpha = 1.05 \sim 1.10$; for a grate-firing boiler, $\alpha = 1.25 \sim 1.60$; for a PC boiler, $\alpha = 1.15 \sim 1.20$; for a BFB boiler, $\alpha = 1.25 \sim 1.30$; for a CFB boiler, $\alpha = 1.25 \sim 1.35$.

D4   COMBUSTION PRODUCTS

D4.1   Theoretical Volume of Combustion Products

For the complete combustion of 1 kg of fuel (at the theoretical air amount), the flue gas volume is V_y^0 (in Nm³/kg fuel), which is the basis for flue gas quantity calculation. Flue gas composition includes CO_2, H_2O, SO_2, N_2.

$$C + O_2 \rightarrow CO_2$$

The CO_2 produced from the complete combustion of 1 kg of carbon is:

$$V_{CO_2} = \frac{22.4}{12} \cdot \frac{C_{ar}}{100} = 1.866 \frac{C_{ar}}{100} \tag{D5}$$

The SO_2 produced from the complete combustion of 1 kg of sulfur is:

$$V_{SO_2} = \frac{22.4}{32} \cdot \frac{S_{ar}}{100} = 0.7 \frac{S_{ar}}{100} \tag{D6}$$

The triatomic gases can be denoted by V_{RO_2}:

$$V_{RO_2} = V_{CO_2} + V_{SO_2} = 1.866 \frac{C_{ar}}{100} + 0.7 \frac{S_{ar}}{100} \tag{D7}$$

or:

$$V_{RO_2} = 0.01866(C_{ar} + 0.375S_{ar}) = 0.01866\,K_{ar} \tag{D8}$$

where $K_{ar} = C_{ar} + 0.375S_{ar}$

Theoretical N_2 $V_{N_2}^0$ comes from the fuel and air:

$$V_{N_2}^0 = \frac{22.4}{28} \cdot \frac{N_{ar}}{100} + 0.79V_0 = 0.8 \frac{N_{ar}}{100} + 0.79V_0 \tag{D9}$$

Theoretical water vapor volume $V_{H_2O}^0$ comes from the moisture in the fuel, combustion of H in the fuel, and moisture in the air. Oil boilers also contain steam from atomization.

The water vapor volume from the evaporation of moisture in 1 kg of fuel is:

$$\frac{22.4}{18} \cdot \frac{M_{ar}}{100} = 0.0124 \, M_{ar} \tag{D10}$$

From $2H_2 + O_2 \rightarrow 2H_2O$, the combustion of H in 1 kg of fuel is:

$$\frac{2 \times 22.4}{2 \times 2.016} \cdot \frac{H_{ar}}{100} = 0.111 H_{ar} \tag{D11}$$

The water vapor in flue gas includes the moisture from theoretical air. Assuming there is d g (generally, $d = 10$ g/kg) in 1 kg of dry air, the density of dry air is 1.293 kg/Nm³, and the density of water vapor is 0.804 kg/Nm³, then the water vapor in 1 Nm³ of dry air is:

$$\frac{d}{1000} \cdot \left(\frac{1}{0.804} \right) \Big/ \left(\frac{1}{1.293} \right) = 0.00161d = 0.0161 \tag{D12}$$

Then the water vapor volume from the theoretical air amount V_0 for 1 kg of fuel can be determined. The total water vapor amount is:

$$\begin{aligned} V_{H_2O}^0 &= 0.0124 M_{ar} + 0.111 H_{ar} + 0.00161 d V_0 \\ &= 0.0124 M_{ar} + 0.111 H_{ar} + 0.0161 V_0 \end{aligned} \tag{D13}$$

To summarize, the theoretical flue gas volume for the complete combustion of 1 kg of fuel in a theoretical amount of air is:

$$V_y^0 = V_{RO_2} + V_{N_2}^0 + V_{H_2O}^0 \tag{D14}$$

The flue gas without water vapor is the dry flue gas, with the following volume:

$$V_{gy}^0 = V_{RO_2} + V_{N_2}^0 \tag{D15}$$

Then:

$$V_y^0 = V_{gy}^0 + V_{H_2O}^0 \tag{D16}$$

where the unit of flue gas is Nm³/kg fuel.

D4.2 Actual Volume of Combustion Products

The actual flue gas volume is necessary for the boiler thermal calculation. When excess air coefficient $\alpha > 1$, the combustion product volumes of complete combustion form the actual flue gas volume which equals the sum of the theoretical flue gas volume, excess air amount, and moisture carried with the excess air.

The excess air amount is:

$$\Delta V = V_a - V_0 = (\alpha - 1)V_0 \tag{D17}$$

The moisture carried wih the excess air is $0.0161(\alpha - 1)V_0$, and the actual flue gas volume V_y is:

$$
\begin{aligned}
V_y &= V_y^0 + (\alpha - 1)V_0 + 0.0161(\alpha - 1)V_0 \\
&= V_{RO_2} + V_{N_2}^0 + V_{H_2O}^0 + (\alpha - 1)V_0 + 0.0161(\alpha - 1)V_0
\end{aligned}
\tag{D18}
$$

The flue gas volume can be divided into dry flue gas volume and water vapor volume:

$$V_y = V_{gy} + V_{H_2O} \tag{D19}$$

where the dry flue gas volume is:

$$V_{gy} = V_{RO_2} + V_{N_2} + V_{O_2} = V_{RO_2} + V_{N_2}^0 + (\alpha - 1)V_0 = V_{gy}^0 + (\alpha - 1)V_0 \tag{D20}$$

For grate-firing furnaces and suspension-firing furnaces, heat transfer is mainly radiative. The diatomic gases O_2 and N_2 are transparent for heat radiation, and the concentration of CO is low; only the triatomic gases RO_2 and H_2O and ash particles in the flue gas participate in radiative heat transfer. The volume fraction of triatomic gases r_{RO_2} and r_{H_2O} and the fly ash concentration μ need to be calculated based on the flue gas composition. See the following:

$$r_{RO_2} = \frac{V_{RO_2}}{V_y} \tag{D21}$$

$$r_{H_2O} = \frac{V_{H_2O}}{V_y} \tag{D22}$$

$$\mu = \frac{A_{ar}a_{fh}}{100G_y} \tag{D23}$$

where G_y is the flue gas mass of 1 kg of fuel (kg/kg), which is calculated as follows:

$$G_y = 1 - \frac{A_{ar}}{100} + 1.306\alpha_f V_0 \tag{D24}$$

where a_{fh} is the fly ash fraction related to combustion type. For grate-firing boilers, $a_{fh} = 0.2\sim0.25$; for dry bottom PC boilers, $a_{fh} = 0.9\sim0.95$; for wet bottom PC boilers, $a_{fh} = 0.6\sim0.7$; for BFB boilers, $a_{fh} = 0.4\sim0.7$; and for CFB boilers, $a_{fh} = 0.45\sim0.95$. μ is the fly ash concentration and α_f is the excess air coefficient at the furnace exit.

D5 ENTHALPY–TEMPERATURE TABLE

The enthalpy–temperature table is related to fuel composition and the excess air coefficient. The table describes the enthalpy of flue gas under certain excess air conditions, reflecting the relationship of flue gas enthalpy and flue gas temperature at the heating surface exit.

D5.1 Enthalpy of Combustion Products

For accurate thermal balance calculation and thermal calculation it is necessary to calculate the heat absorbed by the heating surface, thus the enthalpy of the flue gas and air is required information. Because the enthalpy of air and flue gas varies according to fuel type, excess air coefficient, and temperature, the enthalpy–temperature table for the specific fuel and excess air coefficient α entering before and leaving after the heating surface should be established prior to boiler design.

The enthalpy–temperature table is calculated based on 1 kg as-received fuel and the air enthalpy and flue gas enthalpy of 1 kg of fuel (in kJ/kg fuel). The enthalpy of air and flue gas at 0°C is 0.

The enthalpy of the theoretical air amount ($\beta = 1$) at t (°C) I_k^0 is:

$$I_k^0 = V_0 c_k t \tag{D25}$$

where c_k is the average heat capacity at constant pressure from 0 to t°C, kJ/(Nm^3 °C); at t°C, and the actual air ($\beta > 1$) enthalpy I_k is:

$$I_k = \beta V_0 c_k t = V_k c_k t \tag{D26}$$

At θ (°C), the enthalpy of the theoretical flue gas I_y^0 is:

$$\begin{aligned} I_y^0 &= V_{RO_2} c_{CO_2} \theta + V_{N_2}^0 c_{N_2} \theta + V_{H_2O}^0 c_{H_2O} \theta \\ &= (V_{RO_2} c_{CO_2} + V_{N_2}^0 c_{N_2} + V_{H_2O}^0 c_{H_2O}) \theta \end{aligned} \tag{D27}$$

where c_{CO_2}, c_{N_2}, and c_{H_2O} are the average heat capacity at constant pressure from 0 to θ °C kJ/(Nm^3 °C), the specific values of which are shown in Table D3.

The actual flue gas enthalpy I_y is the sum of the theoretical flue gas enthalpy and excess air enthalpy:

$$I_y = I_y^0 + (\alpha - 1) I_k^0 \tag{D28}$$

If fly ash concentration in the flue gas is high (ie, converted fly ash concentration is larger than 14.33 g/MJ and $10\,000 \dfrac{A_{ar} a_{fh}}{Q_{ar,net,p}} > 14.33$), the enthalpy of fly ash in the flue gas must be accounted for. In such a case, Eq. (D28) can be rewritten as:

$$I_y = I_y^0 + (\alpha - 1) I_k^0 + I_{fh} \tag{D29}$$

TABLE D3 Average Specific Heat Capacity from 0 to $t°C$ of Several Gases $(kJ/Nm^3 °C)$

$t/°C$	CO_2	CO	N_2	O_2	H_2O	Dry air	Wet air
0	1.5998		1.2946	1.3059	1.4943	1.2971	1.3188
100	1.7003	1.2979	1.2958	1.3176	1.5052	1.3004	1.3243
200	1.7873	1.3004	1.2996	1.3352	1.5223	1.3071	1.3318
300	1.8627	1.3075	1.3067	1.3561	1.5424	1.3172	1.3423
400	1.9297	1.3176	1.3163	1.3775	1.5654	1.3289	1.3544
500	1.9887		1.3276	1.3980	1.5897	1.3427	1.3682
600	2.0411		1.3402	1.4168	1.6148	1.3565	1.3829
700	2.0884		1.3536	1.4344	1.6412	1.3708	1.3976
800	2.1311		1.3670	1.4499	1.6680	1.3842	1.4114
900	2.1692		1.3796	1.4645	1.6957	1.3976	1.4248
1000	2.2035		1.3917	1.4775	1.7229	1.4097	1.4373
1100	2.2349		1.4034	1.4892	1.7501	1.4214	1.4583
1200	2.2638		1.4143	1.5005	1.7769	1.4327	1.4612
1300	2.2898		1.4252	1.5106	1.8028	1.4432	1.4725
1400	2.3136		1.4348	1.5202	1.8280	1.4528	1.4830
1500	2.3354		1.4440	1.5294	1.8527	1.4620	1.4926
1600	2.3555		1.4528	1.5378	1.8761	1.4708	1.5018
1700	2.3743		1.4612	1.5462	1.8996	1.4788	1.5102
1800	2.3915		1.4687	1.5541	1.9213	1.4867	1.5177
1900	2.4074		1.4758	1.5617	1.9423	1.4939	1.5257
2000	2.4221		1.4825	1.5692	1.9628	1.5010	1.5328

Note: The data are from Standard Methods of Boiler Unit Thermal Calculation [11].

where I_{fh} is fly ash enthalpy, which can be calculated as follows:

$$I_{fh} = \frac{A_{ar}}{100} a_{fh} c_{fh} \theta \tag{D30}$$

where c_{fh} is the average mass specific heat capacity with fly ash temperature from 0 to $\theta°C$ (Table D4), and a_{fh} is the fly ash fraction.

D5.2   Compiling the Enthalpy–Temperature Table

Actual and theoretical air amounts can be calculated according to fuel composition, the excess air coefficient at the furnace exit can be chosen according to combustion type, and the heating surface air leakage factor can be determined

TABLE D4 Average Specific Heat Capacity of Fly Ash (kJ/kg °C) [11]					
$t/°C$	c_{fh}	$t/°C$	c_{fh}	$t/°C$	c_{fh}
100	0.808	800	0.959	1500	1.172
200	0.846	900	0.972	1600	1.172
300	0.879	1000	0.984	1700	1.214
400	0.900	1100	0.997	1800	1.214
500	0.917	1200	1.005	1900	1.256
600	0.934	1300	1.047	2000	1.256
700	0.946	1400	1.130		

according to heating surface arrangement. The flue gas quantity and composition within the heating surfaces (ie, the average characteristic values of the flue gas) can then be determined. According to the amount of flue gas and composition within the heating surfaces (furnace, slag screen, superheater, etc.), the sum of the enthalpy of individual components in the flue gas can be calculated at specific temperatures–this is the flue gas enthalpy. The flue gas enthalpy at certain intervals (eg, 100°C) can then be calculated to compile the enthalpy–temperature table.

The enthalpy–temperature table is the basis for heat transfer calculation in furnaces and heating surfaces, and for thermal balance calculation. Flue gas enthalpy can be obtained from the excess air coefficient and flue gas temperature, and flue gas temperature can be obtained from the excess air coefficient and flue gas enthalpy based on the enthalpy–temperature table.

D6 SUPPLEMENTARY INFORMATION FOR APPENDIX C

For Chinese boiler desingers, standard thermal calculation methods for boiler units, boiler principles, and all necessary calculations are commonly available in handbooks. Due to page limitations, we were limited to thermal calculation of 113.89 kg/s and thus could only provide a handful of formulae—a few others are provided later (though a few other formulae that are overly complex are not provided here).

In the following captions C##.** is a reference to the cell of a table in Appendix C, where ## denote the sequence number of the table and ** denote the row number of a cell in the table ##. For example, C1.6 means that this information is found in the cell of row 6 of Table C1. C2. Note represents this is for the note of Table C2.

D6.1 C1.6, 4.3, 4.7 Thermal Load and Other Related Data for Solid Slag Pulverized Coal Furnaces

See Table D5.

TABLE D5 Thermal Load and Other Related Data for Solid Slag Pulverized Coal Furnaces

Fuel type	Allowable volume thermal load q_V/(kW/m³) Boiler output D/(kg/s)				Heat loss due to unburnt carbon q_{gt}/% Boiler output D/(kg/s)				Heat loss due to gas incomplete combustion q_{qt}/% Boiler output D/(kg/s)		Excess air coefficient at furnace exit α_l''	Fly ash fraction α_{fh}
	7	10	14	≥20	7	10	14	≥20	<20	≥20		
Anthracite				140				4~6		0	1.2~1.25	0.95
Lean coal				160				2		0	1.2~1.25	0.95
Bituminous coal	250	210	185	175	5	3	2~3	1~1.5	0.5	0	1.2	0.95
Middling				160				2~3		0	1.2	0.95
Lignite	230	245	210	185	3	1.5~2	1~2	0.5~1	0.5	0	1.2	0.95
Oil shale				115				0.5~1		0	1.2	0.95

D6.2 C2.Note The Excess Air Coefficient at the Platen Superheater Exit

See Table D6.

D6.3 C2.Note Air Leakage Factor of the Heating Surface

See Table D7.

D6.4 C3.Note The Basis of Omitting Fly Ash Enthalpy

$$I_y = I_y^0 + I_{fa} + (\alpha'' - 1)I_a^0 \quad \text{(kJ/kg)} \tag{D31}$$

When the ash content of fuel is high, the fly ash content in the flue gas is high. See the following equation:

$$1000\frac{A_{ar}a_{fh}}{Q_{ar,net,p}} > 1.43 \tag{D32}$$

When the above equation is satisfied the enthalpy of fly ash has to be accounted for.

TABLE D6 Excess Air Coefficient at Furnace Exit

Type of combustion chamber		Fuel	Excess air coefficient at furnace exit α_1''
PC boiler	Dry bottom	Anthracite, mean coal, bituminous coal, lignite	1.20~1.25* 1.20
	Wet bottom	Anthracite, mean coal, bituminous coal, lignite	1.20~1.25* 1.20
Heavy oil, gas furnace-fired boiler		Heavy oil, coke oven gas, natural gas, blast furnace gas	1.10**
Grate-firing boiler		Bituminous coal, lignite	1.5~1.6 1.3

*If coal powder is fed by hot air, a larger value should be input here.
**If air is supplied to the sealing furnace wall with positive pressure, 1.05 should be the set value when burning coal gas, 1.02–1.03 when burning oil, the ratio of which to air is automatically controlled and furnace air is less than 0.05.

TABLE D7 Flue Ductwork Air Leakage Factor Under Rated Load

Flue ductwork			Air leakage factor $\Delta\alpha$
Pulverized coal, oil, and gas-fired furnace	Dry bottom furnace with membrane wall		0.05
	Dry bottom furnace with steel frame and casing		0.07
	Dry bottom furnace without casing		0.10
	Wet bottom oil and gas-fired furnace with casing		0.05
	Wet bottom oil and gas-fired furnace without casing		0.08
Grate-firing furnace	Mechanical or semimechanical coal feeding		0.10
Convection heating surface	Slag screen, platen superheater, the first convective evaporator bundle ($D > 14$ kg/s)		0.00
	The first convection evaporator bundle ($D \leq 14$ kg/s)		0.05
	Superheater		0.03
	Reheater		0.03
	Economizer	$D > 14$ kg/s, each stage	0.02
		$D \leq 14$ kg/s Steel tube	0.08
		Cast iron (with casing)	0.10
		Cast iron (without casing)	0.20
	Tubular air preheater	$D > 14$ kg/s, each stage	0.03
		$D \leq 14$ kg/s, each stage	0.06
	Rotary air preheater	$D > 14$ kg/s	0.20
		$D \leq 14$ kg/s	0.25

D6.5   C4.9 Heat Loss Due to Surface Radiation and Convection

See Fig. D2.

D6.6   C4.10 Ash Physical Heat Loss

In a grate-firing furnace or wet bottom PC furnace, physical heat loss q_{ph} due to ash and slag must be calculated; when $A_{ar} \geq 0.0025Q_{ar,net,p}$, however, q_{ph} can be neglected.

Heat loss can be calculated as follows:

$$q_{ph} = \frac{Q_{ph}}{Q_{ar,net,p}}100 = \frac{a_{hz}(ct)_{hz}A_{ar}}{Q_{ar,net,p}}\% \tag{D33}$$

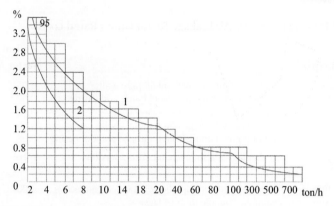

FIGURE D2 Heat loss due to surface radiation and convection. (1) Entire boiler with back-end heating surface. (2) Boiler itself without back-end.

TABLE D8 Furnace Volume Thermal Load q_V (kW/m³)

| Coal type | Dry bottom furnace | Wet bottom furnace | | |
		Open furnace	Half-open furnace	Slagging pool
Anthracite	110~140	≤145	≤169	523~698
Lean coal	116~163	151~186	163~198	523~698
Bituminous coal	140~198	≤186	≤198	523~640
Lignite	93~151			

D6.7 C5.1 Recommended Values of Furnace Volume Thermal Load (kW/m³)

See Table D8.

D6.8 C5.3 Furnace Thermal Load

See Tables D9A, D9B, and D10.

D6.9 C5.13, C5.14, C5.15, C5.39 Length and Dip Angle of Furnace Nose

The length of the furnace nose is generally 1/3 of furnace depth. The up-dip angle is 20–45 degree. When the ash content is low, and the flue gas velocity is high, a smaller up-dip angle can be chosen. The elevation angle is 20–30 degree.

TABLE D9A Upper Limit of Furnace Sectional Thermal Load of Tangential Firing Burner

Q5

Evaporation capacity D	Upper limit of furnace sectional thermal load /(MW/m^2)		
/(t/h)	/(kg/s)	ST≤1300°C	ST≈1350°C	ST>1450°C
25	6.9	1.63	1.95	2.23
35	9.7	1.65	1.98	2.28
50	13.9	1.72	2.05	2.33
65	18.1	1.77	2.09	2.37
75	20.8	1.84	2.12	2.44
130	36.1	2.13	2.56	2.59
220	61.1	2.79	3.37	3.91
410	113.9	3.65	4.49	5.12
500	138.9	3.91	4.65	5.44
1000	277.8	4.42	5.12	6.16
1500	416.7	4.77	5.45	6.63

TABLE D9B Statistic Value of Furnace Volume Thermal Load (MW/m^3)

Boiler capacity D/(t/h)		220	410	670
Tangential firing	Lignite and easy-slagging coal	2.1~2.56	2.9~3.36	3.25~3.71
	Bituminous coal	2.32~2.67	2.78~4.06	3.71~4.64
	Anthracite, mean coal	2.67~3.48	3.02~4.52	3.71~4.64
Front wall or opposite firing		2.2~2.78	3.02~3.71	3.48~4.06

The height of the furnace exit is determined by gas temperature and flow velocity. The gas velocity is generally approximately 6 m/s. The structure of the furnace nose is shown in Fig. D3.

D6.10   C5.18, C10.7, C10.11 Mass Flux of Working Medium in Platen Superheater Tubes

Mass flux is irrelevant to the pressure and temperature of steam; its value remains constant when pressure and temperature change. To this effect, it is

TABLE D10 Allowable Maximum Furnace Sectional Thermal Load Recommended by Former Soviet Union Standards (MW/m²)

Furnace sectional thermal load		Fuel	Front wall firing swirl burner or direct flow burner	Opposite firing swirl burner or direct flow burner	Tangentially fired burners
Multirow arranged burner	Overall q_F	Easy-slagging bituminous coal and lignite	3.48	$D \leq 264$ kg/s, 3.48 $D \leq 444$ kg/s, 4.06 $D > 444$ kg/s, 4.06–4.46	
		Nonslagging coal	4.64	6.38	6.38
		Anthracite	2.32	2.9	
	q_F of each burner zone	Easy-slagging bituminous coal and lignite	1.16	1.51	0.93
		Anthracite	1.74	2.32	1.74
Single-row arranged burner	Overall q_F	Easy-slagging bituminous coal and lignite	1.74	2.32–2.9	
		Anthracite	2.9	3.48	3.48

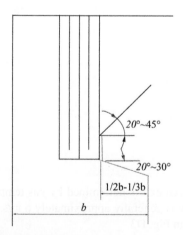

$20°\sim45°$

$20°\sim30°$

1/2b-1/3b

b

FIGURE D3 Structure/size of furnace nose.

convenient to determine the total flow section of the superheater based on mass flux. Mass flux in the superheater of a modern boiler can be determined based on the following parameters.

Medium temperature: $\rho\omega$ =250–400 kg/(m²·s);

High temperature: $\rho\omega$ =400–700 kg/(m²·s).

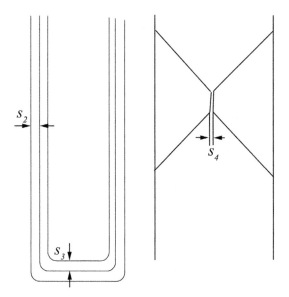

FIGURE D4   Schematic diagram of platen superheater structure.

Higher mass flux is generally adopted when heat transfer intensity is high; a lower mass flow rate is generally adopted when heat transfer intensity is low. In a high-pressure boiler that has a radiation supherheater with high heat transfer intensity, mass flow rate can reach 700–1200 kg/(m²·s).

D6.11   C5.23 Transverse Spacing of Platen Superheater Panels

The transverse spacing of platen superheater panels s_1 is generally 550–1500 mm. When the ash content of fuel is high, slag is easily formed, so s_1 should be larger. When boiler capacity is large, the spacing should be large, usually s_1 = 700–900 mm (see Fig. D4).

D6.12   C5.29, 31 Longitudinal Spacing of Platen Superheater Tubes

The longitudinal spacing s_2, vertical spacing s_3 of bottom tubes, and spacing of supporting tube s_4 of the platen superheater are usually determined as listed in Table D11.

TABLE D11 Tube Spacing of Superheater Tubes

Tube diameter	s_2	s_3	s_4
φ38×4.5	42	55	42
φ42×5	46	60	

D6.13 C5.30 Minimum Bending Radius of Platen Superheater

The minimum bending radius of a platen superheater is usually two times its diameter d. $R = 75$ for $\varphi 38$ tubes; $R = 80$ or 85 mm for $\varphi 42$ tubes.

D6.14 C5.35 Gas Velocity at furnace exit

For platen superheaters, the gas velocity is approximately 6 m/s at rated load. For convective superheaters, the gas velocity is 8–15 m/s. When fuel has low ash content, gas velocity may be higher (and vice versa).

D6.15 C7.4 Air Leakage Factor of Pulverized Coal System

See Table D12.

D6.16 C7.13 Mean Diameter of Ash Particles

See Table D13.

TABLE D12 Air Leakage Factor of Pulverized Coal System

Characteristics of pulverized coal system	$\Delta\alpha_{zf}$	Characteristics of pulverized coal system	$\Delta\alpha_{zf}$
Tubular ball mill:		Impact (or Hammer) coal mill:	
Storage system, hot air used for drying	0.1	Negative pressure	0.04
		Positive pressure	0.00
Storage system, hot air, and flue gas used for drying	0.12	Medium-speed mill:	
Direct fired system	0.04	Negative pressure	0.04
		Fan mill with drying tube	0.20–0.25

Note: Higher values are necessary when the coal moisture is high.

TABLE D13 Mean Diameter of Fly Ash Particles

Combustion equipment	Coal type	Fly ash particle diameter $d_{fa}/\mu m$
Pulverized coal combustion, tubular ball mill	All kinds of coal	13
Pulverized coal combustion, medium-speed, hammer mill	All kinds of coal other than peat	16
	Peat	24
Grate firing	All kinds of coal	20

D6.17   C7.23 Radiation Extinction Coefficient of Coke Particles

$$k = k_q r_\Sigma + k_{fh}\mu_{fh} + k_j x_1 x_2 \tag{D34}$$

where μ_{fh} is the dimensionless concentration of fly ash in the flame (kg/kg) and k_j is the radiation extinction coefficient of coke particles suspended in the flame, set as 10.2 during calculation.

D6.18   C7.24 Dimensionless Value x_1

Considering the influence coefficient of coke particle concentration in the flame, which is related to coal type, $x_1 = 1$ for anthracite and mean coal and $x_1 = 0.5$ for other coals with high volatile content.

D6.19   C7.25 Dimensionless Value x_2

Considering the influence coefficient of coke particle concentration in the flame, which is related to combustion mode, $x_2 = 0.1$ for suspension-firing combustion and $x_2 = 0.03$ for grate-firing combustion.

D6.20   C7.29 Fouling Factor of Water Wall

See Table D14.

TABLE D14 Fouling Factor of Water Wall

Water wall type	Fuel	Fouling factor ζ
Plain tube, fin tube, tube attached to wall	Coal gas	0.65
	Heavy oil	0.55
	Pulverized anthracite, carbon content in fly ash $C_{fh} \geq 12\%$, pulverized meager coal, $C_{fh} \geq 8\%$, pulverized bituminous coal and lignite	0.45
	Pulverized lignite, converted moisture $M_{zs} \geq 35g/MJ$, direct fired pulverized system dried with flue gas	0.55
	All kinds of coal, grate firing	0.60
Dry bottom furnace, coated with refractory material	All kinds of coal, oil, and gas	0.20
Covered with refractory brick	All kinds of coal, oil, and gas	0.10

Note: For pulverized anthracite with $C_{fh} < 12\%$, for pulverized mean coal $C_{fh} < 8\%$ and $\zeta = 0.35$. When combustion occurs with blended fuels, the minimum fouling factor for each fuel should be used.

D6.21 C7.35, 7.37 Δx for Calculation of Relative Coefficient at Flame Center

See Table D15.

TABLE D15 Calculation Method for Flame Center Modification Factor M

Combustion type	Fuel	M	Flame center relative position	Note
Suspension-firing	Coal gas and heavy oil	$M = 0.54 - 0.2x_h$	$x_h = x_r + \Delta x$ 1. When $\alpha_r > 1$, $\Delta x = 0$; 2. When $\alpha_r < 1$, $\Delta x = 2(1 - \alpha_r)$; 3. When D≤10 kg/s, $\Delta x = 0.15$.	When burner tilts 20 degree upward, x_h increases by 0.1; when burner tilts 20 degree downward, x_h decreases by 0.1. Interpolation should be adopted when the tilting is between ±20 degree.
	Pulverized coal	Pulverized coal with high volatile $M = 0.59 - 0.5x_h$ Anthracite and mean coal $M = 0.56 - 0.5x_h$	$x_h = x_r + \Delta x$ 1. For tangential firing, $\Delta x = 0$; 2. Front wall horizontal firing or opposite firing direct flow burner, front wall firing or opposite firing swirl burners with multirows arranged: when D≤116 kg/s, $\Delta x = 0.1$; when D>116 kg/s, $\Delta x = 0.05$. The maximum of M should not be larger than 0.5.	
Grate-firing	All types of coal	Thin coal layer $M = 0.59$ Thick coal layer $M = 0.52$		Spreader stoker is thin coal layer, chain grate and fixed grate-firing is thick coal layer.

Note: $x_r = h_r/h_1$, where h_r is burner height from the horizontal plane passing through half-height of furnace hopper or furnace bottom (m), h_1 is furnace height from the horizontal plane passing through half-height of furnace hopper or furnace bottom to furnace exit center (m), and α_r is the excess air coefficient of the burner.

D6.22   C8.14 Water Wall Configuration Factor

See Fig. D5.

D6.23   C9.17 Coefficient Considering the Effect of Reradiation on the Fouling Factor

See Fig. D6.

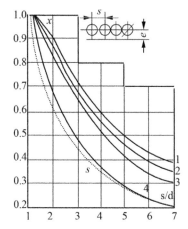

FIGURE D5   **Configuration factors of single-row bare water wall tube.** (1) e ≥1.4d, considering furnace wall radiation (2) e = 0.8d, considering furnace wall radiation (3) e = 0.5d, considering furnace wall radiation (4) e = 0, considering furnace wall radiation (5) e ≥0.5d, without considering furnace wall radiation.

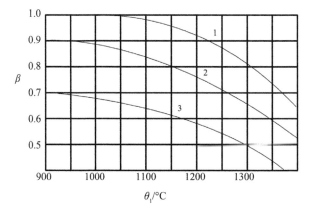

FIGURE D6   **Correction for fouling factor of interface between furnace and platen superheater.** (1) Solid fuel. (2) Heavy oil. (3) Coal gas.

D6.24 C9.19, C9.72 Distribution Coefficient of Thermal Load

See Fig. D7.

D6.25 C9.27 Radiation Fuel Correction Coefficient

ζ_r is the correction coefficient for fuel type. For coal and heavy oil, $\zeta_r = 0.5$; for shale, $\zeta_r = 0.5$; and for natural coal gas, $\zeta_r = 0.7$.

D6.26 C9.41, C11.54, C11.59, C13.33 Correction Factor for Tube Diameter C_d

See Fig. D8.

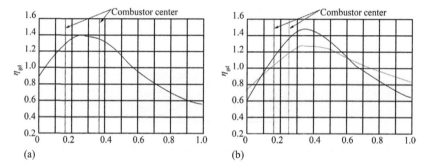

(a)

(b)

FIGURE D7 Nonuniform weight coefficient of the distribution of furnace radiation thermal load along furnace height [in (b) solid line for anthracite, mean coal, bituminous coal, and dry lignite coal; dotted line for lignite and peat]. (a) Coal gas and heavy oil furnace $x = h/h_1$. (b) Dry bottom pulverized coal furnace $x = h/h_1$.

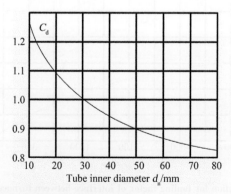

FIGURE D8 Correction factor for tube diameter.

D6.27   C9.47, C9.48, C11.19, C11.20, C11.21, C13.22, C13.23, C18.22, C18.23, C20.21, C20.22, C22.37, C22.38, C24.36, C24.37 Standard Flue Gas Thermal Conductivity

See Table D16.

TABLE D16 Thermal Conductivity, Viscosity, Pr Number of Standard Flue Gas

	Air			Flue gas		
Temperature/°C	$v \times 10^5$ /(m²/s)	$\lambda \times 10^2$ /[W/ (m·°C)]	Pr	$v \times 10^5$ /(m²/s)	$\lambda \times 10^2$ /[W/ (m·°C)]	Pr
0	13.2	2.43	0.70	11.9	2.28	0.74
100	23.2	3.19	0.69	20.8	3.13	0.70
200	34.8	3.90	0.69	31.6	4.01	0.67
300	48.2	4.48	0.69	43.9	4.84	0.65
400	62.9	5.05	0.70	57.8	5.70	0.64
500	79.3	5.62	0.70	73.0	6.56	0.62
600	96.7	6.15	0.71	89.4	7.42	0.61
700	115	6.66	0.71	107	8.27	0.60
800	135	7.14	0.72	126	9.15	0.59
900	155	7.61	0.72	146	10.0	0.58
1000	177	8.05	0.72	167	10.9	0.58
1100	200	8.47	0.72	188	11.7	0.57
1200	223	8.87	0.73	211	12.6	0.56
1300	247	9.27	0.73	234	13.5	0.55
1400	273	9.65	0.73	258	14.4	0.54
1500	300	10.03	0.73	282	15.4	0.53
1600	327	10.39	0.74	307	16.3	0.52
1700	355	10.75	0.74	333	17.3	0.51
1800	384	11.11	0.74	361	18.1	0.50
1900	415	11.46	0.74	389	19.0	0.49
2000	448	11.86	0.74	419	19.9	0.49
2100	478	12.10	0.75	450	20.7	0.48
2200	511	12.44	0.75	482	21.6	0.47

Note: Flue gas with $r_{H_2O} = 0.11$, $r_{CO_2} = 0.13$ is referred to as standard flue gas.

D6.28 C9.49, C11.21, C13.24, C18.24, C20.23, C22.39, C24.38 Average Prantl of Flue Gas and Air

See Table D17.

D6.29 C9.51, C11.23, C13.26 Correction Factor for Tube Rows

$$C_z = 0.91 + 0.0125(z_2 - 2) \tag{D35}$$

D6.30 C9.53, C11.25, C13.28 Correction Factor for the Geometric Arrangement of Tube Bundles

$$C_s = \left[1 + (2\sigma_1 - 3)\left(1 - \frac{\sigma_2}{2}\right)^3 \right]^{-2}, \text{ when } \sigma_1 \leq 1.5 \text{ or } \sigma_2 \geq 2, C_s = 1. \tag{D36}$$

Side spacing is s_1, relative transverse pitch is $\sigma_1 = s_1/d$, back spacing is s_2, and longitudinal relative pitch is $\sigma_2 = s_2/d$.

D6.31 C9.55, 9.59 Ash Deposition Coefficient and Utilization Coefficient

See Fig. D9.

D6.32 C9.56 Tube Wall Fouling Emissivity

At present, the tube wall fouling emissivity of metal heating surface ε_w is generally 0.8.

TABLE D17 Average Prantl of Flue Gas and Air

Flue gas*	Calculation equation $Pr = (0.94 + 0.56 r_{H_2O}) Pr_{pj}$		
	Pr_{pi} value	$100°C \leq \theta \leq 400°C$	$Pr_{pi} = 0.71 - 0.00002\theta$
		$400°C \leq \theta \leq 1000°C$	$Pr_{pi} = 0.67 - 0.00001\theta$
		$1000°C \leq \theta \leq 2000°C$	$Pr_{pi} = 0.68 - 0.00001\theta$
Air**	When $t_k < 400°C$, $Pr = 0.69$		
	When $t_k \geq 400°C$, $Pr = 0.70$		

*Standard flue gas.
**Wet air.

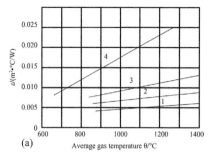
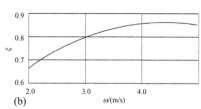

(a)   Average gas temperature θ/°C   (b)   ω/(m/s)

FIGURE D9 (a) Ash deposition coefficient. (b) Utilization coefficient of platen superheater. (1) Nonslagging coal. (2) Half-slagging coal with sootblowing. (3) Half-slagging coal without sootblowing. (4) Oil shale with sootblowing.

D6.33   C11.64, C13.38, C15.17 Ash Deposition Coefficient

For an in-line superheater and a staggered superheater when burning oil, $\varepsilon = 0.0026\,(\text{m}^2 \cdot {}^\circ\text{C})/\text{W}$.

For an in-line arranged superheater when burning solid fuel, $\varepsilon = 0.0043\,(\text{m}^2 \cdot {}^\circ\text{C})/\text{W}$.

D6.34   C11.75, C13.47, C18.40, C20.39 Fuel Correction Coefficient

A is the correction coefficient. When buring heavy oil and coal gas, $A = 0.3$; when buring anthracite, lean coal, and bituminous coal, $A = 0.4$; and when buring lignite and shale, $A = 0.5$.

D6.35   C11.79, C13.51 Effective Coefficient

See Table D18.

TABLE D18 Effective Coefficient of In-line Arranged Convection Heating Surface When Burning Solid Fuels

Fuel type		Sootblow or not?	ψ
Anthracite and lean coal		Yes	0.6
Bituminous coal and middling		Yes	0.65
Lignite	Fly ash with weak viscosity	Yes	0.65
	Fly ash with strong viscosity	Yes	0.6
Wood		Yes	0.6
Oil shale		Yes	0.5

D6.36 C18.28, C20.27, C22.48, C24.47 Correction Factor for the Tube Row Number

C_z denotes the tube row correction factor in bundles, related to longitudinal rows z_2 and relative transverse pitch σ_1. When $z_2 < 10$ and $\sigma_1 < 3.0$:

$$C_z = 3.12z_2^{0.05} - 2.5 \tag{D37}$$

When $z_2 < 10$ and $\sigma_1 \geq 3.0$:

$$C_z = 4z_2^{0.02} - 3.2 \tag{D38}$$

When $z_2 \geq 10$:

$$C_z = 1 \tag{D39}$$

D6.37 C18.42, C20.41 Basic Ash Deposition Coefficient ε_0 Values

See Fig. D10.

D6.38 C18.43, C20.42 Additional Correction Factor $\Delta\varepsilon$ Values for Ash Deposition

See Table D19.

D6.39 C18.44, C20.43 Correction Factor for Tube Diameter

See Fig. D11.

D6.40 C22.22, C24.21 Temperature Difference Correction Coefficient

See Fig. D12.

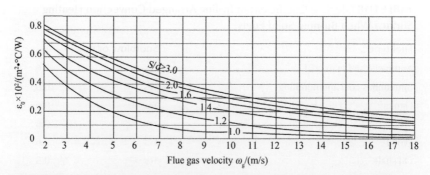

FIGURE D10 Basic ash deposition coefficient values for staggered, plain tube bundles.

TABLE D19 Additional Correction Factor $\Delta\varepsilon$ Values for Ash Deposition $(m^2 \cdot {}^\circ C/W)$

Heating surface	Coal with loose ash deposition	Anthracite		Lignite with soot-blower
		With steel ball soot-blower	Without soot-blower	
First-stage economizer, single-stage economizer, and other heating surfaces with $\theta' \leq 400$	0	0	0.0017	0
Single-stage economizer with $\theta' > 400^\circ C$, second-stage economizer, transitional zone of straight-flow boiler	0.0017	0.0017	0.0043	0.0026
Staggered arranged superheater	0.0026	0.0026	0.0043	0.0034

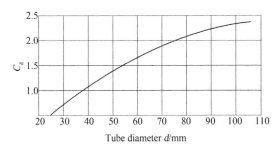

FIGURE D11   Tube correction factor for staggered, plain tube bundles.

D6.41   C22.34, C24.33 Correction Factor for Tube Length

See Fig. D13.

D6.42   C22.35, C24.34 Correction Factor for Temperature/ Composition

See Fig. D14.

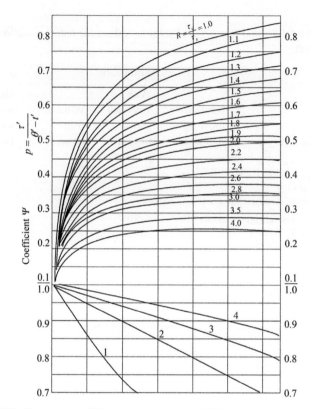

FIGURE D12 Temperature difference correction coefficient of cross-flow arrangement $\psi(\Delta t = \psi\Delta tnl)$**.** (1) Single path. (2) Double paths. (3) Three paths. (4) Four paths.

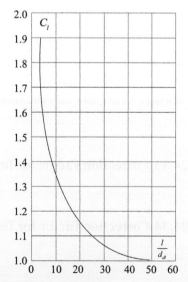

FIGURE D13 Correction factor for tube length.

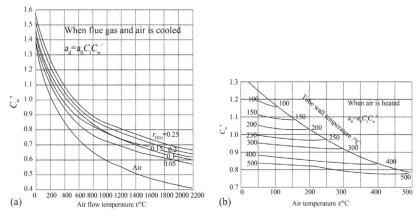

FIGURE D14   Correction coefficient for temperature/composition. (a) Cooling flue gas and air. (b) Heating air.

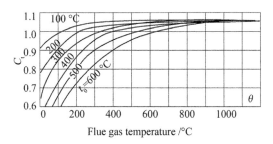

FIGURE D15   Temperature correction factor when working out fluid longitudinal flow through tubes. (a) Cooling flue gas and air. (b) Heating air.

D6.43   C22.36, C24.35 Correction Factor of Flue Gas Temperature and Wall Temperature

C_t is the correction factor for the flue gas temperature and wall temperature of the heating surface. When flue gas is cooled in the heating surface, $C_t = 1$; when the working medium is heated, its value is as shown in Fig. D15.

D6.44   C22.49, C24.48 Correction Factor for Flue Gas Temperature/Composition

See Fig. D16.

D6.45   C22.52, C24.51 Utilization Coefficient

See Table D20.

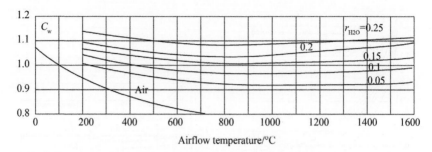

FIGURE D16   Correction factor for flue gas temperature/composition.

TABLE D20 Utilization Coefficient in Tubular Air Preheater ξ

Fuel type	First stage (low-temperature)	Second stage (high-temperature)
Anthracite	0.80	0.75
Heavy oil	0.80	0.85
Other coals and gas fuel	0.85	0.85

References

[1] Zhou P. Quantum mechanics. Beijing: Higher Education Press; 1989.

[2] Zhong Y. Thermodynamics and statistical physics. Beijing: Science Press; 1988.

[3] Guo S. Electrodynamics. Beijing: Higher Education Press; 1986.

[4] Lu D. Engineering radiation heat transfer. Beijing: National Defence Industry Press; 1988.

[5] Eckert ERG, Drake RM. Analysis of heat and mass transfer. New York: McGraw-Hill; 1972.

[6] Sparrow EM, Cess RD. Radiation heat transfer. New York: Hemisphere; 1978.

[7] Siegel R, Howell JR. Thermal radiation heat transfer. 2nd ed. Washington: Hemisphere Publishing Corporation; 1981.

[8] Holman JP. Heat transfer. 5th ed. USA: McGraw-Hill; 1981.

[9] Yang S. Heat transfer. 2nd ed. Beijing: Higher Education Press; 1987.

[10] Wang X, Mei F. Radiation heat transfer. Beijing: Higher Education Press; 1989.

[11] Beijing Boiler Company. Calibration method of boiler unit thermodynamic calculation. Beijing: Mechanical Industry Press; 1976.

[12] Qin Y. Heat transfer in furnace. 2nd ed. Beijing: Mechanical Industry Press; 1992.

[13] Industrial Boiler Design Calculation Standard Method Editorial Board. Industrial boiler design calculation standard method. Beijing: China Standards Press; 2003.

[14] Yu Q. Radiation heat transfer basis. Beijing: Higher Education Press; 1990.

[15] Cen K, Fan J, Chi Z, Shen L. Principles and calculations of boiler and heat exchanger fouling, slagging, wear, and corrosion prevention. Beijing: Science Press; 1994.

[16] Basu P, Fraser SA. Circulating fluidized bed boilers: design and operations. Boston: Butterworth-Heinemann; 1991.

[17] Feng J, Shen Y, Yang R. Boiler principles and calculations. 3rd ed. Beijing: Science Press; 2003.

[18] Yang X, Ma Q. Handbook of radiation heat transfer angular coefficient. Beijing: National Defence Industry Press; 1982.

[19] Ministry of Machine-Building Industry. Stratified Combustion and Industrial Fluidized Bed Combustion Boiler Thermodynamic Calculation Method (JB/DQ1060-82), 1982.

[20] Bian B. Analysis and calculation of radiation heat transfer. Beijing: Tsinghua University Press; 1988.

[21] Chen X, Chen T. Boiler principles. 2nd ed. Beijing: China Machine Press; 1991.

[22] Feng Junkai Selected Papers Editorial Group. Feng Junkai selected papers. Beijing: China Machine Press; 2002.

[23] Greiner W, Neise L, Stock H. Thermodynamics and statistical mechanics. New York: Springer; 2001.

[24] Bao K. Thermal physics basis. Beijing: Higher Education Press; 2001.

[25] Ma T. Plasma physics principles. Hefei: University of Science and Technology of China Press; 1988.

[26] Wang B. Engineering heat and mass transfer (first part). Beijing: Science Press; 1982.

[27] Han C, Xu M, Zhou H, Qu J. Pulverized coal combustion. Beijing: Science Press; 2001.

[28] Wang Y, Fan W, Zhou L, Xu X. Numerical combustion process. Beijing: Science Press; 1986.

Theory and Calculation of Heat Transfer in Furnaces. http://dx.doi.org/10.1016/B978-0-12-800966-6.00019-1

[29] Edwards DK, Weiner MM. Comment on radiative transfer in nonisothermal gases. Combust Flame 1966;10:202–3.

[30] Cess RD, Wang LS. A band absorptance formulation for nonisothermal gaseous radiation. Int J Heat Mass Transfer 1970;13:547–56.

[31] Edwards DK, Balakrishnan A. Thermal radiation by combustion gases. Int J Heat Mass Transfer 1973;16:25–40.

[32] Basu P. An investigation into heat transfer in circulating fluidized beds. Int J Heat Mass Transfer 1987;30:2399–409.

[33] Cen K, Ni M, et al. Circulating fluidized bed boiler theoretical design and operation. Beijing: China Electric Power Press; 1988.

[34] Liu X, Yu Z, Hui S, Xu T. Research on heat pipe heat flux meter. Power Eng 2001;21(4):1335–7.

[35] Hong M, Dong P, Qin Y. Separation calculation method of heat transfer in furnace of large boilers. J Harbin Inst Tech 2000;32(3):90–4.

[36] Zhang Y. The influence of boiler parameters on heat transfer in furnace. Therm Power Gen 2000;1:15–6.

[37] Tang B. A new heat exchange calculation method for large boiler furnaces. Power Eng 1992;6:26–31.

[38] Cheng L, Cen K, Ni M, Luo Z. Thermal calculation of a circulating fluidized bed boiler furnace. Proc CSEE 2002;22(12):146–51.

[39] Lv J, Xing X, et al. Experimental investigation on heat transfer in industrial-scale circulating fluidized bed boilers. J Combust Sci Technol 2000;6:156–60.

[40] Zhou K, Zhao Z, Cao H. Comparison and analysis of calculation methods of furnace outlet temperature. Power Eng 1999;19(5):363–6.

[41] Zhang M, Bie R. Heat transfer coefficient of water wall in circulating fluidized bed boiler. Boiler Manuf 2005;2:32–3.

[42] Xie Z, Gao K. Industrial radiation temperature measurement. Shenyang: Northeastern University Press; 1994.

[43] Liu L. Development status of heat transfer calculation methods in boiler furnaces. Power Eng 2000;20:523–7.

[44] Blokh AG. Heat Transfer in Steam Boiler Furnaces. Berlin: Hemisphere Publishing Corporation; 1988.

[45] Basu P, Kefa C, Jjestion L. Boilers and Burners. New York: Springer-Verlag Inc; 1999.

[46] Kitto JB, Stultz SC. Steam its generation and use. 41st ed. Barberton, Ohio: Babcock & Wilcox Company; 2005.

Subject Index

D

E

F

Printed and bound by CPI Group (UK) Ltd, Croydon, CR0 4YY

08/05/2025

01864799-0003